Learning from Megadisasters

Learning from Megadisasters

Lessons from the Great East Japan Earthquake

Federica Ranghieri and
Mikio Ishiwatari, editors

THE WORLD BANK
Washington, DC

CONTENTS

BOXES

FIGURES

MAPS

TABLES

Foreword

Yoshiki Takeuchi

The Great East Japan Earthquake and Tsunami on March 11, 2011, is a tragic reminder that no country or community is totally safe from natural disasters. The earthquake measuring a staggering 9.0 on the Richter scale hit the Tohoku region along the Pacific coast of Japan. While the damage from the earthquake itself was minimal because people were prepared and had learned from previous disasters, the subsequent tsunami caused extreme devastation to life and property, which shows that even the best prepared country will experience exceptional disasters. We express our sincere condolences to those affected by the Great East Japan Earthquake and Tsunami, and admire the courage and efforts of people for recovery and reconstruction.

At least 80 countries around the world are considered vulnerable to natural disasters. Large-scale natural disasters, once they occur, take a heavy toll on the lives of people. They can also destroy years of development efforts in an instant. Disaster risk management (DRM) should be taken into account as a major development challenge because the poor and the vulnerable are the most exposed to the risks of natural disasters. Therefore, the Government of Japan, in cooperation with the World Bank Group, has repeatedly advocated the importance of integrating DRM into development agenda. We believe it important to take advantage of lessons learned from the disaster and the reconstruction efforts in Japan as global public goods for future development policy.

This report, *Learning from Megadisasters*, consolidates the set of 36 Knowledge Notes, research results of the joint study undertaken by the Government of Japan and the World Bank. It summarizes the lessons learned from the Great East Japan Earthquake and Tsunami and provides guidance to other disaster-prone countries for mainstreaming DRM in their development policies. It is clear that financial resources alone are not sufficient to deal with disasters and to spur development. Technical assistance and capacity building are equally important. In Japan's case, we learned how communities can play a critical role in preparing for and coping with natural disasters. Communities can help prevent damage from spreading, maintain social order, and provide support to the vulnerable. Only through technical cooperation can such know-how be passed on to other countries and be adapted to their local circumstances.

The Sendai Statement, a joint statement on mainstreaming DRM issued by the World Bank Group president and Japan's finance minister in October 2012, emphasized the need to increase both technical and financial assistance for DRM in developing countries. It recognized that DRM is an essential part of enhancing sustainable development. Therefore, we urge the World Bank and other development assistance agencies to mainstream DRM into their operations. Japan, on its part, will spare no effort in building a more disaster resilient world in cooperation with the World Bank and other partners, by leveraging its expertise, technology, and staff. We expect a newly established Disaster Risk Management Hub of the World Bank in Tokyo to play a leading role to serve to match developing countries' needs with our technologies and expertise, and also disseminate the knowledge to the world.

We hope that *Learning from Megadisasters* will help development partners explore how to best integrate DRM into development policies and programs.

Yoshiki Takeuchi *is Deputy Director-General of the International Bureau, Ministry of Finance, Japan*

Foreword

Sanjay Pradhan

Haiyan, the typhoon that struck the Philippines in November 2013, was thought to be the strongest tropical storm ever to have made landfall in human history. It has caused untold damage and suffering. Physical damages from the storm are estimated at $14.5 billion. What the numbers do not show, of course, is the devastation faced by people who have lost their homes, livelihoods, savings, and loved ones. Picking up the pieces is harder when you have little to begin with, and poor communities are often hardest hit and take the longest to recover from disaster.

If the world warms 4 degrees Celsius by century's end, as most scientists predict, the change will mean prolonged droughts and heat waves, intensified precipitation, and the death of coral reefs, nature's barrier against storm surges. Urbanization, too, has increased the poor's vulnerability to disasters, as migrants crowd into unregulated, unsafe housing. Over the past three decades natural disasters worldwide have caused close to $4 trillion in economic losses, much of that in the developing world. Given these trends, disasters of the magnitude of Haiyan can no longer be viewed as once-in-a-century events, but rather as *probabilities*. In the era of climate change and mass urbanization, they will continue to affect in a major way the developing world's long-term prosperity and safety. More than three-quarters of global fatalities from natural disasters occur in developing countries.

Evidence shows that mainstreaming disaster risk management (DRM) into policies, strategies, regulation, and building codes can save lives and assets when adverse natural events hit. While ex post initiatives, such as disaster response strategies, have been formulated in several regions and preparedness in some countries is more advanced than in others, the general level of ex ante initiatives through prevention, mitigation, and preparedness across countries is still low.

The world must shift from a tradition of *response* to a culture of prevention and resilience. While not all natural disasters can be avoided, their impact on a population can be mitigated through effective planning and preparedness. These are the lessons to be learned from Japan's own megadisaster: the Great East Japan Earthquake of 2011, the first disaster ever recorded that included an earthquake, a tsunami, a nuclear power plant accident, a power supply failure, and a large-scale disruption of supply chains—with global consequences for several industries.

Japan has an advanced DRM system that has evolved over nearly 2,000 years as the country has coped with natural risks and hazards. The loss of life and property during the Great East Japan Earthquake might have been much greater if the nation's policies and practices had been less effective. Following the disaster, these policies and practices were reviewed, and recommendations for improvement were proposed to make DRM even more effective.

The World Bank and the Government of Japan jointly created a set of searchable online Knowledge Notes to enable DRM practitioners and policy makers to learn from Japan's experience. This set of 36 Knowledge Notes, which highlight key lessons learned in seven DRM thematic clusters—structural measures; nonstructural measures; emergency response; reconstruction planning; hazard and risk information and decision making; the economics of disaster risk, risk management, and risk financing; and recovery and relocation—have been consolidated in this report, *Learning from Megadisasters*.

This report contains crucial information on DRM and lessons learned from Japan's terrible ordeal in 2011. Our hope is that this experience will help developing countries weather their own megadisasters.

Sanjay Pradhan is Vice President of Change, Knowledge and Learning at the World Bank Group.

Acknowledgments

This book represents the outcomes of the "Learning from Megadisasters" project of the Government of Japan and the World Bank Group. The manuscript was prepared by a team led by Federica Ranghieri (senior urban specialist, World Bank) with Mikio Ishiwatari (senior disaster risk management specialist, World Bank), under the guidance of Bruno Laporte (former director for knowledge and learning, World Bank Institute), Abha Joshi Ghani (director of knowledge and learning, World Bank Institute), Akihiko Nishio (director of operations, South Asia Region, World Bank), and Christine F. Kessides (manager for urban practice, World Bank Institute).

Co-coordination of the project was provided by Japan's Ministry of Finance with support and advice from other agencies. Contributing agencies of the Japanese government include the Cabinet Office; the Ministry of Internal Affairs and Communications; the Ministry of Land, Infrastructure, Transport and Tourism; and the Financial Service Agency. Other contributors include the Japan International Cooperation Agency, the Asian Disaster Reduction Center (ADRC), the International Recovery Platform, CTI Engineering, and prominent academic institutions. Several departments at the World Bank contributed to the work; namely, the Global Facility for Disaster Reduction and Recovery (GFDRR), the Social Development Department (SDV), the Tokyo Development Learning Center, the East Asia and Pacific Region, and the External and Corporate Relations (ECR) Vice-Presidency, all under the coordination of the World Bank Institute.

The authors of the individual chapters are, in alphabetical order and including several organizational authors, as follows: Bianca Adam (World Bank), Masaru Arakida (ADRC), Margaret Arnold (World Bank), Mitsuhiro Fukao (Keio University), Nobuaki Hamaguchi (Kobe University), Kenzo Hiroki (International Centre for Water Hazard and Risk Management), Akihiko Hokugo (Kobe University), Ai Ideta (Kyoto University), Makoto Ikeda (ADRC), the International Recovery Platform, Hideki Ishii (Fukushima University), Mikio Ishiwatari (World Bank), Toshiaki

Keicho (World Bank), Olivier Mahul (World Bank), Japan's Ministry of Internal Affairs and Communications, Satoru Mimura (Fukushima University), Shingo Nagamatsu (Kansai University), Tatuo Narafu (Japan International Cooperation Agency), Yusuke Noguchi (Kyoto University), Kenji Ohse (Fukushima University), Hiroshi Okumura (Kobe University), Makoto Okumura (Tohoku University), Takashi Onishi (University of Tokyo), Takahiro Ono (ADRC), Yasuaki Onoda (Tohoku University), Yukie Osa (Association for Aid and Relief), Brett Peary (Kyoto University), Junko Sagara (CTI Engineering), Keiko Saito (World Bank), Yoko Saito (Disaster Reduction and Human Renovation Institute), Shinichi Sakai (Kyoto University), Kazuko Sasaki (Kobe University), Daisuke Sato (Tohoku University), Motohiro Sato (Hitotsubashi University), Rajib Shaw (Kyoto University), Hironobu Shibuya (Save the Children, Japan), Yoshimitsu Shiozaki (Kobe University), Akira Takagi (Fukushima University), Yukiko Takeuchi (Kyoto University), Yasuo Tanaka (Kobe University), Masato Toyama (CTI Engineering), and Emily White (World Bank and Financial Service Agency).

The document also reflects technical inputs from the World Bank team, which included Abigail Baca (Disaster risk management [DRM] specialist, East Asia and Pacific [EAP]), Sofia Bettencourt (lead operations officer, Africa), Laura Elizabeth Boudreau (DRM analyst, Finance and Private Sector Development [FPS]), Wolfgang Fengler (lead economist, World Bank, Nairobi), Abhas K. Jha (lead urban specialist, EAP), Josef Leitmann (program manager, multidonor trust fund for Haiti), Markus Kostner (sector leader, Social, Environment, and Rural Development, EAP), Daniel Warner Kull (DRM specialist, GFDRR), Olivier Mahul (program coordinator, FPS), Robin Mearns (lead social development specialist, SDN), Niels Holm Nielsen (senior DRM specialist, Sustainable Development Department, Latin America and the Caribbean), Mr. Prashant (senior DRM specialist, GFDRR), Christoph Pusch (lead DRM specialist, GFDRR), Sahar Safaie (DRM specialist, Middle East and North Africa), and Satoru Ueda (lead water management specialist, Sustainable Development Department, Africa).

Among the advisers and reviewers who provided guidance and contribution at various stages of the work were Prof. Yoshiaki Kawata (Kansai University and chair of the project's advisory committee), Prof. Masahisa Fujita (Research Institute of Economy, Trade and Industry), Toshio Arima (Fuji Xerox), Prof. Mitsuhiro Fukao (Keio University), Prof. Fumio Imamura (Tohoku University), Yukimoto Ito (Sendai City), Prof. Toshitaka Katada (Gunma University), Hideaki Oda (United Nations Secretary-General's Advisory Board), Yukie Osa (Association for Aid and Relief, Japan), Prof. Rajb Shaw (Kyoto University), and Prof. Toru Takanarita (Sendai University). Advice and reviews were also received from Prof. Antonis Pomonis (Cambridge Architectural Research, Ltd., and principal investigator of the Global Earthquake Model's Earthquake Consequences Database Project); Prof. William Seimbeida (California Polytechnic State University and Kyoto University); Prof. Costanza Bonadonna, Prof. Chris Gregg, Dr. Franco Romerio, and Dr. Corine Frischknecht (CERG-C [Specialization Certificate for the Assessment and Management of Geological and Climate Related Risk], University of Geneva); Prof. Mehedi Ahmed Ansary (Bangladesh University of Engineering and Technology); Muralee Thummarukudy (United Nations Environment Programme); Karen Sudmeier-Rieux (Commission on Ecosystem Management of the International Union for Conservation of Nature and University of Lausanne); Prof. Reinhard Mechler (International Institute for Applied Systems Analysis); Prof.

Charles Scawthorn (SPA Risk LLC and Kyoto University); and the Earthquake Engineering Research Institute's (EERI) Global Technical Expert Review Group. The EERI team was chaired by Farzad Naeim and included Sergio M. Alcocer, Mohsen Ghafouri-Ashtiani, Jay Berger, Marcial Blondet, Lori Dengler, Marjorie Greene, Polat Gulkan, Kenzo Hiroki, Luxin Huang, Marshall Lew, Eduardo Miranda, Robert B. Olshansky, Mimi Sheller, Alpa Sheth, Gavin Smith, Emily So, Steven L. Stockton, and Balbir Verma.

Steven Kennedy edited the manuscript.

Abbreviations

ACA	Agency for Cultural Affairs
APEC	Asia-Pacific Economic Cooperation
ASEAN	Association of Southeast Asian Nations
B/C	benefit-cost ratio
BCM	business continuity management
BCP	business continuity plan
BOJ	Bank of Japan
BRI	Building Research Institute
BRR	Agency for the Rehabilitation and Reconstruction of Aceh and Nias
CAO	Cabinet Office (Japan)
Cat	catastrophe bonds
CBA	cost-benefit analysis
CBO	community-based organization
CEA	cost-effectiveness analysis
CFW	cash for work
CoP	Community of Practice
CSO	civil society organization
DMAT	Disaster Medical Assistance Team
DRM	disaster risk management
EEW	Earthquake Early Warning
EIRR	economic internal rate of return
EMT	Emergency Mapping Team
ESD	education for sustainable development
FDMA	Fire and Disaster Management Agency
FFW	food for work
FTTH	fiber to the home

FURE	Fukushima Future Center for Regional Revitalization
FY	fiscal year
Gal	galileo [unit of acceleration]
GCNJ	Global Compact Network Japan
GDLN	Global Development Learning Network
GDP	gross domestic product
GEJE	Great East Japan Earthquake
GFDRR	Global Facility for Disaster Reduction and Recovery
GDMS	Geospatial Disaster Management Mashup Service Study
GIS	geographic information system
GoJ	Government of Japan
GPEA	Government Policy Evaluations Act
GPS	global positioning system
GSI	Geospatial Information Authority of Japan
GW	gigawatt
ICOMOS	International Council on Monuments and Sites
ICT	information and communication technology
IFRC	International Federation of Red Cross and Red Crescent Societies
IMF	International Monetary Fund
IRP	International Recovery Platform
IT	information technology
JANIC	Japan NGO Center for International Cooperation
JER	Japanese Earthquake Reinsurance Company
JGB	Japanese Government Bond
JICA	Japan International Cooperation Agency
JMA	Japan Meteorological Agency
JRCS	Japanese Red Cross Society
JRI	Japan Research Institute
JSDF	Japan Self-Defense Forces
JSMCWM	Japan Society of Material Cycles and Waste Management
JWWA	Japan Water Works Association
LPG	liquefied petroleum gas
MAFF	Ministry of Agriculture, Forestry and Fisheries
MCA	multicriteria analysis
MCU	micro control unit
METI	Ministry of Economy, Trade and Industry
MIC	Ministry of Internal Affairs and Communications
MLIT	Ministry of Land, Infrastructure, Transport and Tourism
NAB	National Association of Commercial Broadcasters
NGO	nongovernmental organization
NHK	National Broadcasting Corporation (Japan)
NIED	National Research Institute for Earth Science and Disaster Prevention
NILIM	National Institute for Land and Infrastructure Management
NPO	nonprofit organization
NPV	net present value

OSM	Open Street Map
PCB	polychlorinated biphenyl
POP	persistent organic pollutant
ppm	parts per million
PPP	public-private partnership
RAHA	Refugee Affected and Hosting Areas programme
RC	reinforced concrete
RDC	Reconstruction Design Council
RIA	regulatory impact analysis
SME	small and medium enterprise
SPV	special-purpose vehicle
TEC-FORCE	Technical Emergency Control Force
TEPCO	Tokyo Electric Power Company
TSE	Tokyo Stock Exchange
UKG	Union of Kansai Governments
UNDP	United Nations Development Programme
UNEP	United Nations Environment Programme
UNICEF	United Nations Children's Fund
UN-WB	United Nations–World Bank
VAT	value added tax
VSATs	very small aperture terminals
WBI	World Bank Institute
WTTC	World Travel and Tourism Council

Note: All dollar amounts are U.S. dollars (US$) unless otherwise indicated.

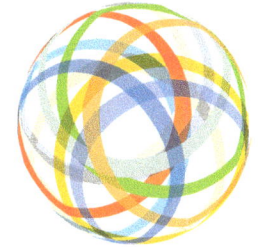

Lessons from the Great East Japan Earthquake

On March 11, 2011, an earthquake of magnitude 9.0 occurred in the Pacific Ocean off the coast of Japan's Tohoku region. The quake shook the ground as far away as western Japan and lasted for several minutes. A half-hour later, a tsunami of unprecedented force broke over 650 kilometers (km) of coastline (map O.1), toppling sea walls and other defenses, flooding more than 500 square kilometers (km^2) of land, and washing away entire towns and villages.

The devastation left some 20,000 people dead or missing, with most of the deaths caused by drowning (table O.1). The tsunami leveled 130,000 houses and severely damaged 270,000 more. About 270 railway lines ceased operation immediately following the disaster, and 15 expressways, 69 national highways, and 638 prefectural and municipal roads were closed. Some 24,000 hectares of agricultural land were flooded. The areas worst hit were the Fukushima, Iwate, and Miyagi prefectures.

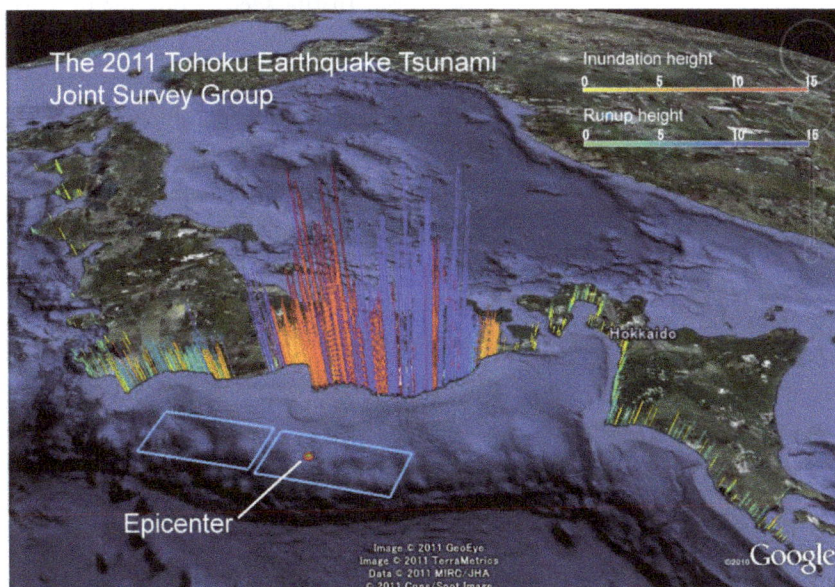

Map O.1 The tsunami struck a wide area of Japan

Source: The 2011 Tohoku Earthquake Tsunami Joint Survey Group, http://www.coastal.jp/tsunami2011/index.php.

Table O.1 The Great East Japan Earthquake of 2011 in figures

CASUALTIES AS OF NOVEMBER 8, 2013	
Dead	18,571
Missing	2,651
Injured	6,150
BUILDING DAMAGE AS OF NOVEMBER 8, 2013	
Total collapse	126,602
Half collapse	272,426
Partial damage	743,089
EVACUEES	
Maximum	470,000 (March 14, 2011)
Current	282,111 (October 10, 2013)
Estimated economic damage	¥16.9 trillion ($210 billion)
Buildings	¥10.4 trillion
Public utilities	¥1.3 trillion
Social infrastructure	¥2.2 trillion
Other (agriculture, forests, fisheries)	¥3.0 trillion
Debris	26.7 million tons (October 2013)

Source: Cabinet Office and Reconstruction Agency.

WHAT THE DISASTER TAUGHT JAPAN—AND WHAT IT CAN TEACH OTHER COUNTRIES

The Great East Japan Earthquake (GEJE) was the first disaster ever recorded that included an earthquake, a tsunami, a nuclear power plant accident, a power supply failure, and a large-scale disruption of supply chains.

Learning from Megadisasters, a knowledge-sharing project sponsored by the World Bank and the government of Japan, is collecting and analyzing information, data, and evaluations performed by academic and research institutions, nongovernmental organizations (NGOs), government agencies, and the private sector—all with the objective of sharing Japan's knowledge on disaster risk management (DRM) and postdisaster reconstruction with countries vulnerable to disasters. The Bank and the Japanese government hope that these findings (see figure O.1) will encourage countries to mainstream DRM in their development policies and planning.

Japan had not foreseen an event of this magnitude and complexity:

- *It was a high-impact event with a low probability of occurrence.* Because of enormous damage from the tsunami and moderate but widespread geotechnical damage, the GEJE event was the costliest earthquake in world history. Japan's Cabinet Office has estimated the direct economic cost at ¥16.9 trillion, or $210 billion.

- *It was a highly complex phenomenon, the effects of which cascaded to sensitive facilities.* The earthquake and ensuing tsunami provoked fires at damaged oil refineries and a potentially catastrophic nuclear accident. The effects of the accident at the Fukushima Daiichi nuclear power station have compromised Japan's energy supply, imperiled its environment, and threatened public health.

- *Direct damage to major Japanese industries rocketed through supply chains around the world.* In the second quarter of 2011, Japan's gross domestic product (GDP) dipped 2.1 percent from the previous year, while industrial production and exports dropped even more sharply—by 7.0 percent and 8.0 percent, respectively. Japan experienced a trade deficit for the first time in 31 years. In the wake of the tsunami, businesses that relied on Japanese electronics and automotive parts faced disruptions and delays in production, distribution, and transportation; they had to scramble to find alternate supply lines and manufacturing partners.

In coping with the GEJE, Japan's advanced DRM system, built up during nearly 2,000 years of coping with natural risks and hazards, proved its worth. The loss of life and property could have been far greater if the country's policies and practices had been less effective. The main elements of that DRM system are

- Investments in structural measures (such as reinforced buildings and seawalls), cutting-edge risk assessments, early-warning systems, and hazard mapping—all supported by sophisticated technology for data collection, simulation, information, and communication, and by scenario building to assess risks and to plan responses (such as evacuations) to hazards

- A culture of preparedness, where training and evacuation drills are systematically practiced at the local and community levels and in schools and workplaces

- Stakeholder involvement, where the national and local government, communities, NGOs, and the private sector all know their role

- Effective legislation, regulation, and enforcement—for example, of building codes that have been kept current

- The use of sophisticated instrumentation to underpin planning and assessment operations.

Certain improvements would have made the Japanese reaction even more effective. Three are particularly important and are singled out here (as well as being included in the section on lessons learned that appears further on):

- *Spreading a better understanding of the nature and limitations of risk assessment* among local authorities and the population at large would improve collective and individual decision making, especially in emergencies. Communication about the unfolding disaster could and should have been more interactive among local communities, governments, and experts. Distributing hazard maps and issuing early warnings were not enough. The magnitude of the tsunami was underestimated, which may have led people to delay their evacuation, if only for a fatal few minutes. If local governments and community members had been more aware of DRM technologies and their margins of error, fewer lives might have been lost.

- *Coordination mechanisms on the ground should be agreed on before the fact.* During the GEJE, coordination among various groups, such as governments (national, prefectural, and local), civil society organizations (CSOs), and private entities was often poor—or at least not optimal. Local governments, whose facilities in some cases were wiped out by the disaster, had little experience working with other organizations on a large scale, and they received insufficient

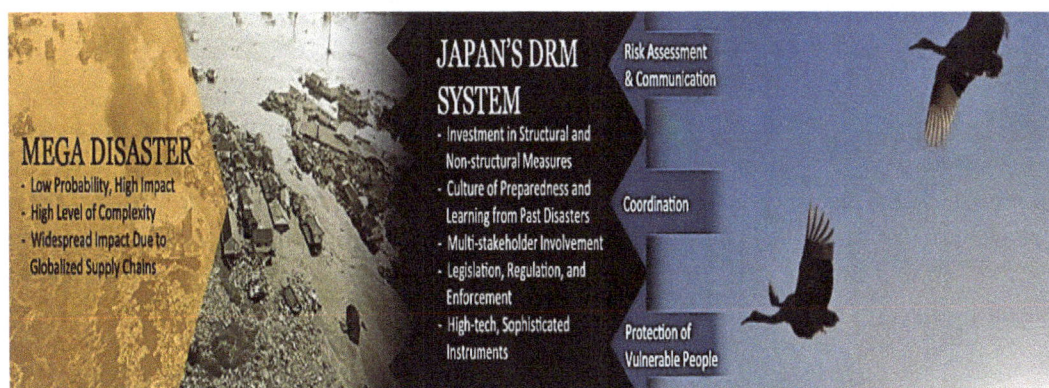

Figure O.1
Summary of findings and lessons learned from the project

support from the central government in managing the new forms of cooperation. As it turned out, coordination with international relief agencies and donors offering exceptional assistance was simply not up to the unprecedented task.

- *Vulnerable groups must be not only protected but also engaged.* Understanding and meeting the challenges of the elderly, children, and women, both during the emergency and in its aftermath, are priorities for effective postdisaster response. Culturally sound solutions that take account of special needs among segments of the population should be planned in advance to enhance resilience and facilitate recovery and reconstruction.

SHARING EXPERIENCES WITH DEVELOPING COUNTRIES

Other countries can protect themselves from major disasters by adopting—and adapting as necessary—some of the measures taken by Japan, and by understanding the strengths and weaknesses of Japan's response to the GEJE. To help them do that, the Learning from Megadisasters initiative provided data, analysis, and insight in printed and Web-based formats (including e-learning), in face-to-face activities, in seminars presented through the good offices of the Global Development Learning Network (GDLN),[1] and through a dedicated community of practice— all designed to build the capacities of government decision makers and other stakeholders in developing countries. A searchable set of online materials at various levels of depth and detail serves as a focal point for this community of learning and practice on DRM. The knowledge base will grow as practitioners from around the world contribute their insights and expertise.

The project delivered a set of 36 "Knowledge Notes" (now chapters) grouped into seven thematic clusters (now parts):

- Structural measures

- Nonstructural measures

- Emergency response

- Reconstruction planning

- Hazard and risk information and decision making

- The economics of disaster risk, risk management, and risk financing

- Recovery and relocation

The notes analyze the response to the March 11, 2011, earthquake and tsunami—and synthesize what worked, what did not, and why, offering recommendations for developing countries that face similar risks and vulnerabilities.

The notes were prepared by more than 40 Japanese and international experts, assisted by 50 advisers and reviewers. The team included developing country practitioners, academic experts, and government officials. The chapters provide a basis for knowledge sharing and exchanges with developing country experts and practitioners.

Key lessons derived from the 36 notes are offered in the four pages that follow, after which the thematic clusters are reviewed in turn.

KEY LESSONS LEARNED FROM THE PROJECT

The successes of Japan's DRM system, as well as the ways in which that system could be improved, are reflected in the lessons drawn from the GEJE and presented in the initial reports from the Learning from Megadisasters project.

Extreme disasters underscore the need for a holistic approach to DRM

Single-sector development planning cannot address the complexity of problems posed by natural hazards, let alone megadisasters, nor can such planning build resilience to threats. Faced with complex risks, Japan chose to build resilience by investing in preventive structural and nonstructural measures; nurturing a strong culture of knowledge and learning from past disasters; engaging in wise DRM regulation, legislation, and enforcement; and promoting cooperation among multiple stakeholders, between government agencies and ministries, between the private sector and the government, and across multiple levels of governance, from local to national to international.

Today, Japan is placing even heavier emphasis on recognizing and respecting complexity and residual risk, designing and managing systems that "fail gracefully"—that is, that mitigate damage to the greatest extent possible before succumbing to overwhelming force. The essence of the approach is to design and maintain resilient infrastructure capable of absorbing damage from natural disasters to some extent, even when an event exceeds all feasible and affordable measures. In the wake of the GEJE, Japan also recognized that additional efforts were required to plan and design measures capable of countering events of low probability but high impact.

Preventive investments pay, but be prepared for the unexpected

Japan's extensive structural precautions were very effective in protecting buildings and people from the earthquake. Although 190 km of the 300 km of dikes in the area collapsed, those dikes decreased the force of the tsunami and, in some areas, delayed its arrival inland. All bullet trains stopped safely without casualty, thanks to a cutting-edge system of detecting the earliest sign of ground movement. The GEJE, however, exceeded all expectations and predictions in the extent of its ensuing tsunami, demonstrating that exclusive reliance on structural measures will ultimately prove ineffective and must be supplemented with nonstructural measures and a basic understanding of the uncertainties surrounding the estimation of events such as earthquakes and tsunamis.

Because it is not practical—from a financial, environmental, or social perspective—to build tsunami dikes 20–30 meters high, Japan's government intends to accelerate the current paradigm shift in its thinking about disaster management, complementing its structure-focused approach to prevention with soft solutions to achieve an integrated approach to disaster risk reduction. Understanding that the risks from natural hazards can never be completely eliminated, the new, balanced approach incorporates community-based prevention and evacuation and other nonstructural measures such as education, risk-related finance and insurance, and land-use regulation.

Learning from disaster is key, as Japan has shown for the past 2,000 years

Japan has used the lessons of past disasters to improve its policies, laws, regulations, investment patterns, and decision-making processes, as well as community and individual behaviors. Investing in preparedness and a strong culture of prevention made all the difference in the Tohoku region when the GEJE struck. The Meiji-Sanriku Tsunami of 1896 killed 40 percent of the population in the affected zone, whereas the GEJE claimed 4 percent.[2] Evacuation drills and DRM education, staples of the country's schools, kept children safe in Kamaishi City. The famous "Kamaishi Miracle" was not really a miracle at all, but rather the result of a sustained effort to instill a culture of resilience and prevention based on continuous learning.

DRM is everyone's business

Japan's disaster management system addresses all phases of disaster prevention, mitigation, and preparedness; emergency response; as well as recovery and rehabilitation. It specifies the roles and responsibilities of national and local government and enlists the cooperation of relevant stakeholders in both the public and private sectors. This comprehensive approach secured a quick and effective mobilization of forces at multiple levels after the 2011 tsunami struck, while also revealing certain problems of coordination that are discussed further on. Since the tsunami, the capacity of local DRM planning systems to prepare for and react to large-scale disasters has been assessed, and revisions have been proposed through new legislation.

Japan's central government plays a leading role in mitigating the risks of disaster across the country, but local governments have the principal responsibility for managing the country's DRM systems. The central government encourages local governments to promote structural measures by providing financial support, producing technical guidelines and manuals, and conducting training for technical staff in planning, design, operation, and maintenance.

Japan's tradition of community participation in preparedness was a key factor in minimizing the number of lives lost to the GEJE. Community-based DRM activities are well integrated into the daily lives of most Japanese, ensuring that awareness of natural hazards is never far from their mind. The national and local governments formally recognize and support the involvement of the community in DRM through laws and regulations that define roles and commitments, through linkages with local institutions (such as *jichikai,* or neighborhood associations), and through participation in meetings at which decisions are made.

Although dikes and communication systems suffered partial failures and forecasting systems underestimated the height of the tsunami, local communities and their volunteer organizations were front and center in responding to the disaster. The GEJE showed that each community needs to explore and identify its best defense, mixing various soft and hard measures, policies, investments, education initiatives, and drills, through sound analysis and stakeholder consultations.

The role of the community goes far beyond evacuation, especially in multihazard DRM (figure O.2). Successful evacuations depend on prior measures such as hazard mapping, warning systems, and ongoing education, all of which proved essential in the evacuation that followed the GEJE. During the GEJE, local governments and communities in affected areas served as first responders, managed evacuation centers, and promptly began post-disaster reconstruction. Partnerships with the private sector were also critical. Rehabilitation could begin the day after the earthquake because agreements with the private sector were already in place. Quick payment of insurance claims allowed individuals and businesses to contribute fully to the rehabilitation effort.

Figure O.2 The many roles of the community in multihazard DRM

Assessing risks and communicating them clearly and widely helps citizens make timely decisions to protect themselves

Accurate risk assessment and interactive communication systems that connect local communities, government agencies, and experts make people less vulnerable and more resilient. But although risk assessments and DRM technologies (including prediction systems) can add enormous value, governments and community members should be aware of their limitations and never stick to a single scenario.

Hazard maps can give the public a false sense of safety, if not properly communicated

Although hazard maps showing risk areas and evacuation shelters had been distributed before the disaster to households in the tsunami-stricken areas, only 20 percent of the people had seen them. Still, 57 percent (which is a relatively high number by international standards) left immediately after the earthquake tremors. In some areas, the tsunami of 2011 proved far greater than indicated on the hazard maps. Warnings that underestimated the size of the earthquake and tsunami may have caused people to delay their evacuation, prolonging their exposure to danger. Because the magnitude of the GEJE and tsunami far exceeded the predisaster estimates, the Japanese government has been revising its methods of assessing earthquake and tsunami hazards, combining historical evidence, topographical and geological studies, and predictions and forecasts based on scenarios for events of low probability but high impact. Manufacturers and other companies are rethinking their strategies for business continuity. Many Japanese companies are already investing in redundancy and diversification within their supply chain, despite the expense of such measures.

Better management of information and communication is crucial in emergencies and recovery operations

The GEJE points to two common information problems: (1) the lack of real-time information on conditions and on coordination among parties (that is, on who is doing what); and (2) the loss of critical public records vital to reconstruction. With regard to the first point, during the GEJE the national government collected information from municipal governments, while additional information was crowd-sourced and channeled through social media and the Internet. On the second point, even though some local governments lacked a formal backup system, data on land ownership were restored fairly quickly, thanks to other official and private backups. Nevertheless, health records in some cities were destroyed, and new policies to avoid a recurrence are needed.

Many postdisaster situations are made worse by the lack of a communications strategy that makes use of appropriate media to deliver critical messages. Good information enables individuals and communities not only to stay safe, but also to contribute more effectively to relief and recovery. It also ensures that citizens have a realistic set of expectations about relief and reconstruction. If communication is to help people stay safe and minimize the disruption to their lives, those people must be able to trust the information and its source. During the GEJE, communication about evacuation, temporary shelters, and emergency food distribution was handled fairly well, but confusion about the scope and extent of the nuclear accident led to public dissatisfaction, as noted in a report from Japan's Nuclear and Industrial Safety Agency.

Coordination mechanisms must be developed and tested in normal times, so that they are ready for use in an emergency

Although the national government established the rescue headquarters very quickly, and interprefectural emergency and rescue units and technical forces were deployed in record time, mechanisms for formal coordination among the various stakeholders (government agencies at all levels, CSOs, and private entities) were inadequate. The GEJE drew an unprecedented level of assistance from 163 countries and 43 international organizations. In all, Japan received $720 million from other countries, almost half of all humanitarian disaster funding dispensed around the globe in 2011. The weakness of coordination observed on the ground during the GEJE demonstrates that coordination mechanisms should be established through advance agreements and clear definitions of responsibility.

Vulnerable groups must be protected— and engaged

Culturally appropriate services and social safety nets for vulnerable groups are needed in times of emergency and during reconstruction. They should be planned in advance. Two-thirds of the deaths during the GEJE occurred among people over the age of 60, who accounted for just 30 percent of the population in the affected areas. At evacuation centers, the needs of women and the disabled were not fully met. New measures are under consideration to assure privacy and security for women, maternal care and gender-balanced policies, and better nursing care for the disabled at evacuation centers. These measures call for empowering marginalized groups for long-term recovery and including a gender perspective in planning and managing shelters, which will require women to be more deeply involved in shelter management. Women should be encouraged to participate in DRM committees, center management, and risk assessment. National and local DRM policies and strategies should be reviewed from a gender perspective.

DETAILED FINDINGS AND RECOMMENDATIONS

The chapters that make up the main body of this report were built around the disciplines employed in the traditional DRM cycle. Grouped into seven thematic clusters that track that cycle, the chapters treat structural measures (part 1) and nonstructural measures (part 2) as preventive options. They also cover the emergency responses put in place after March 11 (part 3) and describe the planning behind the reconstruction process (part 4). The handling of risk assessment and communication before and after the disaster are the subject of part 5. Part 6 deals with risk financing, insurance, and fiscal and financial management; part 7 with the progress of recovery and relocation.

This section of the Overview provides the reader with additional information and details about the main findings of the project and the lessons learned from it, following the scheme of thematic parts used in the chapters. Those chapters may be downloaded from http://wbi .worldbank.org/wbi/megadisasters.

Part 1: Structural measures

Dikes are both necessary and effective in preventing ordinary tsunamis, which are relatively frequent, but they are of limited use against the extreme events that occur less frequently. Japan's Tohoku region built 300 km of coastal defense over the course of 50 years. National and local governments invested a total of $10 billion to build coastal structures and breakwaters in major ports. During the GEJE, the defensive structures along the coast suffered unprecedented damage: 190 of the 300 km of coastal structures collapsed under the tsunami (figure O.3). In some areas those structures did serve to delay the arrival of the waves, buying extra minutes for people to evacuate. Because many tsunami gates designed to reduce

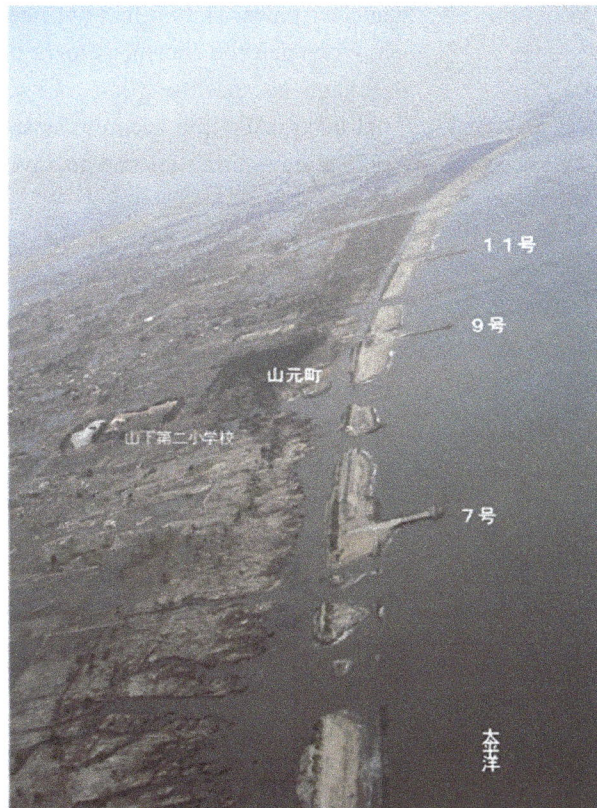

Figure O.3 Dikes in Sendai before and after the tsunami of March 11, 2011

Source: Ministry of Land, Infrastructure, Transport and Tourism, 2011.

flooding along rivers were toppled, the government of Japan launched a structural assessment to better understand the causes of failure. The assessment concluded that construction standards and stability performance under worst-case scenarios should be further investigated. Structures should be able to withstand waves that exceed their design height, reducing the force of the water before they collapse and thereby mitigating damages.

Reinforced infrastructure and buildings erected according to current codes were not seriously damaged. Thanks to Japan's strict and rigorously enforced building codes, earthquake-related losses from the March 2011 disaster were limited, with most of the deaths and economic damage being caused by the ensuing tsunami. Since Japan's first building code was adopted after the Great Kanto

Earthquake of 1923, the government has made regular revisions in light of experiences with a range of natural disasters. During the GEJE, most damage to buildings was caused by phenomena other than the earthquake itself. Liquefaction occurred on building lots that had not been treated against it and in reclaimed lands and on riverbanks, damaging small buildings that lacked pile foundations.

Tsunami damage to crucial facilities, including the Fukushima Daiichi nuclear power station, had cascading effects in several sectors, such as power and energy, petroleum refining, steel production, the automobile industry, fishing, health and medicine, farming, and telecommunications. Critical facilities should be built in safe locations and secured by the most sophisticated disaster management plans. The sea wall protecting the Fukushima Daiichi nuclear power station had not been designed to withstand the enormous force of the GEJE tsunami, because the worst-case scenario had not been taken into account, as stated by the official committee formed to investigate the accident.

The Interim Report of the Government Investigation Committee on the Accident at the Fukushima Nuclear Power Station identified three main causes of failure: (1) DRM plans were focused on earthquakes and not tsunamis; (2) complex scenarios with multiple hazards consisting of earthquakes and tsunamis, compounded by simultaneous transport and communication failures, had not been foreseen; and (3) the complex systems at the nuclear power station had not been managed in an integrated way. The generally accepted myth that nuclear power stations are "safe" had led to an underestimation of certain important risks. The analysis has prompted a reevaluation of risk assessment methods and DRM planning and countermeasures. That reevaluation is likely to shape future policies and procedures.

A multilayered approach to DRM is needed, employing both structural and nonstructural measures. Defensive infrastructure alone is not enough to cope with infrequent disasters of high impact. Nonstructural measures also need to be established, including early-warning systems, rigorous planning and regulation, prompt evacuation of residents, and a variety of institutional and financial measures—among them insurance, rehabilitation funds, and emergency teams.

Part 2: Nonstructural measures

Japan has had a disaster management system in place since the Disaster Relief Act of 1947 and has long used disasters as opportunities to continuously improve that system. The initial emphasis was on disaster response, later complemented by prevention, mitigation, and preparedness; emergency response and recovery; and rehabilitation and rebuilding.

Over the years, the country's investments in disaster preparedness have been wide ranging, covering seismic and tsunami detection, early-warning systems, multichannel systems for disseminating warnings, hazard mapping, evacuation planning (routes and shelters), regular disaster training and drills in schools and at workplaces, and improved signage. Municipal governments have the main responsibility for disaster management, including formulating and implementing local disaster management plans based on the national plan, establishing community-based organizations, distributing hazard maps to the public, raising public awareness, and developing evacuation procedures.

Early warnings and communication

The risk of underestimating a disaster's impact can be extremely costly. The warnings issued on March 11 underestimated the tsunami's height and likely caused people to delay their evacuation. Warning systems were effective in mitigating damage, but experience showed that they have to be better aligned with the communities' evacuation procedures. More than half of the fleeing population evacuated by vehicle, and a third of them got stuck in traffic jams before reaching emergency evacuation shelters. Many people and their vehicles were swept away by the tsunami. Although the general rule is to evacuate on foot, vehicles are also needed, particularly to move the elderly and disabled. New measures to facilitate evacuation by vehicle—for example, rules to mitigate traffic jams and training for drivers on evacuation during disasters—should be considered.

The early earthquake detection system saved thousands of passengers in the Shinkansen. Nineteen bullet trains (Shinkansen) were running when the GEJE occurred, including two at 270 km per hour, almost top speed. All were able to stop safely thanks to early earthquake detection systems. The Japan Meteorological Agency issues earthquake information based on nationwide seismography and observations of seismic intensity. The agency operates an earthquake early-warning system that quickly estimates an earthquake's focus and magnitude and forecasts seismic intensities and the arrival time of ground shaking.

How communities and the private sector saved lives and assets

Community-based organizations saved lives and need to be nurtured. When the tsunami overwhelmed coastal defenses, local communities were forced to use their own knowledge and resourcefulness to survive on March 11. Fortunately, throughout the Tohoku region, communities had been intently engaged in tsunami preparedness. Given the unreliability of predictions and the limitations of defensive structures, community engagement should be put at the center of the disaster-response system.

The "Kamaishi Miracle" was not a miracle at all. Evacuation drills and DRM education are fixtures in Japan's schools. In Kamaishi City, where the tsunami claimed 1,000 members of the population of 40,000, the casualty rate among school children was low: only 5 out of 2,900 primary and junior high school students lost their lives, a rate 20 times lower than for the general public. Regular practice drills, education in the schools, and hazard maps are the keys to preparedness. DRM education saves the lives of children and other members of the community.

Well-prepared business continuity plans prevent disruptions. A business continuity plan (BCP) identifies an organization's critical operations and the potential effects of a disaster, specifying the response and recovery measures the business can take to avoid or minimize disruptions and continue operations at an acceptable level. The GEJE caused 656 private companies to go bankrupt within a year. Fully 88 percent of those businesses were located outside the Tohoku region and failed because of supply-chain problems. A BCP is essential regardless of where a business is based. According to a recent survey, between 80 and 90 percent of medium-sized and large companies indicated that their BCPs had been effective during the response and recovery phase.

Relocation and new regulations

Land-use regulations, including those that relocate houses to higher ground, are successful but sometimes difficult to implement. For that reason, alternative measures need to be considered. Relocation deeply affects the livelihoods and daily lives of many people. Houses that had been relocated after the previous tsunami to hit Yoshihama Village were not affected by the GEJE. But in the coastal village of Taro, identifying suitable relocation sites proved problematic, since its economic activities were situated on the coast. The case of Touni-hongo perhaps best illustrates the benefits of relocation and the challenges of land-use regulation. Houses that had been relocated to higher ground after an earlier tsunami were unharmed by the GEJE tsunami, whereas newly constructed houses in the unregulated lowlands were hard hit. These examples highlight the importance of alternative measures when relocation is not a realistic option—measures such as disaster-preparedness education, evacuation drills, accessible evacuation routes, and appropriately designed structures.

Japan's Basic Disaster Management Plan, as revised after the GEJE, aims to rigorously enforce earthquake and tsunami countermeasures. Addressing a new set of scenarios that take into account the largest possible disaster and multiple simultaneous hazards, the plan calls for the development of disaster-resilient communities, the promotion of disaster awareness, increased research and scientific observation, and stronger systems to warn of tsunamis and deliver evacuation information.

Part 3: Emergency response

Prompt rehabilitation of infrastructure

The Ministry of Land, Infrastructure, Transport and Tourism (MLIT) set up its emergency headquarters at 15:15 (about 30 minutes after the quake). Thanks to the dedicated service of well-trained and experienced government staff, prior agreements with the private sector, and advance financial arrangements, the roads leading to towns on the affected coast were cleared in less than a week. Also, by March 15, all 14 ports were either entirely or partially usable and began accepting vessels delivering emergency supplies and fuel. By April 29, the entire Tohoku Shinkansen line was in operation, as were most of the other railways except for those along the coast. Water supply services were resumed for about 90 percent of residents within a month, while electric power was 90 percent restored within a week.

Governance in time of emergency

The GEJE revealed institutional and legislative features of Japan's governmental system that enabled it to take speedy action toward recovery in coordination with various agencies. In many developing countries, rapid recoveries are more challenging owing to shortages of dedicated agencies and highly skilled and experienced staff. Despite Japan's strengths,

Figure O.4 Otsuchi's mayor was in front of town hall when the tsunami struck

Source: © Mikio Ishiwatari (April 2011). Used with permission. Further permission required for reuse.

local governments in areas hit by the GEJE tsunami have faced difficulties in responding to the disaster. The GEJE affected 62 municipalities in six prefectures in northeastern Japan. Among them, 28 municipalities in the three worst-affected prefectures (Iwate, Miyagi, and Fukushima) suffered serious damage to their office facilities. Computer servers in some of these municipalities were seriously damaged or destroyed, resulting in a loss of data essential for the provision of municipal services. To make matters worse, many municipalities lost their public officials: 221 officials died (see figure O.4) or remain missing from 17 municipalities in the three hardest-hit prefectures.

Fukushima's case was unique. Nine municipalities near the crippled Fukushima Daiichi nuclear power station had to relocate their offices relatively far from the plant (but mostly within the same prefecture) because of concerns about radiation levels in their jurisdictions, even where the physical damage from the earthquake and the tsunami were very limited.

Many prefectures and municipalities outside Tohoku took the initiative to quickly send their own public officials to help the localities deal with postdisaster relief activities and other emergency operations. About 79,000 local government officials were dispatched from all over Japan to the affected prefectures and municipalities until the end of 2011. A year later, many of them were still serving there in capacities ranging from civil engineering and urban planning to social work and finance.

Partnerships to facilitate emergency operations

Twinning arrangements between localities in disaster-affected areas and their counterparts in unaffected areas proved to be effective in dealing with the emergencies. Some of these arrangements were based on formal agreements, while others were based on goodwill. Where local governments are concerned, it is advisable to formalize such mechanisms before disasters strike, obtaining the necessary legal backing and clarifying cost-sharing arrangements. In a large-scale disaster, this kind of counterpart system—in which an unaffected local government provides support to another local government that has been affected by the disaster—allows support and assistance to be provided to all affected areas. For obvious reasons, it is essential that the linked prefectures and municipalities be geographically distant. Support agreements with localities in the same region may not be effective, particularly in a large-scale disaster like the GEJE that affected almost an entire region.

Coordination among government, CSOs, and other stakeholders to deal with the emergency on the ground was an overwhelming challenge. Expert teams, CSOs, volunteers, and military forces from around the world mobilized to help those of Japan, with 163 countries, 43 international organizations, and countless CSOs offering aid and relief. Foreign assistance far exceeded that provided in the wake of the Kobe earthquake in 1995. Considering the difficulties faced by local governments after the GEJE, coordination mechanisms should be established in the central government, or under an umbrella organization.

The system for delivery of relief goods encountered several problems, but measures have been identified to address them. The main problems in the delivery of relief goods were fuel shortages, interruption of telecommunication services, and mismatches between supply and demand that caused goods to be stockpiled in prefectural and municipal depots instead of being delivered promptly to people in need. Several measures can be taken to address these issues, including prior surveys of depots, advance estimation of the quantities of emergency goods that will be required, guidelines on relief goods that are not likely to be culturally acceptable, support from professional logistics specialists, and logistics management support from local governments in unaffected areas.

Evacuation centers and temporary housing

At the peak of the relief effort, more than 470,000 people were housed in evacuation centers. After the disaster struck, nearly 2,500 evacuation facilities were established in the Tohoku region, with additional shelters located outside Tohoku. Most facilities, such as schools and community centers, were publicly owned and had already been designated as evacuation centers. After the GEJE, however, private facilities, such as hotels and temples, were enlisted, because the need for centers far exceeded expectations. Many evacuees stayed with relatives or friends. As construction of temporary housing progressed, evacuees gradually moved out of the centers. Four months after the disaster, about 75 percent of the evacuation facilities had closed, although some in Tohoku stayed open as long as nine months. Because a megadisaster is likely to interrupt essential services such as water and power, it is critical to install alternatives such as portable toilets and power generators. Sendai City plans to equip its shelters with solar panels and other renewable energy options for backup power.

In Fukushima, many had to relocate from one evacuation center to another as the government expanded the mandatory evacuation zone. Some 82 percent of evacuees changed centers at least three times, and one-third changed more than five times. People in Fukushima have continued to migrate to other areas in and out the prefecture. At the end of 2011, more than 150,000 people had been evacuated, at least 60,000 of whom relocated to other prefectures across the country.

At many centers, a self-governing body emerged, with leaders and members of various committees selected by the evacuees themselves. Although managing evacuation centers is a municipal responsibility, most municipalities in the disaster-affected areas suffered staff losses, seriously weakening their capacity to cope with the emergency. At the beginning, most centers were supported by local teachers, volunteers, and other civil society groups. As the evacuation period lengthened, evacuees themselves started taking initiatives to manage their communities.

One of the problems cited at many shelters was lack of gender sensitivity. There was not enough privacy for anyone, but particularly not for women, many of whom did not have private spaces where they could change their clothes or breast-feed their babies. Many shelters eventually installed partitions, but these improvements often were late in coming. It has also been reported that relief goods delivered to the shelters were biased in favor of male evacuees. The lack of gender sensitivity has been attributed to the fact that men were largely responsible for managing the shelters, whether in facilities owned by municipalities or those managed by the evacuees themselves. In Japan, the overwhelming majority of the leaders of community organizations are male.

The special needs of vulnerable groups— including the elderly, children, and the disabled—need to be included in transition-shelter initiatives. The disabled often were not provided with proper care at evacuation

shelters. The earthquake and tsunami left children feeling frightened, confused, and insecure. Following the GEJE, the number of incoming calls to *Childline,* a free counseling service for children, increased fourfold in Fukushima, Miyagi, and Iwate prefectures. The government plans to send some 1,300 mental health counselors to public schools in the affected areas. But the experience points to the importance of bringing in professional staff to care for the disabled and vulnerable.

Japan has learned many lessons about temporary housing from past experience with disaster recovery. In Kobe, for example, large tracts of temporary housing were built too far from the city center. The housing was allocated through a lottery system that created more hardship for those residents (especially the elderly) who wound up far from their old neighborhoods and suffered from the loss of community. The housing should be easily accessible, and complementary care services should be provided. Community-based organizations (such as Japan's *jichikai*) can help community members cope with the stresses of extended stays in transition shelters.

New crowd-sourced information and the use of social media and FM radio

Social media were extensively used for searches, rescues, and fundraising. Social media are Web-based applications that use the Internet to connect people (prominent examples are Twitter and Facebook) as well as websites and computer applications that enable users to collaborate and create content, such as Wikipedia and YouTube. Emergency FM radio also played a crucial role in the aftermath of the GEJE (figure O.5). When the emergency communication systems in many cities broke down because of power failures and lack of emergency backup power, community radio stations were able to send useful information out to residents. In fact, about 20 emergency broadcasting stations dedicated to disseminating disaster information were set up in the Tohoku

Figure O.5. Broadcasting at RINGO Radio

Source: © Kyoto University. Used with permission. Further permission required for reuse.

area. In the immediate aftermath of the disaster, these community radio stations began to provide information about times and locations for the distribution of emergency food, water, and goods. In the following months they gradually shifted to providing other information to help victims in their daily lives or to raise the spirits of people in local communities. Radio was particularly appreciated by the elderly, who were less likely than younger people to have access to Internet information.

With the relatively high levels of mobile-phone penetration in developing countries, social media could be very useful during disasters, at least to the extent that they are already used in normal times. They can also serve to link up with communities outside the stricken areas to facilitate the acquisition and allocation of aid and assistance. In many developing countries, lack of physical accessibility to disaster-affected sites is a key issue. Mobile networks and social media can be used to collect and share localized information to improve access. Reliability or trustworthiness of information is an extremely important factor in the use of social media. Local governments and relevant national government agencies should, therefore, consider using social media in their public relations activities during normal times. When disasters occur, those channels can be used to share disaster-related information with the public.

Part 4: Reconstruction planning

A new law for reconstruction

Based on the recommendations of Japan's Reconstruction Design Council, the national government issued the Basic Act for Reconstruction and the Basic Guidelines for Reconstruction. The Reconstruction Agency, which the prime minister heads, was established under the oversight of the cabinet to promote and coordinate reconstruction policies and measures in an integrated manner. At the prefectural level, the three disaster-affected prefectures developed their own recovery plans.

At the municipal level, most of the disaster-affected municipalities developed recovery plans based on the pertinent policies of the national and prefectural governments. Municipalities have focused on land-use planning to build more resilient communities, including relocation, reconstruction projects, and consensus building among residents on relocation and reconstruction plans. Reports on some outcomes of these planning efforts are offered in cluster 7.

Special reconstruction zones will be identified based on proposals by local governments in the disaster-affected areas, where concessions and incentives (regulatory, fiscal, budgetary, and financial) will be granted to companies that set up new facilities.

Hastening recovery and reconstruction through cooperation between communities and local and national governments

Communities should be involved from the outset in planning reconstruction. In the areas affected by the GEJE, consultations between governments and communities were the rule, and community representatives were invited to serve alongside experts on recovery planning committees from the earliest stages. The most common ways of collecting residents' opinions were surveys and workshops. The central government and local governments outside the disaster-affected area helped affected municipalities plan their recovery by conducting research, seconding staff, and hiring professionals to provide technical support. University faculty members, architects, engineers, lawyers, and members of NGOs participated in the municipal planning process.

Debris and waste management

There was an urgent need to dispose of 20 million tons of debris left behind by the GEJE and tsunami, some of it contaminated by radioactivity. The debris was an enormous obstacle to rescue, and it still impedes reconstruction. The amount of tsunami-related debris in Iwate was

11 times greater than a normal year's waste. In Miyagi, it was 19 times greater. To hasten recovery, local governments across Japan worked together to remove debris. Among the many issues that arose were the availability and selection of storage sites,[3] methods of incineration, decisions about recycling, and waste treatment and disposal. Under Japan's Local Autonomy Act, municipal governments are expected to treat disaster-related waste in accordance with the prefectural government's waste-management plan, and different treatment and disposal methods must be used depending on the composition of the debris. The possibility of recycling should be considered. In general, authorities should prepare for disasters by designating temporary storage sites, traffic routes for transporting waste, and so forth. The role of the private sector in debris management, as well as cooperation with organizations and government bodies outside the affected areas, should be explored.

Livelihood and job creation

Maintaining existing sources of income and creating jobs are crucial during the reconstruction phase. When reconstruction is delayed, income normally generated by neighborhood shops or restaurants will be lost. Under the "Japan as One" work project, local governments in priority areas can avail themselves of job-creation funds. The town of Minamisanriku, for example, received financial support for fiscal year 2011. As of January 2012, it had undertaken 47 job-creation projects employing 460 people. The town will likely receive more financial support for additional employment and livelihood projects.

Part 5: Hazard and risk information and decision making

The limitations of predictive and risk-assessment technologies need to be understood. In Miyagi, the government predicted a high probability of an earthquake occurring but underestimated its size and the ensuing tsunami risk. The official hazard map depicted risk areas that were smaller than the areas actually affected by the GEJE. Given the uncertainties associated with hazard prediction and risk assessment, earthquake and tsunami risks should be assessed based on multiple scenarios, taking into account every conceivable eventuality and utilizing all the tools science has to offer. They should also be informed by historical records going back as far as possible, combined with a thorough analysis of the literature in the field, topographical and geological studies, and other scientific findings.

All districts along the Tohoku coast had prepared tsunami hazard maps prior to the GEJE, but the extent of flooding experienced in some areas far exceeded the maximum extent of inundation predicted on the maps (map O.2). Hazard maps are used by local governments in their disaster-preparedness plans to raise awareness of the risks of disaster among local residents. The hazard map is a crucial tool for communicating information on risks and countermeasures. Involving the community in its preparation helps raise awareness and maximize engagement when a disaster strikes.

The sharing of information among governments, communities, and experts left much to be desired. For example, only 20 percent of the population had seen the hazard maps before the March 11 disaster. Effective risk communication does not necessarily require a sophisticated communication system. Although science-based early-warning systems are important during a disaster, regular sharing of predisaster information at the local level is equally important. The sharing should be accompanied—over time and with the community's involvement—by disaster drills, community mapping, and other measures. In recent years, remote-sensing data has been used around the world to rapidly map the damage resulting from natural disasters. Japan has a well-established track record in disaster mapping: as early as 1995, remotely sensed data were used to map the damage from the Kobe earthquake.

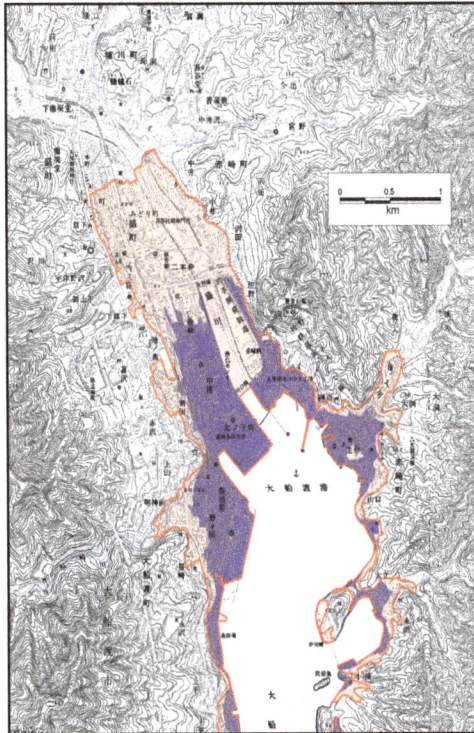

a. Ofunato City, Iwate Prefecture

b. Sendai City, Miyagi Prefecture

☐ Actual inundation

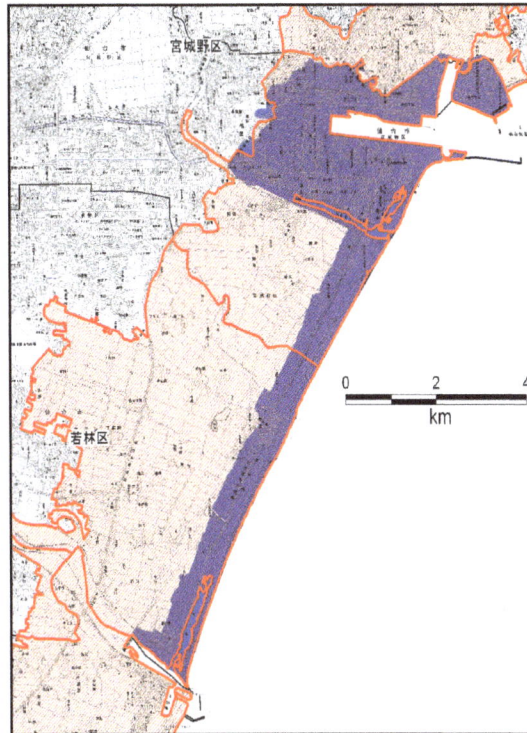

■ Inundation predicted on hazard maps

Map O.2 Actual inundation areas were much larger than predicted

Source: Cabinet Office.

Part 6: The economics of disaster risk, risk management, and risk financing

Prompt government intervention to keep damage from spreading across sectors and countries

In 2011, the GEJE contributed to a 0.7 percent contraction of Japan's GDP. But the full extent of the GEJE's economic impacts will not be known for some time. Manufacturing and services suffered significant direct and indirect impacts. Direct damage to buildings has been estimated at approximately ¥10.4 trillion, or 62 percent of total damages. The amount of damage to the capital stock (asset base) of agriculture, forestry, and fisheries is estimated as ¥2.34 trillion, while damage to the tourism industry amounts to approximately ¥0.7 trillion.

Although the Tohoku and Kanto regions were the most directly affected by the earthquake, the entire manufacturing sector in Japan and some industries abroad were forced to suspend production, as the impact of supply-chain disruptions triggered by the disaster spread through the globe's networked production system. A dense network of supply chains runs throughout Japan, enabling manufacturers to engage in highly efficient production while keeping inventory to a minimum. But this efficiency-oriented management of supply networks backfired in the wake of the earthquake. Although Japanese companies were remarkably responsive, restoring supply chains and getting production almost back to normal by the end of summer 2011, the need remains to create more resilient supply chains both inside and outside Japan.

The auto industry recorded the greatest fall in production but recovered rapidly as facilities reopened and vital transport networks were repaired. After an initial 15.0 percent drop in March, industrial production rebounded from April onward, with growth of 6.2 percent in May and 3.8 percent in June.

Because of the accident at the Fukushima Daiichi nuclear power plant and damages to other power plants, the government had to cut power consumption in the Tohoku and Kanto regions in the summer of 2011. The government ordered large-scale users to cut their consumption by 15 percent and called on smaller electricity users and individual households to curb their consumption voluntarily.

The government played an important role in alleviating the disaster's impact on households and businesses through measures to ensure the stability of the financial system, timely approvals of supplementary budgets, and provisions for rapid disbursement of disaster assistance, all of which helped citizens and firms jumpstart their recovery processes. The financial resources for recovery and reconstruction are being funded by taxes to avoid leaving the cost to future generations.

Earthquake insurance helps people get back on their feet. Dual earthquake insurance programs, consisting of private nonlife insurers and cooperative mutual insurers, cover about four in ten Japanese households. These programs do not provide a one-size-fits-all solution, however. They offer a range of coverage based on level of risk and other factors. Data on natural disasters by country show that both industrialized and developing countries have the same probability of suffering a disaster. The difference is that developed countries tend to have more comprehensive and effective central government policies and better-developed insurance markets, which protect lives and preserve economic assets. A functioning market in catastrophic risk insurance requires major investments in risk models, exposure databases, product design, pricing, and other basic infrastructure of the system. Governments can play an important role in fostering the growth of this kind of infrastructure, thereby enabling the private insurance industry to offer cost-effective and affordable insurance solutions.

Part 7: Recovery and relocation
Relocation and new regulations for land use in at-risk areas in the wake of megadisasters

Since the GEJE, the Japanese government has strengthened DRM systems based on lessons learned from that event. One of those lessons is that relocation is effective in mitigating disaster damage. However, managing relocation projects—and consulting with affected communities—is challenging. It is difficult to achieve a consensus among community members on any rehabilitation plan. For example, while some prefer to rebuild their hometowns on the original sites, others want to move to safer areas.

Governments should examine various recovery schemes, such as relocation to safer areas, and reconstruction at the original sites. When planning a recovery scheme, it is crucial to consider community needs. But there is a trade-off between speed and quality in the recovery process. A government can promptly rehabilitate towns by taking a top-down approach. On the other hand, community consultation requires more time.

Local governments should establish a participatory mechanism, since community participation is essential in promoting recovery. One lesson from the humanitarian response systems used after the Indian Ocean tsunami in 2004 is the importance of striving to understand local contexts and working with and through local structures. Experts and CSOs are expected to play a role in assisting recovery, for example, by organizing and facilitating workshops or consultation meetings and working with government and other experts (see figure O.6).

A cross-sectoral approach is required to rehabilitate people's daily lives. Organizations should harmonize recovery plans among all sectors concerned, such as roads, DRM, and urban planning. Coordination among local governments, the ministries of the central government, and reconstruction

Figure O.6 Community rehabilitation facilitator

Source: Japan International Cooperation Agency.

agencies is crucial for effective planning and implementation.

Local governments should lead recovery, but support from the national government is essential. Since local governments can more closely respond to the varied needs of affected people on the ground, they should take the principal responsibility for recovery planning and implementation. The national government should support local governments' efforts by creating legislation and new project schemes, providing subsidies, and providing technical support (such as conducting tsunami simulations and dispatching technical staff).

The relative merits of "self-reconstruction" and public housing in postdisaster reconstruction

An essential task of government is to help people affected by natural disasters, particularly the most vulnerable groups, to reconstruct permanent housing. Local governments must strive to identify the best way to provide such assistance. Close communication between government and affected communities is an essential aspect of any effective response.

Governments should establish support mechanisms for housing reconstruction, in particular, for vulnerable and low-income groups. Wherever possible, local governments should encourage affected community members to assume responsibility for rebuilding their lost dwellings. This approach is desirable because it allows people to rebuild to suit their needs and because it lightens the load on government. Some groups, however, such as the low-income and the elderly, cannot rebuild on their own because of financial constraints. Local governments in the Tohoku area are providing these groups with public housing.

Support from experts and private sector involvement are useful. Because completing large tracts of public housing in a short time is a difficult task, local governments responsible for reconstruction works should accept assistance from other organizations and experts, and through public-private partnerships. Local governments are well advised to take advantage of the private sector's experience with project management.

Local governments should formulate a plan to operate and maintain public housing. While the central government provides financial support for construction, local governments and the affected population will have to operate and maintain public housing. Local governments should consider operation and maintenance at the design stage.

Preserving cultural heritage

In Japan, earthquakes and tsunamis have damaged an enormous number of cultural properties—for example, 744 designated cultural properties were damaged by the GEJE.

A country's cultural heritage is fundamental for national and community pride and for social cohesion. Historical monuments are regarded as national and community treasures. Since these properties are deeply connected to people's lives and communities' history, their disappearance is equivalent to losing part of a nation's identity.

Governments should embrace the importance of preserving cultural heritage. Protecting and preserving cultural properties and historical buildings are often considered low priorities in disaster management. The DRM plans of governments rarely cover the preservation of cultural heritage.

It is important to prepare for disasters by conducting collaborative activities with local communities during normal times. The owners of historical records, local residents, government officials, and experts should be involved in creating a mechanism for preservation. Without systems for preserving historical records, records in private collections are at a high risk of disappearing during disasters. Digital copies should be made of original historical records. These copies are crucial when original records are lost to disaster for their contribution to the preservation and rehabilitation process.

Museums should produce a database of properties. Information on properties is crucial in conducting preservation work after disasters. At a museum in Rikuzentakata City, it was quite difficult for experts to address the property and materials they encountered, since the staff had died in the disaster and all information was lost.

Recovering from damage to the Fukushima Daiichi Nuclear Power Station

The recovery process following the nuclear accident at the Fukushima Daiichi Nuclear Power Station on March 11, 2011, presented challenges different from those faced in the recovery of areas damaged by the tsunami waves and tremors of the GEJE. The nuclear accident left communities concerned with the serious effects of radiation exposure, relocation, the dissolution of families, the disruption of livelihoods and lifestyles, and the contamination of vast areas.

Following the nuclear accident, people in Fukushima were removed to municipalities and prefectures outside their home communities. There they faced difficulties finding housing, jobs, and schooling in unfamiliar places. Many families separated in the process of seeking employment and uncontaminated places to

live. To date, those affected do not have a clear vision of when they can return to their original communities. Even in areas where living restrictions have been lifted, there are few job opportunities, educational opportunities, and medical and other social services. In addition, the fear of radiation has not yet dissipated. Many displaced people continue to reside in transition shelters, perpetuating the possibility of conflict between the host community and temporary residents.

Nuclear disaster can divide a society. The affected population of Fukushima has been divided by differences in radiation exposure, risk perception, age, and income. Following adjustments to evacuation zoning, some affected people have begun to return to their hometowns. More than 20,000 people in four municipalities, however, will not be able to return to their communities for at least five years because of high levels of radiation. Some groups, in particular families with children, are seriously concerned about radiation and have moved outside the prefecture, while others stay on. In general, younger people tend to move away and start new lives, while older people seek to return to their home communities. People with higher incomes are more likely than poorer people to relocate voluntarily.

Prolonged evacuation causes conflict between communities. Conflicts have emerged between evacuees and host communities. Towns and villages in the prefecture that suffered from the earthquakes and tsunamis are hosting evacuees from areas affected by the nuclear accident. Because the evacuees occupy housing and use public services (such as health, education, and transport facilities) in the host communities, natives encounter shortages, leading to resentment, which is exacerbated by the fact that evacuees from the nuclear accident are being compensated by the operator of the nuclear plant.

Developing "temporary towns" is an enormous challenge. Developing temporary sites for evacuees in other municipalities is more complicated than the normal practice of building resettlement shelters in the disaster-affected area. It is necessary to clarify responsibilities and cost-sharing arrangements among the affected and the host municipalities and with the national and prefectural governments. The question of how to use the facilities and buildings of the temporary towns after evacuees return to their hometowns will have to be studied and resolved.

Radiation monitoring requires participation from various stakeholders (such as communities, governments, and academia) to produce an accountable database. Merely providing risk information on radiation is not enough to prevent rumors or to overcome their influence.

about the risks and consequences of devastating events, and by making informed decisions to better manage both. Disaster management is increasingly important as the global economy becomes more interconnected, as environmental conditions shift, and as population densities rise in urban areas around the world. As the GEJE showed, proactive approaches to risk management can reduce the loss of human life and avert economic and financial setbacks. To be maximally effective, and to contribute to stability and growth over the long term, the management of risks from natural disasters should be mainstreamed into all aspects of development planning in all sectors of the economy.

CONCLUSION

The global cost of natural hazards in 2011 has been estimated at $380 billion—resources that could have been used in productive activities to boost economies, reduce poverty, and raise the quality of life. No region or country is exempt from natural disasters, and no country can prevent them from occurring. But all can prepare by learning as much as possible

NOTES

1. The GDLN is a network of video-conferencing facilities in many locations around the world that can be mobilized on short notice for real-time meetings and workshops.
2. The Meiji tsunami occurred at night, whereas the GEJE struck during the day.
3. Waste treatment outside the affected area is usually required but difficult to arrange. Previous experience in Tohoku suggested that finding dumping sites would be a problem.

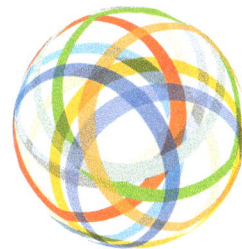

PART I

Structural Measures

CHAPTER 1

Structural Measures Against Tsunamis

Structures such as dikes play a crucial role in preventing disasters by controlling tsunamis, floods, debris flows, landslides, and other natural phenomena. But structural measures alone cannot prevent all disasters because they cannot mitigate damages when the hazard exceeds the level that the structures are designed to withstand. The Great East Japan Earthquake demonstrated the limitations of Japan's existing disaster management systems, which relied too heavily on dikes and other structures. Damage can be kept to a minimum by multilayered approaches to disaster mitigation that include structural and nonstructural measures and that ensure the safe evacuation of residents.

Dikes, dams, and other structures are regarded as core measures in disaster risk management (DRM) in Japan. Japan has constructed dikes to mitigate flooding for nearly 2,000 years. The first dike system was constructed in the Yodogara River in Osaka in the 4th century. The Japanese used dike systems to protect crucial areas, such as castles and residential areas, in the middle and early modern periods. The government established after the Meiji Revolution in the late 19th century has promoted structural measures to control floods, high tides, landslides, and tsunamis by employing modern technology introduced from the Netherlands and other Western countries. Disaster damage had substantially decreased because of concentrated investment in structural measures (chapter 28).

Surrounded by seas, Japan has an extremely long, geographically complex coastline of approximately 35,000 kilometers (km). People, productive assets, and social capital are concentrated on small coastal plains over a limited land area. Not only are Japan's coastal areas situated where earthquakes are exceptionally common, but they are also subject to harsh natural events, such as typhoons and winter ocean storms. Historically, the country has suffered severe damage from tsunamis, storm surges, ocean waves, and other natural phenomena. To protect life

BOX 1.1

The enormous tsunami walls of Taro, Miyako City, Iwate Prefecture

The people of the Tohoku region have built and maintained tsunami dikes for decades. Following the Meiji Sanriku Tsunami of 1896, the village of Taro was hit by a 15-meter tsunami that washed out 285 houses and killed 1,447 people. The 7.6-meter Showa Sanriku Tsunami of 1933 also hit Taro, washing out all 503 houses and killing 889 of the village's 2,950 residents. Because insufficient high ground could be found for 500 houses, the village chose to build dikes. Construction began in 1934 using borrowed money and took more than three decades to complete. The largest dike was 2,433 meters long and 7 meters high (10.65 meters above the sea level). It was 3 meters wide at the top and as much as 25 meters wide at the base. The March 11 tsunami swept over this dike before destroying it, leaving a path of death and destruction across the community.

Source: © Mikio Ishiwatari/World Bank. Used with permission. Further permission required for reuse.

and property concentrated near its coastline, the country has been developing coastal and port facilities for the last half century.

FINDINGS

Coastal structures in the region affected by the Great East Japan Earthquake

When the tsunami hit eastern Japan in March 2011, 300 km of coastal dikes, some as high as 15 meters high, had been built (map 1.1). Prefectural governments, which have the main responsibility for building the dikes (supported by national subsidies that cover two-thirds of

the cost), built 270 km of the total, with the national government building the remaining 30 km. The national government also had developed technical standards, guidelines, and manuals for use in the design and construction of coastal structures. In response to the economic damage caused by the Great East Japan Earthquake (GEJE)—¥300 billion ($3.75 billion) in destroyed dikes—the government has invested several hundred billion yen in dike construction in the Iwate, Miyagi, and Fukushima prefectures. It has also invested ¥400 billion ($5 billion) in constructing bay mouth breakwaters in major ports, such as Kamaishi, Kuji, and Ofunato, to protect them from tsunamis. A cost-benefit analysis of these investments appears in chapter 28.

The disaster-affected region had frequently sustained devastating damage from tsunamis, including the Sanriku tsunamis of June 1896 and March 1933, and a tsunami caused by a massive earthquake off the coast of Chile in May 1960. The 1933 Showa Sanriku Tsunami was the first disaster to provoke modern tsunami countermeasures at the initiative of the central and prefectural governments. Those countermeasures included mainly relocation to higher ground and the building of dikes, albeit at just five sites (box 1.1).

The Chilean Earthquake Tsunami of 1960 prompted extensive construction of coastal dikes in the region. The dike height was initially based on the height of the 1960 tsunami but was revised several times thereafter to take into account other major tsunamis that had occurred in the previous 120 years, as well as predictions of future storm surge levels. These dikes are designed to withstand the largest of the predicted tsunami heights and storm surge levels. In Iwate and northern Miyagi, the heights were based on historical records, whereas in southern Miyagi and Fukushima they were based on the predicted storm surges. Methods of risk assessment are explained in chapter 25.

How structures performed against the GEJE tsunami

Some towns in the region were well protected by the structures in place, even though the tsunami caused by the earthquake far exceeded their design height. In Iwate's Fudai Village, the 15.5-meter floodgate, built in 1984, protected the village and its 3,000 inhabitants. The village was severely damaged by the Meiji Sanriku Tsunami of 1896 (height 15.2 meters), the Showa Sanriku Tsunami of 1933 (11.5 meters), and the Chilean Earthquake Tsunami of 1960 (11.5 meters). The mayor of the village in the early 1980s was convinced that a 15-meter tsunami would hit the village again at some point, and built the 200-meter-wide floodgate about 300 meters inland from the mouth of the Fudaigawa River, which runs through the village. Although the 20-meter-high GEJE tsunami did top the floodgate, the gate kept the water from reaching the town center (figure 1.1). The topography of Fudai Village, being surrounded by cliffs with a narrow opening to the sea, was a major factor in enabling the construction of such a high gate.

The dikes also served to protect communities in areas where the tsunami was lower than the dike (northern Iwate, Aomori, Ibaraki, and others), as shown in the example of Hirono Town (figure 1.2).

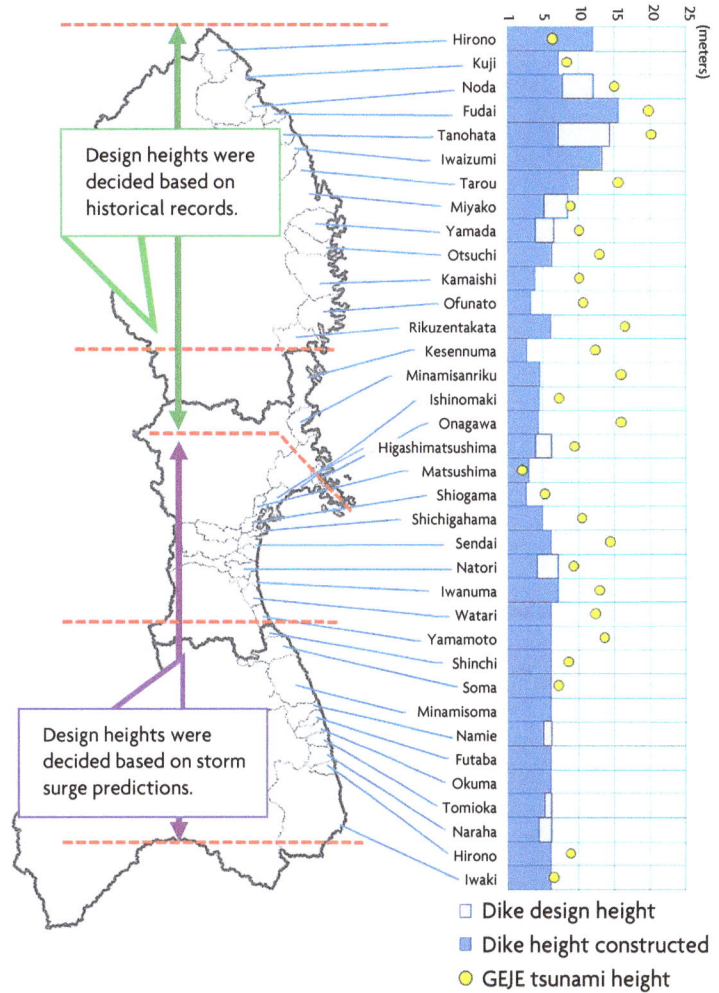

Design heights were decided based on historical records.

Design heights were decided based on storm surge predictions.

□ Dike design height
■ Dike height constructed
○ GEJE tsunami height

Map 1.1 Determining dike height

Source: Ministry of Land, Infrastructure, Transport and Tourism (MLIT).

Figure 1.1 Inundation area in Fudai Village, Iwate

Source: MLIT.

Levee height T. P.+12.0m
堤防天端高 T.P.+12.0m

Tsunami height
T. P.+9.5m

Figure 1.2 No tsunami inundation in Hirono Town, Iwate

Source: MLIT.

Note: T.P. = tidal plane.

27

Certain breakwaters were also effective in mitigating damage from the tsunami. The breakwater at the mouth of Kamaishi Bay in Kamaishi City, Iwate, was completed in 2009, at a total cost of some ¥120 billion ($1.5 billion). It was the world's deepest breakwater. Although destroyed by the GEJE tsunami, the breakwater reduced the tsunami's force, and therefore its height, by about 40 percent and

delayed its arrival by some six minutes, allowing more time for people to evacuate to higher ground (figure 1.3).

The GEJE tsunami destroyed many coastal structures. Of the 300 km of dikes along the 1,700 km coast of the Iwate, Miyagi, and Fukushima prefectures, 190 km were destroyed or badly damaged. In many cases the tsunami was twice the height of the dikes (map 1.1). All 21 ports along the Pacific coast in the Tohoku region (from Aomori to Ibaraki) sustained extensive damage to their breakwaters, quays, and other coastal facilities, suspending all port functions.

Run-up from the tsunami caused significant damage along major rivers in the region. Traces of the run-up were found as far as 49 km upstream from the mouth of the Kitakami River. Ishinomaki City in the Miyagi Prefecture, where the Kitakami flows out to the sea, experienced severe tsunami run-up in addition to the direct attack along the coast. Approximately 73 square kilometers (km^2) along the river, or about 13 percent of the entire city, were inundated (map 1.2). The city suffered badly, with 3,280 dead and 539 missing (as of March 11, 2012); 20,901 houses were completely destroyed, and 10,923 houses badly damaged (as of October 21, 2011).

New thinking about structural measures in light of the GEJE

The GEJE exposed the limitations of DRM strategies focused disproportionately on structural measures. Dikes and breakwaters built before the GEJE were designed to protect against relatively frequent tsunamis, and were effective in preventing damage from those of limited height. In the GEJE, however, the height of the tsunami far exceeded predictions. Although the structures helped to reduce water levels, to delay the arrival of the tsunami, and to maintain the coastline, many of them were breached, resulting in enormous inland damage.

Planning for the largest possible event is a significant policy shift in Japan's thinking

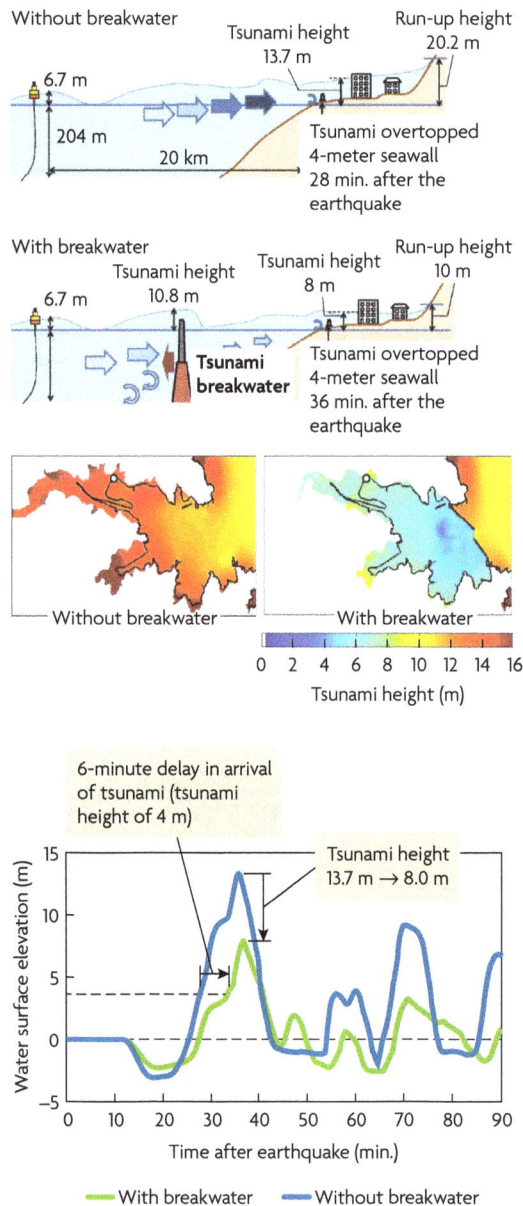

Figure 1.3 Effectiveness of the Kamaishi tsunami breakwater

Source: MLIT.

about DRM. Building 20- or 30-meter tsunami dikes is neither realistic nor financially, socially, or environmentally practical. But lives can and must be protected by other means, notably multilayered approaches that combine structural and nonstructural measures to ensure the safe evacuation of residents (chapter 32). Nonstructural measures are discussed in the chapters of cluster 2. Planning for the new generation of multilayered DRM approaches is based on a comprehensive assessment of historical records, documents, and physical traces of past tsunamis, and by drawing on the latest seismological research and simulations.

Since the GEJE, the Japanese government has taken a two-level approach. Level 1 includes tsunamis that occur as frequently as every 100 years and that cause significant damage, whereas level 2 covers the largest possible tsunami, which has an extremely low probability of occurrence (once every 1,000 years) but has the power to cause devastating destruction (figure 1.4).[1] Conventional structural measures such as dikes and breakwaters protect human lives and property, and stabilize local economic activities, in the face of level 1 tsunamis. To withstand level 2 tsunamis, however, coastal structures must be improved to be more resistant to collapse and to reduce the likelihood of their complete destruction through scouring (figure 1.5). Some 87 percent of dikes that had been reinforced against scouring were not damaged in the GEJE, although the tsunami spilled over them.

The government has issued new guidelines for rebuilding river and coastal structures, taking into consideration their appearance as well as local characteristics, ecosystems, sustainability issues, and financial feasibility.

Operation of floodgates and inland lock gates

Although floodgates and inland lock gates can protect against tsunamis, their operation posed problems during the GEJE. Such gates should be closed before the tsunami arrives, but in the case of the GEJE tsunami this operation could not be completed in time, and a number of volunteer firefighters and other workers were killed in the process. In addition, many gates were left open because equipment failed or because operators were caught in traffic

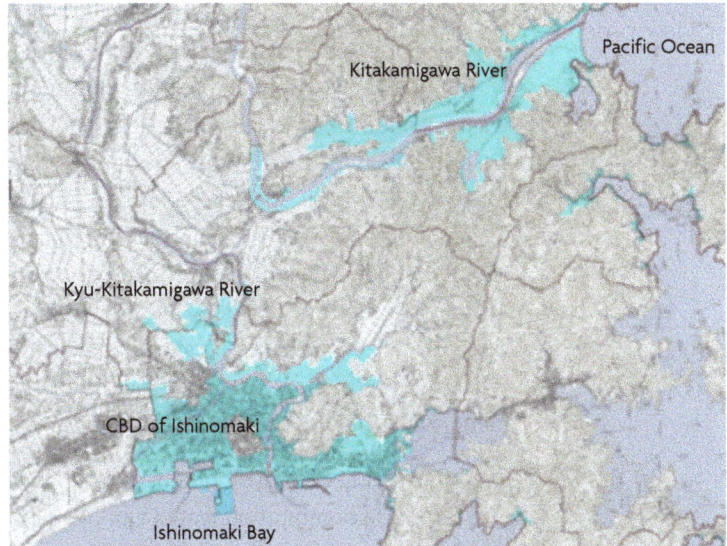

Map 1.2 Tsunami inundation area along the Kitakami and Kyu-Kitakami rivers
Source: MLIT.

Note: CBD = central business district.

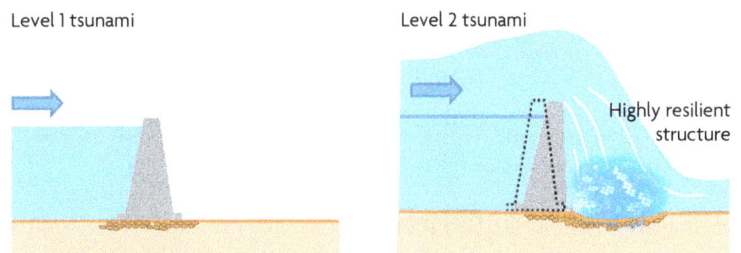

Figure 1.4 Countermeasures against level 1 and level 2 tsunamis
Source: MLIT.

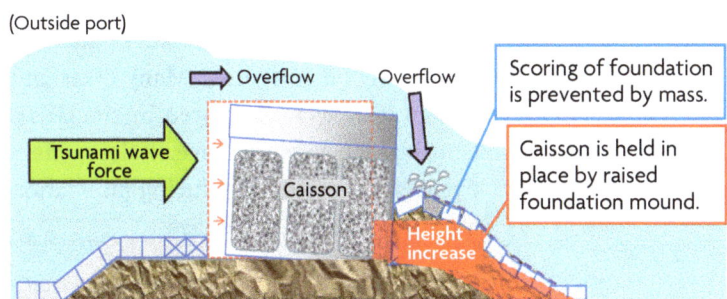

Figure 1.5 Structure of a highly resilient breakwater
Source: MLIT.

jams and could not reach the site. Other gates became nonfunctional owing to power losses.

In December 2011, the Flood Prevention Act was amended to require local governments to ensure the safety of volunteer firefighters and other workers who operate floodgates, inland lock gates, and similar facilities. In March 2012, the Ministry of Land, Infrastructure, Transport and Tourism (MLIT) and the Fire and Disaster Management Agency issued the following recommendations to local governments and other concerned organizations:

- Remove unnecessary floodgates and ensure that the remaining floodgates can be operated automatically, semiautomatically, or by remote control.

- Keep inland lock gates closed at all times. Introduce automatic floating gate systems or install ramps or steps.

- Install emergency power supplies and make facilities earthquake-resistant.

LESSONS

- *Structural measures alone cannot prevent tsunami disasters.* The enormous tsunamis experienced in the GEJE have revealed the limitations of DRM measures that rely too heavily on structures. Many dikes and breakwaters were destroyed by the GEJE tsunami. They were nevertheless effective to some extent in reducing inundation areas and mitigating damage.

- *It is important to learn from past disasters and to revise countermeasures accordingly.* In the GEJE-affected areas, various structural measures had been implemented in light of historical disasters, and they were successful in mitigating damage until the GEJE. Scenarios that envision the greatest possible hazard should be taken into consideration when designing DRM measures.

An appropriate combination of structural and nonstructural measures is required in order to achieve maximum mitigation of damage. Structural measures should be designed to prevent damage to human lives and property caused by level 1 events and to mitigate damage from level 2 events.

- *Building design can mitigate damage if not prevent it.* Though it is unrealistic to build structures large enough to protect against the largest conceivable events, the resilience of conventional structures must be enhanced. These should be built to mitigate damage even when the hazard level exceeds their design specifications. It is possible for structures to "fail gracefully" (meaning that they do not fail completely or collapse), thereby delaying the onslaught and reducing the energy of tsunamis. The concept of failure should be incorporated into the design to take into account unforeseen events.

- *Power failure and other emergency conditions need to be considered in structure design.* Coastal facilities such as floodgates should be designed so that they can be properly managed even in the event of power failure and in the absence of operators. Standardized guidelines should be established for their safe operation in emergencies.

RECOMMENDATIONS FOR DEVELOPING COUNTRIES

Prepare for disasters by integrating structural and nonstructural measures. DRM measures should account for two levels of hazard. Level 1 events are *relatively frequent and produce major damage*; level 2 events, *the largest possible disasters, have an extremely low probability but produce a devastating impact*. Every possible structural and nonstructural measure should be employed to protect against level 2. Structural measures should be designed to protect people, property, and socioeconomic activities against level 1 and to mitigate damages at level 2.

Provide technical and financial support for local governments. The central government plays a crucial role in reducing disaster risks across the country. The central government should encourage local governments to promote structural measures by providing financial support and guide them in meeting minimum requirements for structures by producing technical guidelines and manuals. Also, the central government should provide the local governments with technical support, such as conducting training for technical staff in planning, design, operations, and maintenance.

Consider designs and improvements to enhance the resilience of structures and to prevent sudden and complete failure. Extraordinary external loads caused by earthquakes, floods, and other events should be considered in designing structures such as dams and dikes, which should be designed in such a way so that they will mitigate damage even when the hazard level exceeds their design levels. Their effectiveness in mitigating damage should be ensured even in the event of their technical failure.

Raise dike levels in a phased manner, considering the country's financial and social conditions. Safety standards and structural design upgrades against level 2 events should reflect the concentration of population and economic assets in the protected areas. Although it may not be possible to build dikes capable of withstanding level 2 disasters, appropriate and feasible targets for dike design safety should be identified.

Assure reliable operation of key facilities during emergencies. The safe and reliable operation of infrastructure must be ensured in emergency situations. Structural measures such as floodgates cannot provide reliable protection if they cannot be operated under extreme conditions, such as power failures and the absence of operators. Multiple layers of operation should be assured. A sufficient number of qualified operators should be available during disasters, but not necessarily onsite. Developing manuals and conducting regular drills are required during normal times. The danger to which operators are exposed should be minimized.

NOTES

Prepared by Mikio Ishiwatari, World Bank, and Junko Sagara, CTI Engineering.

1. The two-level approach has already been adopted in the design of other key infrastructure, such as dams and flood-prevention dikes. Dams typically consider the maximum probable flood or a flood with a 10,000-year return period when designing structural safety, and a 100- to 200-year return period for flood-control operations. For flood-prevention dikes to protect some critical areas of Tokyo and other locations, the government has increased design standards beyond the norm of 100- to 200-year floods.

BIBLIOGRAPHY

Ishinomaki City. 2011. *Ishinomaki wo osotta ootsunami to fukkoukeikaku* [Ishinomaki City's mega tsunami attack and reconstruction plan]. http://www.thr.mlit.go.jp/iwate/kawa/seibi_keikaku/dai6/image/dai6_02.pdf.

Ishiwatari, M. 2012. "Review of Countermeasures in the East Japan Earthquake and Tsunami." In *East Japan Earthquake and Tsunami: Evacuation, Communication, Education and Volunteerism,* edited by R. Shaw and Y. Takeuchi. Singapore: Research Publishing.

MLIT (Ministry of Land, Infrastructure, Transport and Tourism). 2011a. *Kowan ni okeru sougouteki na tsunami taisaku no arikata* [Comprehensive tsunami countermeasures in ports]. Interim report. http://www.mlit.go.jp/common/000149434.pdf.

——. 2011b. "Higashinihon daishinsai wo fumaeta seki suimon nadono sekkei sousa no arikatanitsuite" [Design and operation of weirs and gates in light of the lessons of the Great East Japan Earthquake]. http://www.mlit.go.jp/river/shinngikai_blog/kakouzeki_suimon/arikata/arikata110930.pdf; http://www.jice.or.jp/sonota/t1/pdf/02arikata.pdf.

———. 2011c. *Kasen heno sojou tsunami taisaku ni kansuru kinkyu teigen* [Emergency proposal on countermeasures for tsunami run-up along rivers]. http://www.mlit.go.jp/common/000163992.pdf.

———. 2011d. *Kasen kaigan kouzoubutsu no fukkyu ni okeru keikan hairyo no tebiki nit suite* [Guidelines for preserving landscapes during rehabilitation of river and coastal structures]. http://www.mlit.go.jp/river/shinngikai_blog/hukkyuukeikan/tebiki/tebiki.pdf.

———. 2011e. *Tohokuchiho taiheiyo oki jishin wo humaeta kakouzeki suimon tou kentou iinkai* [Technical committee on weirs and gates following the earthquake off the pacific coast of Tohoku]. Documents of the 4th meeting. http://www.mlit.go.jp/river/shinngikai_blog/kakouzeki_suimon/dai04kai/dai04kai_siryou3-1.pdf; http://www.mlit.go.jp/river/shinngikai_blog/kakouzeki_suimon/dai04kai/dai04kai_siryou3-2.pdf.

Technical Committee on Tsunami Countermeasures. 2011. *Kaigan teiboutouno fukkyunikansuru kihontekina kangaekata* [Basic approach for rehabilitating dikes damaged by GEJE]. http://www.mlit.go.jp/common/000182993.pdf.

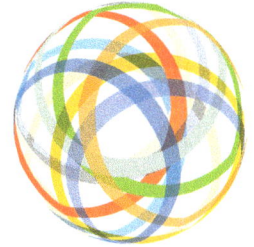

CHAPTER 2

Building Performance

The strong main shock of the Great East Japan Earthquake of March 11, 2011, caused little damage to buildings. Buildings designed under the current building code and those with base isolation fared well. However, seismic design guidelines for nonstructural members had not been considered adequately, which resulted in problems such as the collapse of ceiling panels. Soil liquefaction occurred in reclaimed coastal areas along Tokyo Bay and riverside areas. The key lessons of the event are that seismic-resistant building design prevents collapse of buildings and protects human lives, that retrofitting vulnerable buildings is essential to reduce damage, that seismic isolation functioned well, and that nonstructural building components can cause serious damage. When applying these lessons to developing countries, local technical and socioeconomic conditions should be taken into account.

FINDINGS

History of building codes in Japan

The world's first national seismic design code

Due to its location and tectonic settings, Japan is prone to large earthquakes. The Great Kanto Earthquake in 1923 caused some of the most serious damage in Japanese history, as fires consumed a large part of Tokyo, killing more than 100,000 people (table 2.1). Based on the lessons learned from the disaster, a seismic design code was introduced in the building code of 1924, the first national seismic design code applied anywhere in the world.

Building code updates following major earthquakes

After every major earthquake, Japan's national government and academic community carry out detailed surveys of building damage, and the building code is revised accordingly. Technical recommendations are based on the most recent lessons. The Tokachi-oki earthquake in 1968 caused serious damage to reinforced concrete (RC) buildings and inspired a major revision of the building code in 1981. Until 1981, the building code required buildings to withstand a lateral force of 20 percent of the total weight of the building without damage to structural

Table 2.1 Comparison of three major disasters in Japan

DISASTER	GREAT KANTO EARTHQUAKE	GREAT HANSHIN-AWAJI (KOBE) EARTHQUAKE	GREAT EAST JAPAN EARTHQUAKE AND TSUNAMI
Year	1923	1995	2011
Magnitude	7.9	7.3	9.0
Location	Tokyo and surrounding area	Kobe and surrounding area	Extended area. Tsunami affected 1,000 km of coastline
Casualities (dead and missing)	105,385	6,437	19,845 (as of September 26, 2011)
Main cause of deaths	Fire	Collapse of old houses	Tsunami (drowning)
Conditions	Noon. Residents were using stoves to cook lunch. Strong winds spread fire, which burned for three days. Fire created tornados and whirlwinds.	Before dawn. Sleeping residents were killed when their houses collapsed. Few were killed on trains or highways.	Mid-afternoon. People were at school or work, where evacuation protocols were put into effect.

members. The revised code, part of which is still in use, requires that buildings be strong enough to withstand a lateral force equal to 100 percent of the building's weight. Damage to the building is permissible as long as human lives are not threatened.

Current building code (1981) in Japan

The main aspects in the current building code of 1981 are as follows:

- Within their lifetime, buildings should be able to withstand several large earthquakes without structural damage.

- Building should be able to endure, without collapse or other serious damage, an extremely large earthquake with a return period of 500 years.

Technical guidelines for assessing and retrofitting existing RC buildings constructed under building codes in effect prior to 1981 were produced.

Initiative to retrofit buildings following the Great Hanshin-Awaji Earthquake (Kobe earthquake) in 1995

The 1995 Kobe earthquake caused heavy damage, 6,437 casualties, and economic losses estimated at more than $120 billion. Of the buildings that collapsed in the Kobe quake,

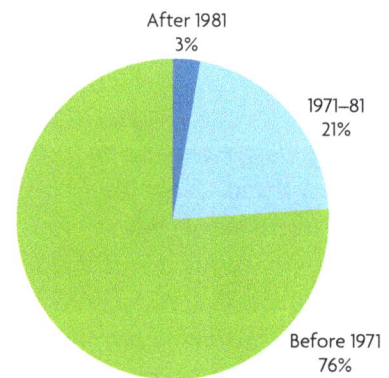

Figure 2.1 Share of houses that collapsed in the 1995 Kobe earthquake, by year of construction

Source: Ministry of Land, Infrastructure, Transport and Tourism (MLIT).

97 percent were built before 1981 (figure 2.1). Based on this finding, the government implemented a new law in 1995 to promote retrofitting of old buildings.

Under the Act for Promoting Seismic Retrofitting of Existing Buildings (1995), the national and local governments offer incentives to private homeowners, such as

- Subsidies for assessments of structural soundness

- Subsidies for the cost of retrofitting

- Reductions in income tax and property tax

- Low-interest loans to cover the cost of retrofitting

Some 80 percent of local governments have established subsidy programs to encourage owners to assess the structural integrity of their homes, and, as of April 2011, some 64 percent of the local governments had programs that subsidized retrofitting work. The government's target is to increase the ratio of earthquake-resistant houses to 95 percent before 2020. In 2008 the ratio was 79 percent, with some 10.5 million houses still requiring retrofitting. In spite of efforts to promote retrofitting, only 300,000 houses were retrofitted between 2003 and 2008. These numbers show that it is difficult to motivate homeowners to retrofit.

Damage to buildings from the Great East Japan Earthquake

Minimal damage from earthquake

Table 2.2 shows the summary of the damage caused to buildings following the Great East Japan Earthquake (GEJE). Most of the collapsed residential buildings were washed away or destroyed by the tsunami rather than the earthquake. The death toll from the earthquake itself is estimated to be less than 200.

The earthquake produced violent shaking over a very wide area. The strongest peak acceleration of 2,933 galileo (Gal) was recorded in Tsukidate, Miyagi Prefecture, but 18 observation stations in six prefectures observed acceleration greater than 1,000 Gal. In spite of the strong acceleration, damage from shaking was minimal, owing partly to the characteristics of the ground motion (the dominant frequency was relatively high). Damage to buildings constructed under the 1981 and later building codes was limited and within the range anticipated by the design code.

Serious damage from the tsunami

The cause of most of the damage to houses was the tsunami that followed the main shaking. Most wooden houses in deeply inundated areas were washed away or totally destroyed (figure 2.2). Many steel structures were also severely damaged (figure 2.3). By contrast, buildings of RC performed well against the tsunami. Although many were completely

Figure 2.2 Houses and cars were washed away by the tsunami

Source: © Yamada-machi. Used with permission. Further permission required for reuse.

Table 2.2 Damage to buildings following the GEJE

CATEGORY	NUMBER
Residential buildings	
Total collapse	107,779
Partial collapse	117,019
Burned	263
Partial damage	434,327
Nonresidential buildings	32,445

Source: NILIM (National Institute for Land and Infrastructure Management) and BRI (Building Research Institute) 2012.

Figure 2.3 The tsunami destroyed the outer walls of steel structures

Source: NILIM and BRI 2012.

Figure 2.4 Reinforced concrete building withstood tsunami even though submerged (note car on roof)

Source: NILIM and BRI 2012.

Figure 2.5 Reinforced concrete building damaged by buoyancy

Source: NILIM and BRI 2012.

Figure 2.6 Reinforced concrete building scoured by the tsunami current

Source: NILIM and BRI 2012.

Figure 2.7 Overturned building of reinforced concrete with pile foundation

Source: NILIM and BRI 2012.

submerged, they did not suffer structural damage (figure 2.4). Those RC buildings that were damaged tended to be small and without a pile foundation (figures 2.5 and 2.6). Figure 2.7 shows a damaged building where the probable causes of the damage were a combination of weak connections between piles and footings, strong water pressure from the tsunami current, and liquefaction.[1]

Effectiveness of building countermeasures

Good performance of seismic base isolation system

Japan's Building Research Institute (BRI) reported that the seismic base isolation[2] systems in all 16 buildings in Miyagi Prefecture performed well, reducing lateral motion by 40–60 percent. No damage was observed to the structures or to mechanical and electrical facilities inside the buildings. No fittings or furnishings fell. The dampers and the cover over the slits between the isolated and nonisolated parts were damaged as expected.

Enhanced seismic design and retrofitting of transportation infrastructure facilities

A major campaign to reinforce key infrastructure such as bridges following the Kobe earthquake in 1995 was undertaken by highway and railway companies and governmental agencies. As a result, serious structural collapses of infrastructure were avoided following the GEJE. The East Japan Railway Company had reinforced more than 17,000 bridge piers under the Shinkansen (bullet train) lines, and the central government had retrofitted 490

Figure 2.8 Fallen ceiling panels in school gymnasium
Source: NILIM and BRI 2012.

Figure 2.9 Subsidence of houses from liquefaction
Source: NILIM and BRI 2012.

bridges in the Tohoku region. Because of these works, some 1,500 bridges on national routes in the region were spared serious damage. Five bridges collapsed under the force of the tsunami. Because damage was generally limited, it was possible to repair the main highways and roads to the affected areas within one week of the event. However, serious damage in the coastal areas affected by the tsunami took longer to repair. Shinkansen service to the Tohoku region resumed after 49 days (chapter 20), a huge improvement over the situation after the Kobe earthquake, when reconstruction of the roads required more than 18 months and repair of the Shinkansen line took 82 days.

Areas for improvement

Damage to nonstructural building components

Much of the damage observed in buildings following the GEJE involved nonstructural components attached to structures, such as ceiling panels, nonstructural walls, and finishing materials (figure 2.8). To date, no guidelines or codes cover the wide variety of materials and designs used on nonstructural components. In Japan, few engineers have devoted attention to the matter.

Liquefaction

Liquefaction occurred on reclaimed lands and river banks over a wide area. Small buildings without pile foundations built on plots that had

not been treated for liquefaction were affected (figure 2.9). Existing building codes cover countermeasures against liquefaction for RC and other buildings, but not for the detached wooden houses owned by most ordinary people. The Ministry of Land, Infrastructure, Transport and Tourism has now produced technical guidelines to fill these gaps. Some local governments have provided liquefaction risk maps to encourage building owners to take countermeasures.

Damage from failure of retaining walls

In Sendai City, more than 4,000 houses were damaged by landslides caused by the strong ground shaking (figure 2.10). Since 1961, to

Figure 2.10 Houses damaged by failure of retaining walls
Source: NILIM and BRI 2012.

prevent landslide disasters, the city government has regulated housing in hilly areas under the Act on the Regulation of Housing Land Development. Most locations that experienced landslides following the GEJE were developed before the act came into effect. In 2009, in response to landslides caused by earthquakes since 2000, the central government established a subsidy mechanism whereby local governments were tasked to carry out geotechnical work to stabilize the ground for large-scale housing projects in high-risk areas. However, stabilization work had not started by the time the March 2011 disaster struck.

Effect of ground motion of long periods on skyscrapers

The potentially devastating effect of quaking and tremors over long periods on skyscrapers and seismically isolated buildings has been recognized in recent years. New skyscraper designs take this into account. Some skyscrapers had been retrofitted before the GEJE with devices to control deformation or absorb energy. On March 11 strong and sustained ground motion of long periods reached Tokyo (approximately 400 kilometers [km] from the epicenter) and even Osaka (800 km), affecting the skyscrapers in both of these metropolitan areas. Recognizing the importance of countermeasures against the risks of sustained ground motion, the Japanese government released a draft of a new technical guideline that revises structural design procedures, safety measures for furnishings and fittings, and a screening method to identify skyscrapers that need to be examined in detail.

Technical guideline for tsunami evacuation shelters

Japan's first technical guideline for tsunami shelters was published in 2004. A revised guideline was released in November 2011, based on detailed surveys of the areas affected by the GEJE. Where the risks from tsunami pressure are less serious, the tsunami load can

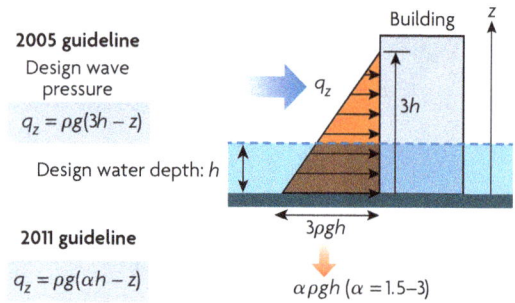

2005 guideline
Design wave pressure
$$q_z = \rho g(3h - z)$$

Design water depth: h

2011 guideline
$$q_z = \rho g(\alpha h - z)$$

$3\rho g h$

$\alpha \rho g h$ ($\alpha = 1.5$–3)

Figure 2.11 Revised design load requirements against tsunamis
Source: BRI and NILIM.

be smaller under the revised guideline than under the previous guideline (figure 2.11).

LESSONS

- *The importance of retrofitting buildings* is demonstrated by the fact that buildings designed under the 1981 building code and retrofitted buildings performed well in the GEJE, whereas most of the damaged buildings were constructed before 1981 and had not undergone any retrofitting. Further efforts to retrofit are required, including more attractive incentives for those who cannot afford to invest in safety or are reluctant to do so (as are many elderly people). More affordable retrofitting methods should be developed. Partial retrofitting, safety shelters inside the home, and beds covered by safety frames are examples of affordable options.

- *The GEJE demonstrated the need to consider nonstructural elements* when thinking about earthquake safety. The materials, design, and construction of nonstructural components vary greatly. Technical guidelines are needed to ensure that such components are earthquake-resistant.

- Even when structures withstood ground shaking and saved the lives of their inhabitants, inhabitants could not reoccupy their dwellings because of deformation of walls and doors. Substantial sheer cracks in nonstructural walls also made inhabitants wary of returning. In addition to ensuring structural safety, it is recommended that efforts to achieve the functional continuity of buildings—with minimum disruption to everyday lives—are made.

- *Countermeasures against liquefaction and landslides need to be enhanced in Japan.* Following the GEJE, the Japanese government has reviewed the method of assessing the risk of liquefaction. Developing more effective and affordable anti-liquefaction treatments is needed. The government is considering a requirement that homebuyers be notified of the risk from liquefaction. The government is also providing subsidies for projects to stabilize slopes with landslide potential near houses.

- *Increasing buildings' capacity to absorb energy reduces structural deformation.* The GEJE demonstrated the possibility of a gigantic earthquake occurring as a result of three large earthquakes (Tokai, Tonankai, and Nankai) occurring in short succession. Such a series of earthquakes would be likely to produce strong ground motions of long periods. Structural and retrofitting measures should be performed according to the new guideline, lowering the risk of long-period ground motions by preventing their amplification. Increasing buildings' capacity to absorb energy reduces structural deformation.

- *Buildings with isolated bases performed well during the GEJE,* enabling them to be used without interruption even immediately after the main shock of the earthquake.

RECOMMENDATIONS FOR DEVELOPING COUNTRIES

Improving the seismic resilience of buildings is the most effective risk mitigation measure. One of the most basic and effective measures

Figure 2.12 Collapsed school building in which furniture is still standing (Yogyakarta province, following Central Java Earthquake, 2006)

Source: Japan International Cooperation Agency (JICA).

to mitigate risks from earthquakes is to build structures that are resilient to ground shaking. Many buildings in developing countries are extremely vulnerable to collapse (figure 2.12).

Use appropriate technologies. Various seismic design guidelines have been developed around the world. Direct application of such guidelines may not be appropriate in developing countries because of their costs, the limited knowledge and skills of builders, and limited tools and facilities on construction sites. What are needed are seismic design guidelines that are suited to local conditions and capable of enhancing the resilience of buildings.

Knowledge and lessons should be adapted and customized to local conditions. In Indonesia a simple technical guideline that is consistent with local technical capacities and other conditions was developed and is being disseminated with help from the Japan International Cooperation Agency (JICA) (box 2.1). Knowledge based on detailed surveys of construction sites and motivation on the part of engineers, workers, government officials, and owners of buildings can improve safety. Pilot buildings may include emergency centers, fire stations, hospitals, or evacuation shelters. These can demonstrate the benefits of advanced seismic resilience while enhancing the knowledge and skills of technicians through on-the-job training.

A simple technical guideline and its dissemination through the building permit process in Indonesia

The Central Java Earthquake in 2006 caused heavy damage and killed some 6,000 people, mostly as their houses collapsed. During reconstruction, the provincial government developed a technical guideline for small, one-story houses. The guideline, simple enough to be illustrated in a poster, has been well accepted by the population. The central government decided to apply it across the country through the building permit system.

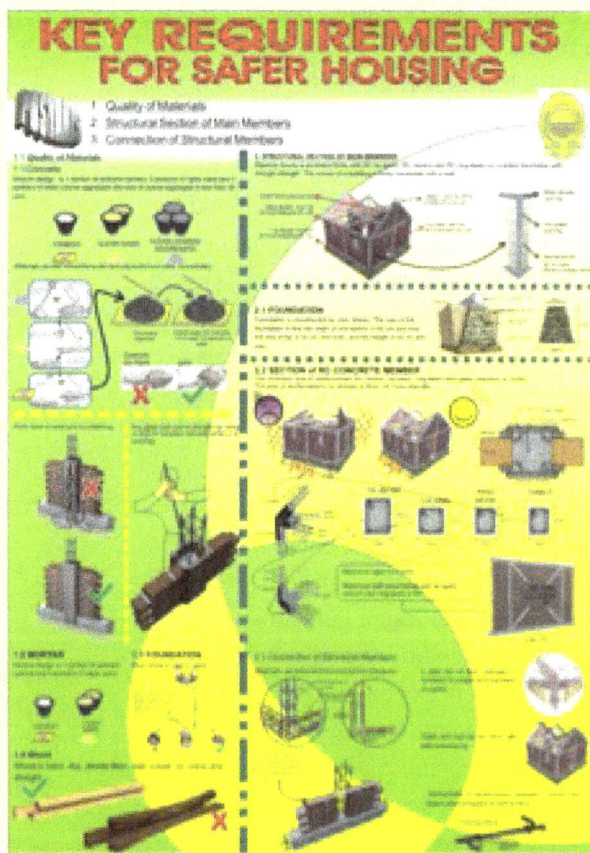

Source: JICA.

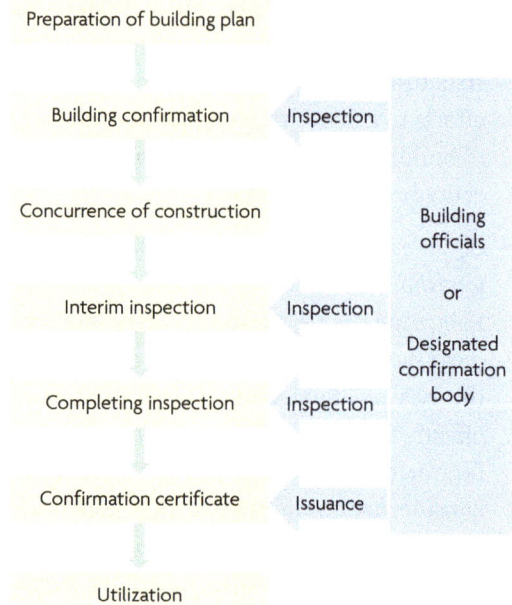

Figure 2.13 Flowchart illustrating the Japanese building permit process
Source: MLIT.

officials and inspectors with access to technical information.

Japan's Building Standard Law mandates the implementation systems shown in figure 2.13. Local government officials (or designated confirmation bodies) conduct examination/inspections before, during, and after construction. If conformity with building standards is confirmed, a confirmation certificate is issued. An interim inspection is performed on buildings that have certain structural characteristics or purposes. Multifamily dwellings, multistoried buildings, and public buildings are generally subject to this type of inspection.

Retrofit historical buildings. In countries with many vulnerable historical buildings, retrofitting is a priority. Retrofitting should be considered in the context of striking a balance between affordable and effective retrofitting methods, a balance that motivates both private owners and government officials and politicians.

Secure the safety of nonstructural components. The issue of nonstructural building components is common in developing countries, although the critical elements may be

Implement building codes. Another important issue is how best to implement building codes and how to monitor their implementation. Legislation should include provisions related to the issuance of building permits, inspection of construction, and enforcement of building codes. Enforcement requires sufficient numbers of trained and equipped

different. Nonstructural walls, roofing materials, and ornamental attachments such as pediments and signs are examples observed in field surveys in affected areas. Complicating this issue are the large variety of materials and designs and the scarcity of engineers. Materials that provide shelter and the curtain walls of outside buildings must be regulated first, given the risks they pose to pedestrians. To resolve the issue of roofing materials, manufacturers and engineers should be involved in improving construction methods and materials. Also, construction workers should be trained to install such materials in safer ways.

Prevent large deformation of structures. Japanese experts are examining ways to minimize structural deformation. This could be useful to countries whose seismic design codes allow larger deformation than Japan's.

Prepare for tsunamis. Japan's experience and knowledge with tsunami evacuation shelters is useful to other countries exposed to tsunamis, such as Indonesia. The tsunami evacuation shelter in Banda Aceh is an example of Japanese technical cooperation (box 2.2).

Promote seismic base isolation. Buildings with seismic base isolation features suffered very little damage from the GEJE. More key public buildings, particularly those that will be used for emergency relief activities and emergency response—that is, evacuation shelters and fire stations—should be built using base isolation. Simple and affordable techniques for base isolation should be developed for use in developing countries.

NOTES

Prepared by Tatsuo Narafu, Japan International Cooperation Agency, and Mikio Ishiwatari, World Bank.

1. In an earthquake, soil behaves like a liquid, losing its strength and bearing capacity.
2. Isolated structures damp the effects of earthquake ground motion through decoupling of horizontal components. Isolation systems may be laminated steel with high-quality rubber pads, or other energy-absorbing materials.

BOX 2.2

Tsunami evacuation shelters applying the Japanese technical guideline

Banda Aceh was severely damaged by the Indian Ocean Tsunami of 2004. Despite the devastation wrought by the tsunami, local people are returning to coastal areas because their livelihoods are tied to the sea. Because no suitable evacuation areas are found along the coast, evacuation shelters are being constructed. The Japan International Cooperation Agency (JICA) is supporting the construction of vertical evacuation shelters that embody Japanese technical guidelines. The shelter shown below was used for emergency evacuation in 2012.

Source: JICA.

BIBLIOGRAPHY

Architectural Institute of Japan. 2011. *Preliminary Reconnaissance Report of the 2011 Tohoku-Chiho Taiheiyo-Oki Earthquake* [in Japanese]. Ministry of Land, Infrastructure, Transport and Tourism. http://www.mlit.go.jp/.

BRI (Building Research Institute). http://www.kenken.go.jp/english/index.html.

Cabinet Office. http://www.bousai.go.jp/.

JMA (Japan Meteorological Agency). http://www.jma.go.jp/jma/indexe.html.

Mizutani, T. 2012. *Emergency Evacuation and Human Losses from the 2011 Earthquake and Tsunami off the Pacific Coast of Tohoku* [in Japanese]. Research Report on the 2011 Great East Japan Earthquake Disaster, Natural Disaster Research Report (48), National Research Institute for Earthquake Science and Disaster Prevention, Japan.

Narafu, T., H. Imai, S. Matsuzaki, K. Sakoda, F. Matsumura, Y. Ishiyama, and A. Tasaka. 2008. "Basic Study for Bridge between Engineering and Construction Practice of Non-Engineered Houses." In *Proceedings of the 14th World Conference on Earthquake Engineering*, Beijing, China.

National Research Institute for Earth Science and Disaster Prevention. http://www.bosai.go.jp/e/.

NILIM (National Institute for Land and Infrastructure Management). http://www.nilim.go.jp/english/eindex.htm.

NILIM and BRI. 2012. "Summary of Field Survey and Research on the 2011 Earthquake off the Pacific Coast of Tohoku." Technical Note of NILIM No. 647/BRI Research Paper No. 150. NILIM and BRI, Tokyo, Japan.

CHAPTER 3

Hydrometeorological Disasters Associated with Tsunamis and Earthquakes

Earthquakes and tsunamis increase the risks of hydrometeorological disasters. After the Great East Japan Earthquake, disaster-prevention structures such as coastal and river dikes were quickly rehabilitated. A phased process of rehabilitation work made it possible to address urgent needs for protection against frequently occurring floods and storm surges, while at the same time meeting longer-term targets for protection against megadisasters. The deterioration of levels of protection against hydrometeorological disasters was quickly assessed after the event in order to identify priority areas for rehabilitation, revise standards for the issuance of warnings, and raise public awareness about the increased risks of hydrometeorological disasters.

FINDINGS

The Great East Japan Earthquake and tsunami increased the risks of hydrometeorological disasters

The Great East Japan Earthquake (GEJE) caused extensive damage to coastal and river infrastructure and diminished the level of protection they provided against floods and storm surges, thereby increasing the risk of hydrometeorological disasters. Countermeasures against these risks have been successfully put in place (figure 3.1). According to the Ministry of Land, Infrastructure, Transport and Tourism (MLIT), 426 coastal units (including coastal dikes and revetments extending along 190 kilometers [km]) out of a total of 515 units with a total length of some 300 km, sustained damage in the Iwate, Miyagi, and Fukushima prefectures.

The MLIT began, on the day of the earthquake, to assess the safety of dams and structures in some 30 rivers. Slope failure and subsidence of dikes were observed at 2,115 sites

	APRIL	MAY	JUNE	JULY	AUG	SEPT	OCT
Flooding period		Snow melt/flood		Rainy season		Typhoon season	
			Period when spring tide is relatively high				

Assessment and announcement of secondary disaster risks	**Overview assessment** ▮▮▮ Assessment and announcement of risks * Subsidence in the Sendai Plain, Miyagi, and Iwate coastal areas has already been announced
Discharge of water from inundated area	**Emergency discharge** ▮▮▮ * River, coastal, agriculture, and sewerage departments collaborate to implement emergency protection of coastal lowlands and continue necessary measures (water removal by discharge pump vehicles, etc.)

Measures against storm surges — Restoration of coastal dikes, etc.

Emergency measures (stacking sandbags to high tide level) **Temporary measures** ▮▮▮▮ (reinforcement of foreside of sandbags, etc.)

	3 prefectures of Iwate, Miyagi, and Fukushima			
	Total	Fully/partly destroyed	Temporary measures *	Completed
Coastal dike protection (km)	approx. 300	approx. 190	approx. 50	approx. 19

* Section where important public facilities exist

- - - - Implement as many measures as possible

Full restoration (restoration of damaged embankments and dike protection)

Measures against heavy rains and floods

Warning/evacuation measures

Lowering of standards for call-out of flood fighters, or announcement standards for river flood forecasting warnings, communication to residents, etc.

- - - - (emergency rehabilitation of dikes by embankments) - - - - Implement as many measures as possible

River dikes/structures (weirs/gates)

Temporary measures

	Damaged sites (state managed)	Temporary measure	Completed
Tohoku	1,195	29	29
Kanto	920	24	24
Total	2,115	53	53

Full restoration	Damaged sites (state managed)	Full restoration	
		by end of June 2011	by end of June 2012
Tohoku	1,195	993	202
Kanto	920	733	187
Total	2,115	1,726	389

Measures against sediment disasters

Warning/evacuation measures

Detailed assessment ▮▮▮ (inspection of sediment disaster risk areas)

	Total	Inspected	Deformation identified
Inspection sites	33,301	32,302	1,143

* Inspection completed in areas other those inaccessible for inspection (no transportation, nuclear accident affected areas, etc.)

Strengthen warning level by lowering of announcement standards for sediment disaster warning information or installation of mudslide sensors

Sediment management facilities

Temporary measures ▮▮▮ (stacking sandbags, etc.)

Construct sediment control dams as emergency measures in area where failure occurred ▮▮▮

	Before rainy season *1	Before typhoon season *2
Construction of sediment control dams	18	24

*1 Areas where failure was caused by the earthquake
*2 Areas where there are risks of failure

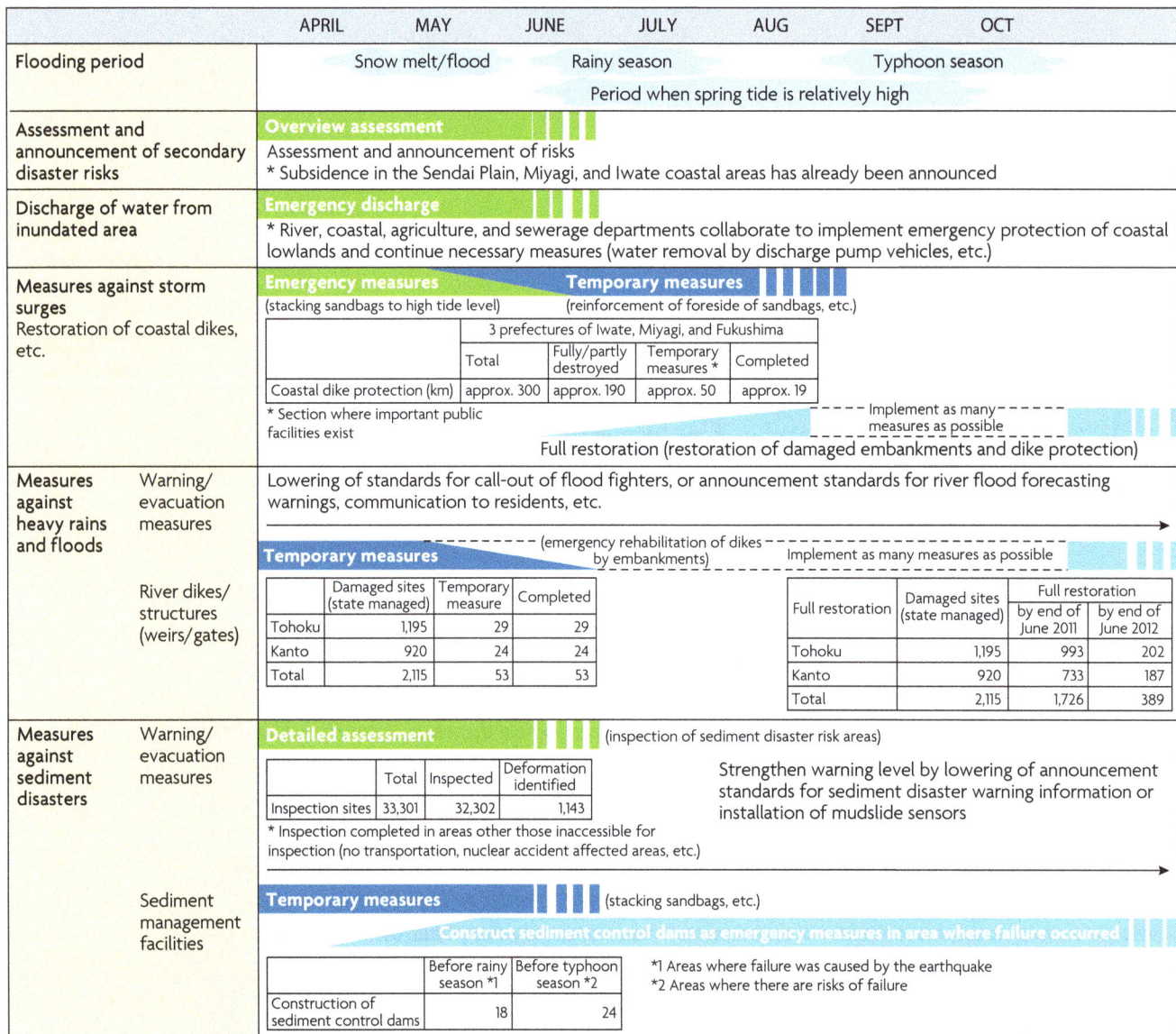

Figure 3.1 Countermeasures taken against hydrometeorological disasters following the GEJE

Source: Ministry of Land, Infrastructure, Transport and Tourism (MLIT).

Figure 3.2 Damage to river dikes at Narusegawa

Source: MLIT.

in eight rivers managed by the MLIT, mainly in the Tohoku and Kanto regions (figure 3.2). Local governments reported damage to a total of 1,627 sites in the rivers they manage. Many river dikes were also damaged by liquefaction caused by earthquakes. The MLIT confirmed that none of the country's dams suffered structural problems, except for minor leaks and cracks. One irrigation dam failed, killing seven and leaving one person missing in Fukushima Prefecture.

Increased inundation risks from subsidence

The earthquake caused extensive subsidence in some areas. Rikuzentakata City in Iwate Prefecture, for example, saw subsidence of 84 centimeters, which led to flooding of coastal areas and roads at high tide, often hampering recovery and rehabilitation efforts.

The level of protection against storm surges and flooding was significantly diminished in the Sendai Plain. The area below mean sea level more than tripled (from 3 square kilometers [km^2] to 16 km^2) after the earthquake (map 3.1), as revealed in the MLIT's laser profiling survey. The MLIT produced subsidence maps and revised downward the water levels at which it issues flood warnings. For management of spatial data and their use in mapping, see chapter 26.

Landslides caused by the earthquake

The earthquake caused 141 landslides, as a result of which 19 people lost their lives (as of February 2012). Immediately after the earthquake, the MLIT began inspecting 1,952 sediment control facilities managed by the ministry, while the prefectural governments inspected 4,324 facilities. The MLIT conducted emergency inspections of about 32,000 sites with potential risks of sediment disasters such as debris flows and landslides in 220 municipalities where the Japan Meteorological Agency (JMA) had observed seismic intensity of 5+ or larger. Significant deformation was found at 66 locations; minor deformation at 1,077. The MLIT shared this information with municipalities so that they could take the necessary measures.

With the higher risk of sediment disaster since the earthquake, triggers for the issuance of sediment disaster warnings were temporarily lowered. Local meteorological observatories and prefectural governments jointly issue warnings about such disasters. Prefectural governments and the JMA reviewed the standards for the issuance of warnings by investigating

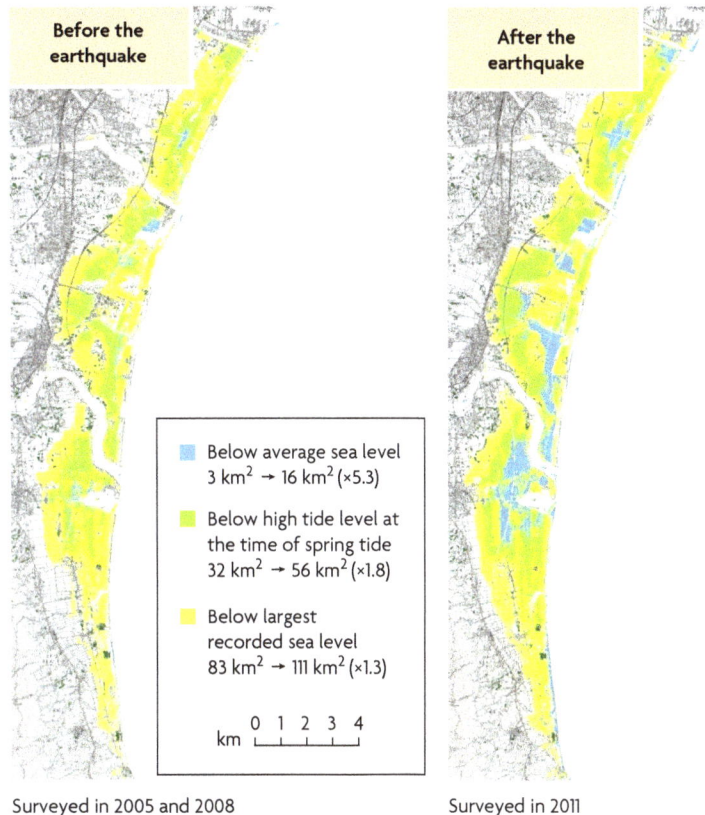

Before the earthquake

After the earthquake

Below average sea level
3 km^2 → 16 km^2 (×5.3)

Below high tide level at the time of spring tide
32 km^2 → 56 km^2 (×1.8)

Below largest recorded sea level
83 km^2 → 111 km^2 (×1.3)

km   0 1 2 3 4

Surveyed in 2005 and 2008   Surveyed in 2011

Map 3.1 Subsidence caused by the earthquake increased inundation risks
Source: MLIT.

the relationship between the amount of precipitation after an earthquake and the probability of a sediment disaster.

Rehabilitating coastal and river dikes to prevent secondary disasters

After the GEJE, emergency measures were implemented to restore coastal dikes to prevent coastal flooding from storm surges. Emergency rehabilitation was first implemented along about 50 of the 190 km of damaged coastline. Those 50 km were selected because of the important facilities and properties in the area, or because of the urgency of restoring livelihoods, industrial activities, transportation, and agricultural activities.

The emergency rehabilitation work was implemented in three phases determined by climatic conditions and the seasonal

Step 1: Emergency rehabilitation 1 completed before onset of flood season

Large sandbags (weatherproofed)

Levee washed out by tsunami

∇ high-tide protection level (T.P. + 2.0m)

Existing levee

Step 2: Emergency rehabilitation 2 completed before onset of typhoon season

Large sandbags (weatherproofed)

∇ high-wave protection level (T.P. + 3.8–6.2m)

∇ high-tide protection level (T.P. + 2.0m)

Existing levee

Step 3: Full rehabilitation to be completed in about five years

∇ high-wave protection level (T.P. + 3.8–6.2m)

∇ high-tide protection level (T.P. + 2.0m)

Existing levee

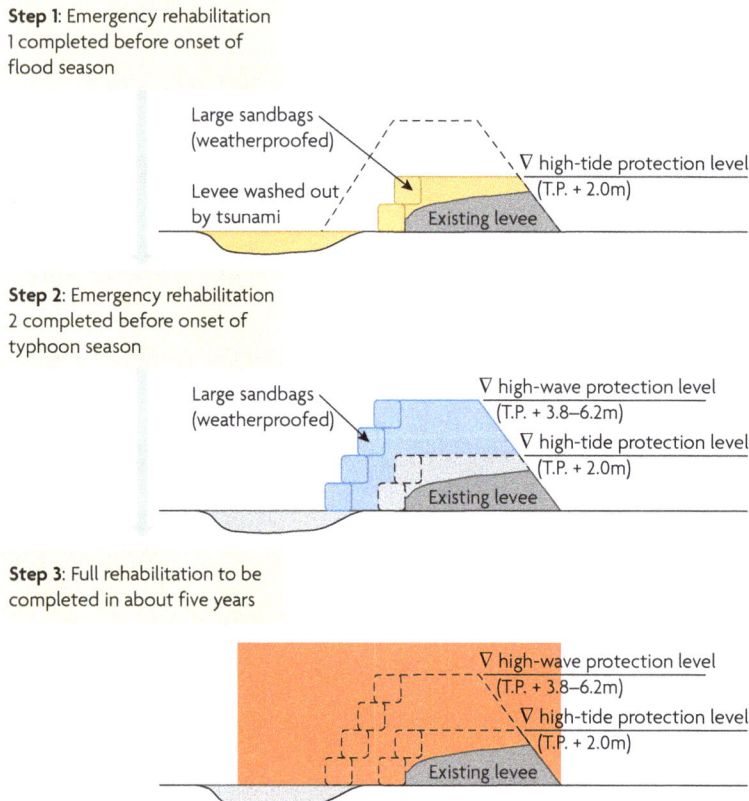

Figure 3.3
Rehabilitation of coastal dikes

Source: MLIT.

Note: T.P. = Tokyo Peil.

occurrence of natural disasters (figure 3.3). The first step was to reinforce and raise the height of the damaged dikes up to the high-tide protection level. This work was done before the June–July flood season. The second step was to raise the dike height to the high-wave protection level, which was completed by September, before the typhoon season.

Full-scale restoration, the third step, started in fiscal 2012 in accordance with reconstruction plans and other rehabilitation projects. The works will be carried out over about a five-year period so as not to disrupt community development and industrial activities. On the Iwanuma Coast and in other coastal areas with facilities that are critical to recovery and reconstruction, such as wastewater treatment plants, full restoration was completed by the end of fiscal 2013, in March 2014.

Rehabilitation of river dikes began directly after the earthquake as the first step in preparing for heavy rain and floods. One of the most urgent tasks was to reconstruct the dikes to

their predisaster height before the rainy season began in June. Emergency rehabilitation work was conducted at the 53 heavily damaged sites: 29 in the Tohoku region and 24 in the Kanto region. These works were completed by July 11, 2011. The standard for flood warnings was lowered during the flood seasons. The MLIT and the prefectural governments measure rainfall and the water level in rivers, using automatic monitoring equipment and telemeter systems. The ministry and the governments then issue flood forecasts and warnings through the mass media, the Internet, and mobile phones.

Complete restoration of the river dikes to their predisaster condition began after the typhoon season and was completed by the time the 2012 rainy season began in June. Countermeasures against liquefaction have also been implemented. The final step will be to improve dikes on the major rivers in the Tohoku region—the Abukumagawa, Narusegawa and Kitakamigawa—to protect against floods and tsunamis.

Measures to mitigate inundation risks in disaster-affected areas

Inundation risks from heavy rain have increased in the disaster-affected lowlands of the Sendai Plain, where river dikes and drainage pump stations were damaged or destroyed and where extensive subsidence occurred. Temporary emergency measures were taken to reduce the risk of flood damage. Thirty-three drainage pump vehicles, provided by other regional bureaus of the MLIT around the country, were deployed in the disaster-affected area. A risk map showing inundation levels from daily precipitation of 100 millimeters and 200 millimeters provided information for local residents and municipalities. Inundation sensors were installed in areas with a high risk of flooding, and the information they collect is published on the Internet. Measures have been taken to send timely notifications automatically to relevant municipalities and local residents when there is a high risk of flooding.

LESSONS

- *Disaster prevention structures such as coastal and river dikes need to be rehabilitated quickly* to prevent secondary disasters. Rehabilitation work should ideally be completed before the next rainy season and typhoon season.

- *In the aftermath of a disaster, it is important to identify the priority areas for rehabilitation* and for protection against hydrometeorological disaster. Priorities can be determined based on the existence of important facilities or commercial production centers and their significance for recovery and reconstruction activities.

- *Rehabilitation work should take place in phases.* This is an effective way of meeting communities' most urgent needs for protection against frequently occurring floods and storm surges, while at the same time meeting longer-term targets for protection from megadisasters.

- *Deterioration in levels of protection against hydrometeorological disasters needs to be quickly assessed,* and the relevant agencies, organizations, and the public should be informed. Damage information should be collected and disseminated as soon as possible (chapter 26). Warning standards should be revised according to the assessment.

RECOMMENDATIONS FOR DEVELOPING COUNTRIES

Following any disaster, protective measures against collateral damage and secondary disasters are essential. The following actions are recommended.

Conduct an assessment immediately following the disaster. Damage to disaster-prevention facilities and the risk of ensuing disasters should be assessed immediately after a disaster by quickly collecting relevant information. To make the most efficient use of resources, the areas to be rehabilitated should be dealt with in order of priority. Expert emergency teams should be formed during normal times by drawing on national networks (chapter 14). Advance agreements can be made to allow the organizations concerned to mobilize private sector resources without going through the usual procurement processes (chapter 20).

Rehabilitate crucial structures before the next disaster. A staged approach is appropriate, taking into account time constraints before the onset of the next season susceptible to hydrometeorological disaster. Rehabilitation works should be prioritized. Practical works, such as temporary structures made of sand bags or gabion boxes, need to be set up quickly.

Consider financial mechanisms. Financial arrangements, in particular the responsibilities of the central and local governments, should be made in advance during normal times (chapter 20).

Share risk information with the community. "Post-disaster risks" should be shared with local communities that may be affected. Nonstructural measures such as warnings should be strengthened in at-risk areas, since the effectiveness of countermeasures will have been diminished by the disaster.

NOTE

Prepared by Junko Sagara, CTI Engineering.

BIBLIOGRAPHY

MLIT (Ministry of Land, Infrastructure, Transport and Tourism). 2012a. *Doshasaigaiheno taioujyokyo, zokuho* [Follow-up report on countermeasures for sediment disasters]. *River* 788: 59–61.

——. 2012b. "Kasenteibotouno fukyujyokyo" [Rehabilitation of river dikes]. *River* 788: 50–54.

Sato, Y. 2012. "Kaiganteibotouno fukyufukojyokyonitsuite" [Rehabilitation of sea dikes damaged in the Great East Japan Earthquake]. *River* 788: 55–58.

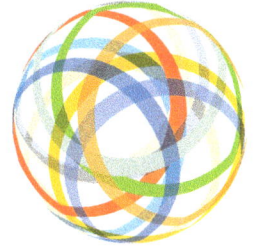

CHAPTER 4

Multifunctional Infrastructure

Public facilities and infrastructure can be built in such a way as to reduce disaster risks and serve as disaster risk management facilities. Roads, expressways, and other public facilities helped reduce damage and loss in the Great East Japan Earthquake by providing protection against flooding, and by serving as evacuation routes and base stations for emergency operations. Organizations for disaster management and other public sector organizations should coordinate to ensure that their public works are multifunctional whenever possible, and cost-sharing mechanisms should be developed to ensure that the financial burden is shared equitably.

FINDINGS

Expressways served as disaster management facilities

Expressways and roads mitigated damage resulting from the Great East Japan Earthquake (GEJE). The East Sendai Expressway, a 24.8-kilometer (km) toll road running through the Sendai Plain, about 4 km off the coast and at an elevation of 7 to 10 meters, acted as a secondary barrier or dike and prevented tsunamis from penetrating further inland (figure 4.1). It also prevented debris from flowing into the inland urban areas. The embankment served as an evacuation shelter for nearby residents, and about 230 people escaped the tsunami by running up to the expressway.

Many expressways were built on high ground, providing routes for evacuation as well as for rescue operations. Many coastal towns and communities were isolated immediately after the disaster because roads were flooded or covered with debris. Expressways built on higher ground served to connect otherwise isolated towns and communities (figure 4.2).

The Sanriku Expressway, a 224-km expressway that runs along the Pacific coast through the Miyagi and Iwate prefectures, is still under construction. About 51 percent of the expressway was open for public use when the area was hit by the GEJE; it helped save many lives.

Expressways constructed on higher ground were not damaged by the tsunami. In the aftermath of the GEJE, they provided an

Figure 4.1 East Sendai Expressway

Source: Ministry of Land, Infrastructure, Transport and Tourism (MLIT).

When the tsunami hit the area, about 60 residents managed to escape from the tsunami by climbing up the expressway embankment.

The Kamaishi–Yamada Road, a 23-km section of the Sanriku Expressway that was opened only six days before the GEJE, served as a disaster management road. It was built to ease traffic congestion on Route 45, the main road connecting the coastal communities. Since Route 45 was prone to flooding from typhoons and tsunamis, the new road was expected to provide an alternative route if Route 45 were cut off in an emergency. In the Unosumai District of Kamaichi City, about 570 residents and school children escaped the tsunami. Because the road that led to the evacuation shelter had been destroyed, they climbed up to the Kamaishi-Yamada Road and managed to reach the evacuation shelter safely.

Service stations and parking areas along highways served as disaster management bases

Roadside service stations, service areas, and parking areas along highways also helped in the disaster management effort, providing bases of operation for rescue teams and evacuation shelters for local residents (table 4.1). The roadside service stations and rest areas along roads and highways, called *Michi-no-eki* (road stations), are equipped with toilets, restaurants, and shops and are also intended to promote local tourism and business. These facilities are developed jointly by the Ministry of Land, Infrastructure, Transport and Tourism (MLIT) in cooperation with local municipalities. In April 2012, there were 987 such stations nationwide. During the GEJE, road stations were turned into disaster management bases equipped with electric power. They were available to the public around the clock when the neighboring area experienced power failures (figure 4.3).

In Minamisanriku City, sports facilities near a highway exit were used as a disaster management center, evacuation shelter, drop-off site

evacuation route for residents and enabled the self-defense forces and other emergency relief teams to get to the coastal municipalities that had been heavily affected. It also served as an important emergency route for transporting food, medical supplies, fuel, and other relief materials going to local disaster management bases and evacuation centers.

Miyako Road, a 4.8-km section of the San-riku Expressway, opened in March 2010.

for emergency supplies, and operating base for the local government, medical institutions, police, and volunteer workers. The local government even moved its office to the site, because its official building had been destroyed by the tsunami.

Evacuation stairs to expressways saved school children

When Iwaizumi Town in the Iwate Prefecture was severely hit by the massive tsunami, an evacuation stairway constructed at the Omoto Elementary School two years before saved the lives of 88 children (figure 4.4). Because there was no escape route from the school, since it was surrounded by steep cliffs, some of the children suggested how improvements might be made during a tsunami evacuation drill. In response to their suggestions and those of local residents, a MLIT field office completed the approximately 30-meter evacuation stairway with 130 steps along Route 45, which runs right behind the school.

Figure 4.2 The Sanriku Expressway was built with tsunamis in mind
Source: MLIT.

Table 4.1 "Road stations" used in the aftermath of the GEJE

ROAD STATION	LOCATION	SERVICES DURING GEJE
Sanbongi	Osaki, Miyagi	Open for 24 hours with power. Supplied food to evacuees.
Tsuyama	Tome, Miyagi	Used as a base for self-defense forces and rescue teams and as an evacuation center.
Fukushima-Touwa	Nihonmatsu, Fukushima	Provided food, water, and toilets for evacuees. Used by 1,500 evacuees.
Kita-no-sato	Kitakata, Fukushima	Provided water and food. The hot-spring facility was made available to the affected residents.
Minamisoma	Minamisoma, Fukushima	Used as an evacuation center and emergency support base.
Hirata	Hirata, Fukushima	Provided power and water to evacuees and food to local hospitals and evacuation centers.

Source: MLIT.

Figure 4.3 Self-defense force at a roadside station
Source: MLIT.

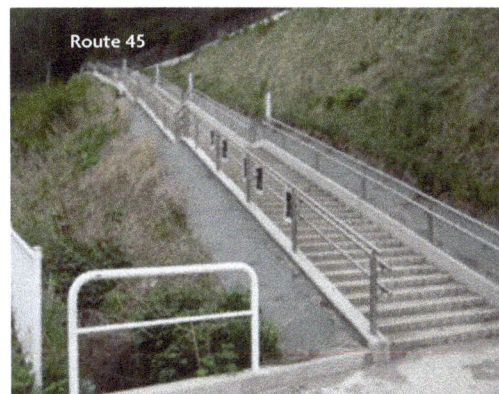

Figure 4.4 Evacuation stairway at the Omoto Elementary School
Source: MLIT.

LESSONS

- *Embankment structures used to raise the elevation of highways and expressways can effectively prevent* penetration of tsunami water and debris further inland. They can also be used as disaster management facilities (box 4.1).

- *Roads, highways, and expressways provided safe evacuation sites and escape routes* because they were designed with earthquakes and tsunamis in mind. It pays to take disaster reduction into account when designing transport and other infrastructure.

- *Public facilities such as roadside stations and highway parking areas* were used by various teams and organizations as base stations for rescue and emergency operations. They were also used as evacuation centers because they were equipped with electricity, food, and water supplies.

BOX 4.1

Evacuation stairs to the East Sendai Expressway

Recognizing that the embankment of the East Sendai Expressway had served as an effective evacuation site for local residents, evacuation stairs were temporarily installed at five locations along the embankment in May 2011. They are intended to facilitate evacuation in case of a tsunami.

Source: MLIT.

RECOMMENDATIONS FOR DEVELOPING COUNTRIES

Infrastructure and public facilities such as roads, highways, and railways can be used as disaster management facilities in the event of floods, tsunamis, mudflows, and landslides. Facilities that are multifunctional are a particularly cost-effective approach to disaster management.

Integrate various facilities into planning for disaster risk management (DRM). DRM plans should include a range of public facilities. For example, playgrounds and parking areas can become rescue team bases or spaces for transition shelters. Expressway embankments can become evacuation sites in the event of cyclones, floods, and tsunamis.

Develop cost-sharing mechanisms. Cost-sharing mechanisms should be established between DRM organizations and public works organizations. The latter cannot be expected to bear all the DRM-related costs of a project, since those costs affect the project's financial feasibility. In Japan the cost of adding height to an expressway is shared by the DRM organizations (chapter 12).

Coordinate with other sectors. Coordination with other sectors, such as transportation, is required to develop multifunctional facilities. Platforms to coordinate planning, construction, and operation and maintenance should be established. In Japan, prefectural governors designate the multifunctional facilities, allowing concerned organizations to initiate coordination under a new tsunami DRM law (chapter 12).

Consider negative effects. High structures such as bridges and highways may have negative effects, such as water logging. They may isolate or separate communities and impose obstacles to the passage of people and animals. These effects should be assessed, and countermeasures or diversion channels and routes developed. In Japan, permission from DRM organizations is required before highways and bridges can be built.

NOTE

Prepared by Junko Sagara, CTI Engineering.

BIBLIOGRAPHY

MLIT (Ministry of Land, Infrastructure, Transport and Tourism). 2011a. *Higashinihon daishinsai ni oite fukujiteki na bousaikinou wo hakkishita jirei* [Cases where structures provided added functions during the Great East Japan Earthquake]. http://www.mlit.go.jp/road/ir/ir-council/hw_arikata/teigen/t01_data04.pdf.

MLIT, Shikoku Regional Development Bureau. 2011b. *Higashinihon daishinsai kara manabu mono* [Lessons of the Great East Japan Earthquake]. http://www.skr.mlit.go.jp/kikaku/senryaku/pdf/1-kaigi/110609%E3%80%80siryou-1.pdf.

MLIT, Tohoku Regional Development Bureau. 2011c. *Higashinihon daishinsa no taiou ni tsuite* [Efforts during the Great East Japan Earthquake]. http://www.thr.mlit.go.jp/road/ir/shouiinkai/pdf/110825/03_siryou1.pdf.

Okumura, Y. 2011. *Higashinihon daishinsai deno tori-kumi* [Efforts in the Great East Japan Earthquake of the Ministry of Land, Infrastructure, Transport and Tourism]. http://www.jari.or.jp/resource/uploads/Symposium2012-01.pdf.

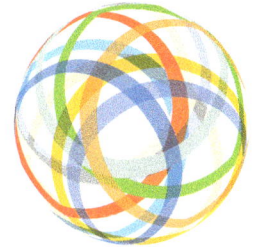

CHAPTER 5

Protecting Significant and Sensitive Facilities

The Great East Japan Earthquake was a multihazard event. A massive quake triggered a series of tsunamis of unprecedented dimensions, as well as the subsequent nuclear accident. Sensitive facilities need to be protected against low-probability and complex events because damage to such facilities can have a cascading effect, multiplying the destruction and leading to irreversible human, social, economic, and environmental impacts.

FINDINGS

Important facilities were seriously damaged by the complex disaster

The Great East Japan Earthquake (GEJE) was a massive disaster triggered by the largest earthquake ever recorded in the history of Japan. But it was not only an earthquake disaster. The quake triggered a series of hazards and events including tsunamis of unprecedented dimensions, as well as a subsequent nuclear accident. Damages to critical disaster-response facilities—such as public buildings, hospitals, and schools—hindered local capacities for response and recovery. Furthermore,

destruction of sensitive facilities—such as a nuclear power station and industrial facilities—led to cascading damages and serious social, economic, and environmental impacts. The cascading effects of the GEJE revealed the weakness of Japanese disaster risk management (DRM) systems in the face of low-probability, high-impact events, and highlighted the importance of protecting sensitive facilities against disasters of any scale.

Government buildings

Local municipalities in Japan have the primary responsibility of saving and assisting people in the event of disasters. However, in the GEJE, many coastal towns and villages were devastated by the earthquakes and tsunamis, suffering great damage to their buildings, facilities, and personnel, and losing their capacity to take response measures promptly.

Based on a survey by Japan's Cabinet Office, of the 237 municipalities that responded and that experienced seismic intensity of 6- or more, about 28 percent had to relocate their buildings either fully or partially (figure 5.1). In Otsuchi Town in Iwate Prefecture, a massive tsunami swallowed up the municipality building, destroying it and taking the lives of town officials, including the mayor, who was at the time directing the disaster-response

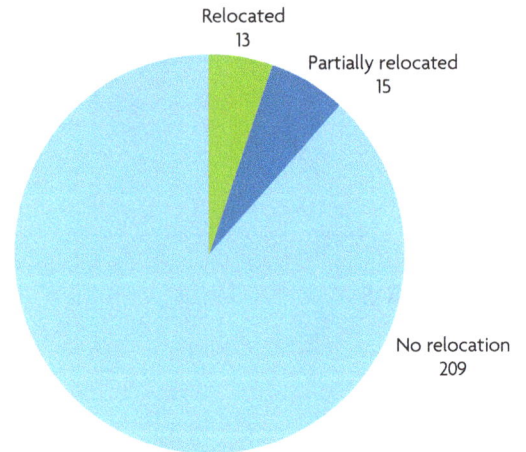

Figure 5.1 Relocation of municipal buildings after the GEJE

Source: Cabinet Office.

An angel's voice

A woman on the municipal staff in Minamisanriku City was urging residents over the radio to evacuate to higher ground. Although tinged with fear and apprehension, her voice gave people courage and helped save countless lives. She continued broadcasting to the very end before being engulfed by the tsunami. She never returned home. She had planned to be married in September 2011. In all, 39 staff members were declared dead or missing. The 12-meter-high building was located in a risk area that was submerged by 2.4 meters of water during the 1960 Chilean Tsunami.

Source: Prime Minister's Office and Fire and Disaster Management Agency.

Figure 5.2 Otsuchi Town Hall

Source: © Mikio Ishiwatari/World Bank. Used with permission. Further permission required for reuse.

operations (figure 5.2). The town was without a mayor for five months.

Disaster management and evacuation facilities

Disaster management and evacuation facilities are critical to protecting people in times of disaster. Many of these facilities were devastated by tsunamis (box 5.1). In the 11 coastal municipalities of Iwate Prefecture, 48 out of 411 emergency evacuation shelters (designated shelters to which people are to evacuate immediately after an earthquake, as distinct from evacuation centers) were inundated by tsunamis; and in Rikuzentakata City, one of the cities

Figure 5.3 The Rikuzentakata City gymnasium

Source: © CTI Engineering. Used with permission. Further permission required for reuse.

with the highest casualty rates, more than half the evacuation shelters were inundated. The city's gymnasium was designated as a primary evacuation shelter, and more than 80 people were there when the tsunami hit (figure 5.3). Only a few survived.

Health and social welfare facilities

Hospitals and social welfare facilities also need to be protected, because without medical response capabilities the number of casualties will increase and health hazards will spread. According to the Ministry of Health, Labour and Welfare, almost 80 percent of hospitals were either destroyed or severely damaged by the earthquakes and tsunamis (figure 5.4). Furthermore, more than 12 percent of social welfare facilities—such as homes for the elderly, children, people with disabilities, and

other vulnerable groups—were damaged by the disaster (see chapter 16).

Industrial facilities

Six out of nine oil refineries in the Tohoku and Kanto regions had to suspend operations; fire broke out at two of the nine facilities. At an oil refinery in Chiba, the structure holding one of the liquefied petroleum gas (LPG) tanks failed, and the tank collapsed, leading to LPG leakage. The leaked LPG caught fire and caused an explosion, spreading the fire from one tank to another (figure 5.5). Six people were injured and all 17 LPG tanks were damaged, along with pipelines and roads. The fire and debris from the explosions damaged the surrounding buildings and vehicles. Nearby residential areas also suffered as the blasts damaged windows, shutters, slate roofs, and more. The explosions at the oil refineries are believed to have been among the factors that accounted for the fuel shortage immediately after the disaster, which disrupted people's lives and hindered emergency recovery operations in the disaster-affected areas.

The collapsed tanks had met all the requirements for earthquake-proof structures; however, at the time of the earthquake the tank was temporarily filled with water, instead of the lighter-weight LPG, in preparation for a

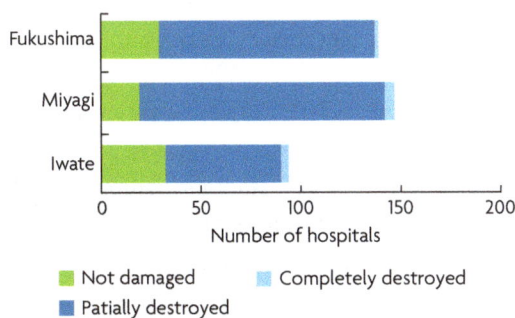

Figure 5.4 Hospitals affected by the GEJE in Fukushima, Miyagi, and Iwate prefectures

Source: Ministry of Health, Labour and Welfare.

Fire broke out near tank No. 364

Figure 5.5 Leaked LPG catches fire at a refinery

Source: © CTI Engineering. Used with permission. Further permission required for reuse.

Figure 5.6 Broken braces led to collapse of LPG tank

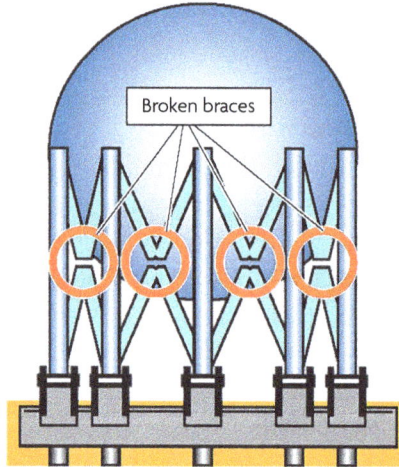

Broken braces

regular inspection. The braces supporting the legs that held the tank up could not bear its weight during the earthquake, leading to its collapse (figure 5.6).

In light of this accident, a government committee that conducted a technical review of LPG facilities recommended:

1. Revision of the technical guideline for the tank braces

2. Confirmation of the facilities' safety by private companies, and government monitoring of the confirmation

3. Risk assessment and countermeasures against liquefaction to be undertaken by private companies

4. Reassessment of earthquake risks following the government review.

Figure 5.7 Retrofitting Jokoji Temple

Source: Agency for Cultural Affairs.

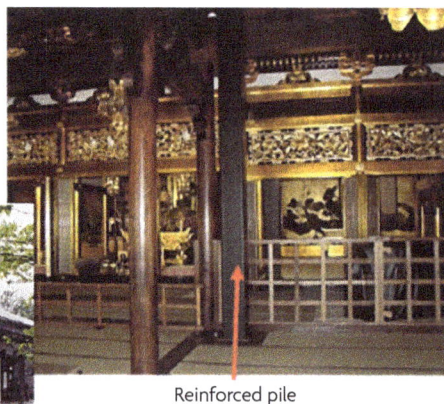

Reinforced pile

Cultural properties

According to the Agency for Cultural Affairs, more than 700 nationally designated cultural properties (such as monuments and historic buildings and landscapes) were heavily damaged by the earthquake and tsunami (see chapter 35). Many national treasures, important cultural properties, and special historic sites were also affected. Fortunately, few cultural properties of national importance were damaged. However, several properties will take a long time to recover, and some have been lost forever.

Disasters that result in irreversible damage or losses of important cultural properties can have a severe negative effect on local businesses, such as those that depend on the tourism industry, and can also undermine people's pride in their communities. A culture-sensitive approach to restoration, in which original or local materials are used, is required to maintain the cultural value of historical buildings (figure 5.7). Retrofitting work should not be carried out in a way that destroys the historic value of a monument or building. If retrofitting cannot be carried out without compromising the structure's cultural value, the area should be closed to visitors rather than altered in a way that changes its character. Following the Great Hanshin-Awaji Earthquake in 1995, the Japanese government established guidelines for protecting cultural properties against earthquakes and began implementing seismic assessments and retrofitting structures associated with national treasures and important cultural properties.

The cascading effect of the accident at the Fukushima Daiichi Nuclear Power Station

Four nuclear power stations comprising 14 units were located close to the epicenter of the March 11 earthquake (map 5.1). The earthquake caused all operating units to shut down automatically (box 5.2). Large tsunamis hit all sites within an hour of the main shock, damaging several of them. The worst affected sites were Fukushima Daiichi and Fukushima Daini.

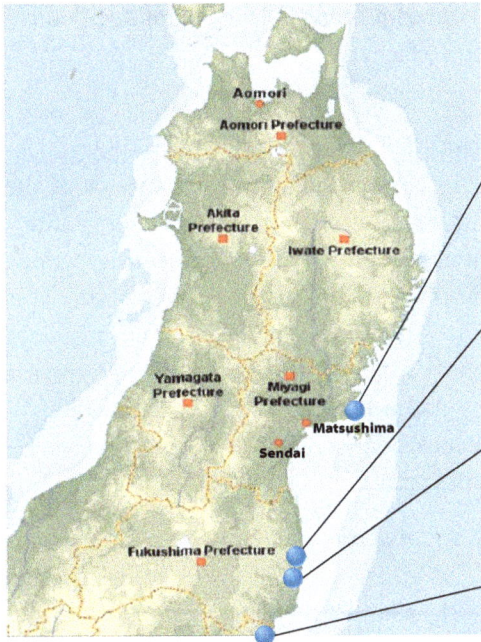

Unit	Output, commission year	Automatic shut down	Cold shut down
Onagawa			
Unit 1	524 MW, 1984	✓	✓
Unit 2	825 MW, 1995	✓	✓
Unit 3	825 MW, 2002	✓	✓
Fukushima Daiichi			
Unit 1	460 MW, 1971	✓	
Unit 2	784 MW, 1974	✓	
Unit 3	784 MW, 1976	✓	
Unit 4	784 MW, 1978	Periodic inspection	
Unit 5	784 MW, 1978		✓
Unit 6	1,100 MW, 1979		✓
Fukushima Daini			
Unit 1	1,100 MW, 1982	✓	✓
Unit 2	1,100 MW, 1984	✓	✓
Unit 3	1,100 MW, 1985	✓	✓
Unit 4	1,100 MW, 1987	✓	✓
Tokai Daini			
Unit 1	1,100 MW, 1978	✓	✓

Map 5.1 Nuclear power stations near the epicenter and their emergency shutdown modes

Source: Office of the Prime Minister of Japan.

Fukushima Daini lost some safety-related equipment, but off-site and on-site power remained available, although not at optimal levels. On the other hand, Fukushima Daiichi lost much of its safety-related equipment because of the tsunami and almost all off-site and on-site power (figure 5.8). This led to a loss of cooling to the operating reactors, and the ensuing nuclear meltdowns and release of radioactive materials.

The failure of the Fukushima Daiichi Nuclear Power Station has had severe social consequences (see chapter 36). About 160,000 people in Fukushima were evacuated, of whom more than 60,000 were taken outside Fukushima Prefecture. Many were unable to return to their homes for a long time because of unsafe levels of radioactivity.

Some agricultural products were found to contain high levels of radiation, resulting in local products being stigmatized as unsafe. There was also an incident in which radioactive gravel from Fukushima was mixed into the concrete used for construction of a new apartment building, exposing the residents to radiation.

The Japanese government has taken decisive steps to clean up contaminated areas

BOX 5.2

The tsunami's impact on the Onagawa nuclear power station

The Tohoku Electric Power Company's Onagawa Nuclear Power Station is located about 120 kilometers west of the epicenter of the March 11 earthquake. Although the tsunami was about 13 meters high at the Onagawa nuclear power station, the station's structures and equipment were not severely damaged.

When the first unit was built in the 1970s, the site elevation of the station was set as 14.8 meters above sea level. A literature review and interview surveys revealed that the maximum tsunami height at the Onagawa site was estimated to be about 3 meters, but the 14.8-meter site elevation was considered appropriate.

Since then, the tsunami hazard assessment has been reviewed many times, using up-to-date findings and cutting-edge tsunami simulations and, every time, the safety of the facility against tsunamis has been confirmed. The most recent tsunami design standard was set as 13.6 meters. Even though the Onagawa site experienced a subsidence of 1 meter, the March 11 tsunami did not submerge the main facility.

At the second unit, however, the intake unit for the seawater pump station was built as a pit-structure, and the pump was situated below the rest of the facility. This caused the seawater to enter the pump room through the tide gauge, submerging an emergency generator and rendering it inoperable.

In the aftermath of the disaster, the main building of the nuclear power station was used as an evacuation center for about 400 local residents whose houses had been washed away. These people stayed at the power station for about three months.

Source: Matsuo 2012 and Tohoku Electric Power Co., Ltd. 2011.

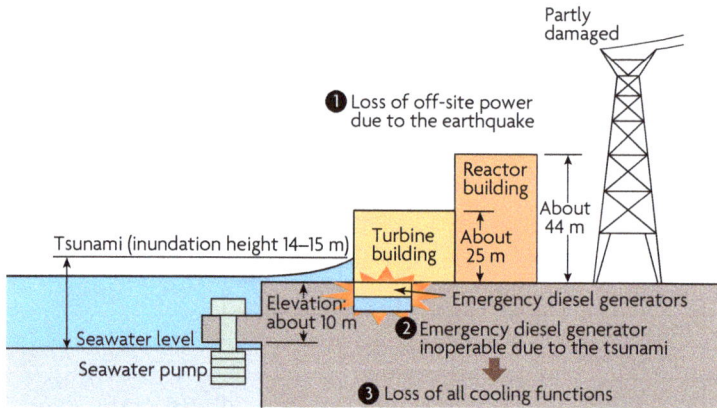

Figure 5.8 Cause of the accident at the Fukushima Daiichi Nuclear Power Station

Source: Office of the Prime Minister of Japan.

Figure 5.9 Fukushima Daiichi Nuclear Power Station

Source: TEPCO (Tokyo Electric Power Company).

The interim report of the committee pointed to three factors:

- *Lack of preparedness for serious accidents caused by tsunamis.* Neither Tokyo Electric Power Company, the operator of the nuclear stations, nor the regulatory authorities had prepared for accidents as serious as those caused by the enormous tsunamis that followed the GEJE. Countermeasures must be put in place to address high-impact events, even those with low probability. All concerned organizations must recognize these risks.

- *Lack of appreciation for the effects of complex disasters.* Securing nuclear stations and ensuring the safety of people in the neighboring communities against unforeseen complex disasters is a serious issue. Existing countermeasures for dealing with complex disasters must be reviewed and revised.

- *Lack of a holistic understanding of complex disaster scenarios.* Existing countermeasures to address nuclear power accidents do not reflect a thorough understanding of the complexity of nuclear power station systems. The excuse that the event was "beyond assumption" is unacceptable. Serious problems existed in the DRM system for nuclear accidents.

around Fukushima and to minimize health risks. It has set aside about ¥1.15 trillion for decontamination and disposal of contaminated waste between fiscal years 2011 and 2013. The long-term environmental and health effects of the nuclear incident are unknown; the Japanese government will be monitoring the health status of residents of Fukushima Prefecture over the next 30 years.

The Government Investigation Committee on the Accident at the Fukushima Nuclear Power Stations stressed that a paradigm shift is required in DRM for catastrophic events.

LESSONS

- *Important facilities were in most cases well protected against large-scale earthquakes* thanks to seismic reinforcement and other measures.

- *Crucial facilities or facilities sensitive to disasters should be designed to withstand extreme events.* Although tsunami hazards were taken into account in the site evaluations and design of facilities, the hazard level had been underestimated.

- *Nuclear power stations and other disaster-sensitive facilities should be carefully evaluated against the risks of all natural hazards, and these assessments should be periodically revised* based on the latest knowledge and technologies. The failure of a sensitive facility, as in the case of the Fukushima Daiichi Nuclear Power Station, can cause not only short-term consequences but also long-term social, economic, and environmental problems.

RECOMMENDATIONS FOR DEVELOPING COUNTRIES

The cascading effects of the GEJE disaster highlight the importance of protecting sensitive facilities against disasters of any scale. The following are recommended as important steps to lower risks for crucial facilities and to prevent high and irreversible impacts of complex disasters.

Identify critical facilities. Critical facilities need to be identified and well protected against extreme events. These include hospitals, government offices, evacuation shelters, schools, and other facilities to be used for rescue operations, evacuations, and other disaster management activities. Also, facilities, such as nuclear power stations and oil refineries that may cause cascading effects in various sectors should be identified. Disaster management plans should include information on the functions of these facilities and the risks they may pose.

Assess critical facilities. Facilities that are required to function as bases for disaster-response activities should be "stress tested" for disaster resistance. Even simple assessments, such as confirming a facility's safety against recorded disasters, is useful in preparing for disaster. The risk of all natural hazards, including that of multihazard events, should be carefully evaluated. Risk assessment should incorporate not only statistics

on recent hazards but also historical records of past disasters as well as future projections, if possible. Such assessments and assessment methodologies should be periodically updated.

Protect critical facilities. Critical facilities should be protected against the risks of all natural hazards. The possibility of multihazards should be considered in their design. Enforcement of building codes should be a high priority for buildings and other important structures.

Prepare for complex disasters. High-risk plants and facilities need to be included in disaster management plans. Plans for quick recovery and rehabilitation after a disaster of unexpected scale should be made. Evacuation drills should be conducted based on various disaster scenarios.

Establish enforcement mechanisms. Regular inspections of critical facilities by firefighters and other disaster management organizations should be established. Responsibility for safety guidelines, monitoring, and enforcement needs to be clearly established in land-use procedures, building codes, fire inspections, and so on. Effective enforcement requires appropriate legislation, organization, and human resources.

NOTE

Prepared by Mikio Ishiwatari, World Bank, and Masato Toyama and Junko Sagara, CTI Engineering.

BIBLIOGRAPHY

Agency for Cultural Affairs. 2011. *Damages to Cultural Properties in the Great East Japan Earthquake.* http://www.bunka.go.jp/english/pdf/2011_ Tohoku_ver14.pdf.

Central Disaster Management Council. 2011. *Report of the Committee for Technical Investigation on Countermeasures for Earthquakes and Tsunamis Based on the Lessons Learned from the "2011 off the Pacific Coast of Tohoku Earthquake."* http://www .bousai.go.jp/kaigirep/chousakai/tohokukyokun /pdf/Report.pdf.

Investigation Committee on the Accident at the Fukushima Nuclear Power Stations. 2011. *Interim Report*.

Matsuo, A. 2012. "Genshiryoku anzenwo sasaeru dobokugijutsunituite" [Civil engineering technologies that support nuclear safety based on a site survey of Onagawa Nuclear Power Station of Tohoku Electric Power Co., Ltd.]. *Journal of Japan Society of Civil Engineers* 97 (4): 95–97.

Nuclear Emergency Response Headquarters. 2011. *Report of Japanese Government to IAEA Ministerial Conference on Nuclear Safety— Accident at TEPCO's Fukushima Nuclear Power Stations*. http://www.kantei.go.jp/foreign/kan /topics/201106/iaea_houkokusho_e.html.

Tohoku Electric Power Co., Ltd. 2011. *Ongawa genshiryoku hatsudensho ni okeru tsunami hyoka taisaku no keii nitsuite* [Tsunami assessment and measures at Onagawa Nuclear Power Station]. http://www.nsc.go.jp/senmon/shidai/jishin /jishin4/siryo4-2.pdf.

PART II

Nonstructural Measures

CHAPTER 6

Community-Based Disaster Risk Management

Local communities play a key role in preparing for disastrous events such as the Great East Japan Earthquake, and are normally the first responders to take action. On March 11, 2011, community-based organizations were active in the disaster response and saved countless human lives. Recognizing the role of communities and providing them with central and local government support is critical to maintaining and strengthening important community-based functions.

Local communities have been responding to and managing disaster risk for centuries. Before the creation of Japan's formal state system, local communities carried out disaster-related activities as volunteers; community-based organizations (CBOs) have existed for centuries. They include: *Suibo-dan* for flood risk dating from the 17th century, *Syobo-dan* for firefighting from the 18th century, and *Jisyubo* for earthquake disasters from the 1970s (see table 6.1).

In addition, various nongovernmental organizations (NGOs) and nonprofit organizations (NPOs) are involved in disaster risk management (DRM) activities at the community level. Many of them collaborate with neighborhood associations (*Jichikai*) and local governments, and sometimes with local academic institutions.

How the government and CBOs coordinate around DRM has evolved over two centuries, shaped by major events and trends. These include the Meiji Restoration at the end of the 19th century, which prompted modernization and centralization; democratization following World War II; and the miracle of economic development in the 1960s. Traditional community structures were eroded over time as

Table 6.1 Community-based organizations engaged in disaster risk management in Japan

ORGANIZATION	HAZARD	LEGAL ACT	SUPERVISING GOVERNMENT ORGANIZATION	DATE ESTABLISHED	NUMBER OF STAFF OR GROUPS
Suibo-dan	Flood	Flood Fighting (*Suibo*) Act	Ministry of Land, Infrastructure, Transport and Tourism	17th century	900,000 staff in two organizations
Syobo-dan	Fire	Fire Defense Organization Act	Fire and Disaster Management Agency (FDMA)	18th century	
Jisyubo	Earthquake	Basic Act on Disaster Reduction	Cabinet Office, FDMA	1970s	140,000 staff
NPO	All	Act to Promote Specified Nonprofit Activities	Cabinet Office	After the Kobe earthquake in 1995	over 2,000 groups

Source: Ishiwatari (2012).

Japanese society modernized and urbanized. As depicted in figure 6.1, this has resulted in a decrease in spontaneous and autonomous community-based engagement in DRM with a corresponding increase in government support to these activities. The government's recognition of and support to community-based DRM has been key to keeping these efforts alive and well.

FINDINGS

The role of community-based organizations in the Great East Japan Earthquake

A key factor in reducing the number of lives lost in the Great East Japan Earthquake (GEJE) was the long tradition of CBOs around risk reduction and preparedness. The tsunami waves

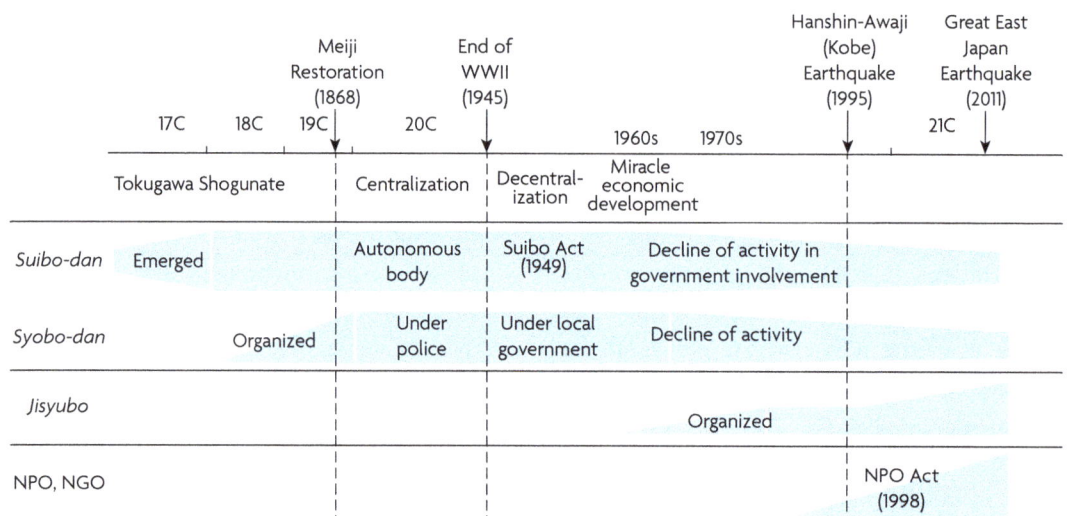

Figure 6.1 Historical timeline of community-based organizations

Source: Ishiwatari (2012).

brought on by the GEJE overwhelmed coastal defenses, and warning systems underestimated the height of the waves. CBOs played critical roles in responding to the event.

The volunteer fire corps (*Syobo-dan*)

The volunteer fire corps traces its history to the 18th century. Corps members have regular jobs but, when disaster strikes, they take part in disaster management activities in their own communities, such as firefighting, issuing warnings, assisting evacuations, conducting search and rescue operations, and operating facilities. There are currently some 890,000 active volunteers across Japan, which is almost six times the number of career firefighters. The Fire Defense Organization Act and its bylaws stipulate the corps's roles, organizational structures, members' status as part-time government staff, and compensation and allowances. The local government has principal responsibility for the corps, while the central government subsidizes their facilities.

The *Syobo-dan* responded to the GEJE at the risk of their own lives. Some 250 members were killed or are missing, including 51 in Rikuzentakata City. Some examples follow:

- A corps member quickly guided all the people in a community to an evacuation shelter preventing any casualties. Corps members supported the evacuation of 30 handicapped and elderly persons, and persuaded three other people to move who were insisting on staying at home (Shiogama City).

- Members closed the tsunami gates by hand, since they could not be operated automatically because of power failures (Miyako and Ofunato cities).

- Members died closing the tsunami gates in Kamaishi and Ishinomaki cities.

- One member died ringing a fire bell to warn people of the tsunami right up until the tsunami hit (Otsuchi Town).

- Six members, on the way back from closing gates, tried to save a bedridden elderly woman from her residence. Five of the six members and the woman died in the tsunami (Otsuchi Town).

Based on lessons learned from the GEJE, the Fire and Disaster Management Agency requested local governments to reinforce the volunteer fire corps in October 2011 with equipment, increased allowances up to the level stipulated by law, and the recruitment of new members.

Neighborhood associations (*Jichikai*)

Communities were generally very well prepared for the GEJE. Most had participated in regular disaster drills and knew what to do when the tsunami warning was issued.

For example, in Kesennuma City, a television program broadcast in 2007 urged neighborhoods to prepare themselves. The program provided a detailed simulation of a tsunami hitting Kesennuma. This simulation was shown to the local residents, and the neighborhood associations (*Jichikai*) subsequently undertook to identify key evacuation routes. Regular disaster drills were also conducted. These preparations helped local residents to evacuate safely and quickly to higher ground immediately after the GEJE, thus saving many lives (figure 6.2).

In the Toni village of Kamaishi City, community members participate in annual disaster evacuation drills conducted by the *Jichikais*.

Figure 6.2 Left: Damaged Hashikami area of Kesennuma; right: Kesennuma Fukkou Yatai Mura (community recovery restaurant)

Source: © Kyoto University. Used with permission. Further permission required for reuse.

The drills are conducted every year on March 3 to mark the anniversary of the Meiji Sanriku Tsunami of 1896. Participation rates in the disaster drill vary from neighborhood to neighborhood, with more people participating in the smaller, more cohesive communities. According to the head of the *Jichikai*, the participation rate in Kojirahama is low, while in Kerobe most people participate in the drill. In Kerobe and Oishi, community members have a strong sense of solidarity, as the population is much smaller than in Kojirahama and they have lived there for decades. Toni residents have written books about the effects of past tsunamis, which are used by the communities as an awareness-raising tool. In addition, there are two tsunami maps: one issued by the Kamaishi City government and the other developed by the community members themselves. The former includes the expected flood area, expected height of the tsunami, and expected arrival time. The latter

includes local information about which areas were flooded in the Meiji Sanriku and Showa-Sanriku tsunamis, evacuation sites, evacuation routes, and dangerous areas. These maps are distributed to all families in the town of Toni (chapter 27). Finally, a number of community festivals are used as opportunities to engage local schools in disaster awareness and preparedness activities (figure 6.3).

In the Wakabayashi ward of Sendai City, the local community forged a very strong relationship with the elementary school to educate people in disaster preparedness. At the initiative of the *Jichikai*, regular drills were conducted in cooperation with the school. A handbook was prepared on managing the evacuation centers. After the 2010 Chilean earthquake, a tsunami warning was issued for the Tohoku coast, and tsunami waves of up to 1.5 meters reached some areas. This prompted communities in Wakabayashi to reexamine their evacuation plans. They found that it would take longer than expected for school children in the coastal school, Arahama Elementary, to evacuate to the designated school, which was 4 kilometers from the coast. The local community therefore decided to take shelter in Arahama Elementary School, and emergency food supplies were increased to feed 800 instead of 300 people and were stored on the top floor of the school building. During the GEJE, Arahama Elementary served as the shelter for more than 300 adults from local communities, in addition to 70 school children. They remained in the school overnight, and the food supplies were well protected on the top floor during the emergency (figure 6.4).

LESSONS

The GEJE experience yielded several important lessons about the need to empower communities to understand and reduce the risks they face, to be prepared, and to act as first responders to hazard events. It also pointed

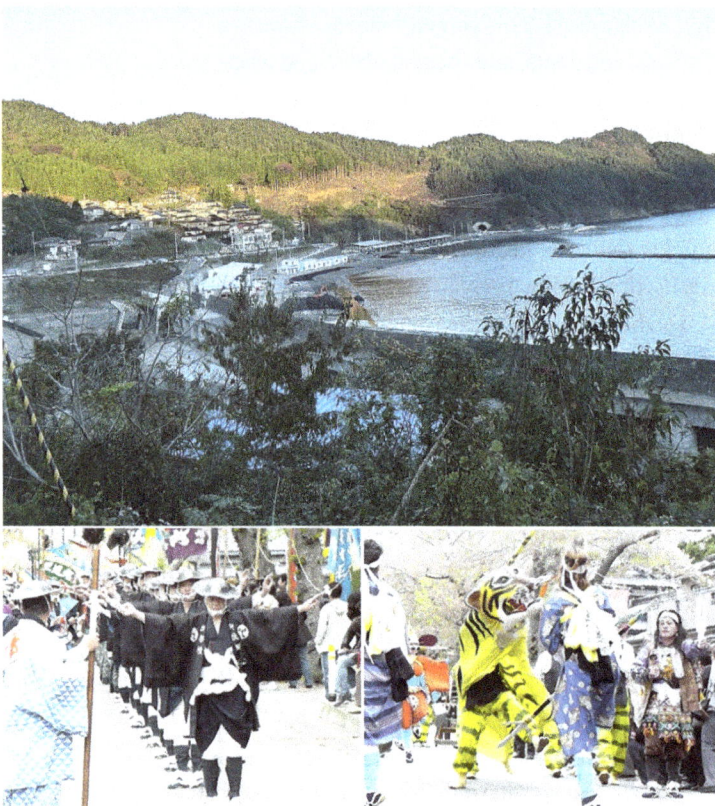

Figure 6.3 Toni Bay area of Kamaishi City (top), Sakura Festival (bottom)

Source: © Kyoto University. Used with permission. Further permission required for reuse.

to important ways that their roles can be strengthened. Specific lessons include:

- *The volunteer fire corps plays a critical role in DRM* for several reasons:
  - Since the volunteers come from the community, they have local knowledge of the context and are familiar with those residents who may need special assistance to evacuate, such as the disabled or bedridden.
  - The total number of volunteers is some six times that of the professional fire-fighting staff, providing a cost-effective way of mobilizing large-scale emergency response capacity.
  - The members receive regular training and can respond immediately because they are locally based.

- *Community-based DRM activities are well integrated in the daily lives* of the residents, ensuring that awareness of natural hazards is maintained, for example, by marking the anniversary of a large catastrophe with disaster drills, and linking awareness-raising activities with local festivals.

- *The role of communities in DRM is formally recognized and supported* by local and national authorities through linkages with local institutions.

RECOMMENDATIONS FOR DEVELOPING COUNTRIES

Empower community members. Most people saved from major disasters are rescued by relatives and neighbors within the first 24 hours—before professional responders can get there. Statistics show that in the 1995 Hanshin-Awaji (Kobe) earthquake, 80 percent of those rescued were saved by their neighbors. So, while local and national authorities have key responsibilities for civil protection in hazard events, communities are always the first responders and should be empowered in that role.

Figure 6.4 Wakabayashi ward (top), and local community activities (bottom)

Source: © Kyoto University. Used with permission. Further permission required for reuse.

Raise awareness. Strong and effective community-based DRM requires grassroots support and linkages to the day-to-day life of the community. Linking disaster risk awareness and preparedness activities to local cultural events can be extremely effective in maintaining a culture of preparedness.

Support community organizations. In addition to grassroots support, building effective and sustainable capacity for community-based DRM requires the formal recognition and support of local and national authorities. In addition to providing financial and technical assistance, local and national governments should develop legislation on and institutionalize the role of CBOs.

NOTE

Prepared by Rajib Shaw, Kyoto University, and Mikio Ishiwatari and Margaret Arnold, World Bank.

BIBLIOGRAPHY

Ishiwatari, M. 2012. "Government Roles in Community-Based Disaster Risk Reduction." In *Community-Based Disaster Risk Reduction: Community, Environment and Disaster Risk Management,* edited by R. Shaw. U.K.: Emerald Group Publishing.

MAG (Neighborhood Disaster Volunteers Foundation). http://www.mag.org.tr/eng/mag.html.

CHAPTER 7

Disaster Management Plans

Following its devastating experience with recent disasters, Japan has been strengthening and drawing up new disaster management plans at the national and local levels. The Great East Japan Earthquake revealed a number of weaknesses in planning for complex and extraordinary disasters. Central and local governments have been revising their plans to reflect what they learned from the disaster.

FINDINGS

Japan's disaster management system addresses all phases of disaster prevention, mitigation and preparedness, and emergency response, as well as recovery and rehabilitation. It specifies the roles and responsibilities of national and local governments, and enlists the cooperation of relevant stakeholders in both the public and private sectors. Following the Great East Japan Earthquake (GEJE), assessments have been made of the capacity of existing disaster risk management (DRM) planning systems to prepare for and react to large-scale disasters.

Revisions have been proposed, based on the lessons learned on March 11.

Disaster management systems in Japan

Disaster Countermeasures Basic Act. In the 1940s and 1950s Japan was repeatedly ravaged by typhoons and earthquakes. In particular, the Isewan Typhoon in 1959 caused tremendous damage; in 1961 the Disaster Countermeasures Basic Act was passed.

The act established the following:

- The Central Disaster Management Council was established to formulate the overall policy for DRM and to function as the national coordinating body for disaster management. The council was chaired by the prime minister, and its members came from line ministries, semipublic organizations (such as Public Broadcasting, the Bank of Japan, the Japanese Red Cross, and a telecommunications company) and representatives from academia (figure 7.1).

- Roles and responsibilities regarding disaster reduction were clearly defined at the national, prefectural, and municipal government levels, as well as for community organizations and citizens; and the three levels of governments were required to draw up master plans for DRM. Also, all the ministries and semipublic organizations were asked to prepare disaster management plans for their sectors.

- The cabinet submitted an annual report to the National Diet covering the status of DRM, and specifying the budgetary allocations for DRM programs. The National Diet formed special committees for disaster management in both its lower and upper

houses, which have continued to monitor governmental DRM initiatives.

In 1995 the occurrence of the Hanshin-Awaji (Kobe) Earthquake forced a revision of the 1961 Act to focus more on countermeasures and prevention, resulting in a new Disaster Countermeasures Basic Act in 1995.

The Central Disaster Management Council retained its leading role in conducting the following activities:

- Formulating and coordinating the implementation of the Basic Disaster Management Plan

- Formulating and coordinating the implementation of contingency plans for emergencies

- Advising the prime minister or the minister of state for disaster management on important issues relevant to disaster management

- Fostering consultations on important issues surrounding disaster management, in response to inquiries from the prime minister or the minister of state for disaster management

The Cabinet Office is the secretariat for this council. The Minister of State for Disaster Management, who is assisted by the staff of the Cabinet Office, has a mandate to oversee the planning and central coordination of basic DRM policy and large-scale disaster countermeasures. The minister is also responsible for integrated information gathering and other disaster emergency measures.

The Basic Disaster Management Plan is the master plan and the basis for DRM activities in Japan. It is prepared by the Central Disaster Management Council in accordance with the Disaster Countermeasures Basic Act. The plan clarifies the duties of the central government, public corporations, and local governments in implementing measures. The plan also describes the sequence of disaster countermeasures such as preparation, emergency

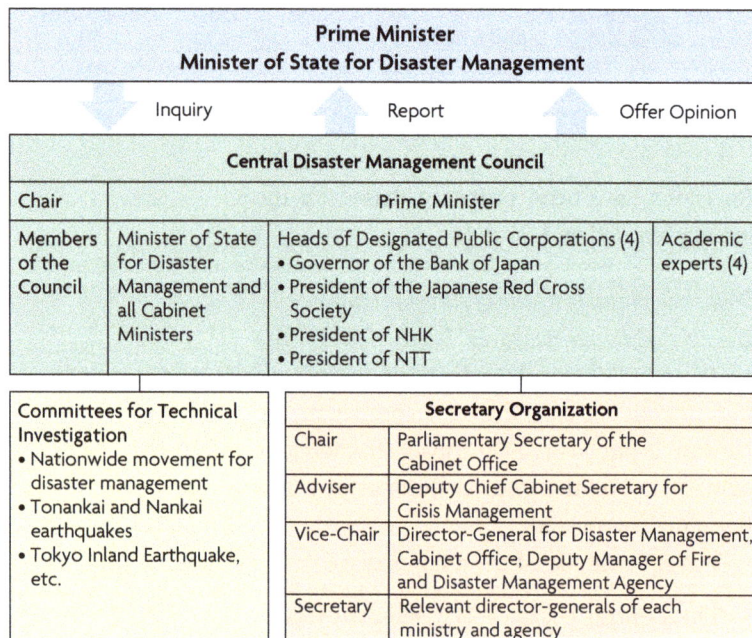

Prime Minister Minister of State for Disaster Management			

Inquiry　　　　　　Report　　　　　　Offer Opinion

Central Disaster Management Council			
Chair	Prime Minister		
Members of the Council	Minister of State for Disaster Management and all Cabinet Ministers	Heads of Designated Public Corporations (4) • Governor of the Bank of Japan • President of the Japanese Red Cross Society • President of NHK • President of NTT	Academic experts (4)

Committees for Technical Investigation • Nationwide movement for disaster management • Tonankai and Nankai earthquakes • Tokyo Inland Earthquake, etc.	Secretary Organization	
	Chair	Parliamentary Secretary of the Cabinet Office
	Adviser	Deputy Chief Cabinet Secretary for Crisis Management
	Vice-Chair	Director-General for Disaster Management, Cabinet Office, Deputy Manager of Fire and Disaster Management Agency
	Secretary	Relevant director-generals of each ministry and agency

Figure 7.1 Structure of the Central Disaster Management Council

Source: Cabinet Office.

Note: NKH = Japan Broadcasting Corporation; NTT = Nippon Telephone and Telegraph.

response, recovery, and reconstruction for various types of disasters.

Based on the Basic Disaster Management Plan, every designated government organization and public corporation draws up a Disaster Management Operation Plan, and every prefectural and municipal disaster management council prepares a Local Disaster Management Plan.

The council has the right to establish technical committees to study technical matters. After the GEJE, the council recommended specific evaluations to identify whether any revisions or additions to the 1995 Disaster Countermeasures Basic Act were required.

The Expert Committee on Earthquake and Tsunami Disaster Management prepared a report to document facts and findings from the GEJE experience. In response to this report, the Japanese government amended the Basic Disaster Management Plan on December 27, 2011, aiming to enhance countermeasures against multihazard, high-impact events.

Major revisions to the plan included:

- Addition of a new section on tsunami disaster management

- Fundamental improvements in disaster management for tsunamis and earthquakes in the light of the GEJE:
  - Requirements to prepare for low-probability and large-scale earthquakes and tsunamis
  - More careful consideration of multihazard and multilocation disasters
  - Mandatory inclusion of DRM in urban land use
  - Raising of public awareness about evacuation, DRM measures, and hazard maps
  - Additional investments nationwide for capacity building of each countermeasure
  - More resources to be invested in understanding disaster risk, and developing innovative systems for monitoring earthquakes and tsunamis
  - Communication tools such as tsunami early warning systems to be strengthened
  - Additional reinforcement and retrofitting of homes and buildings to reduce earthquake damage

- Revision of countermeasures, such as taking gender into consideration at evacuation shelters, and improvement of warning messages

Revising local disaster management plans

The municipal government plays a fundamental role in disaster management: according to the Disaster Countermeasures Basic Act, it is responsible for establishing a local disaster management plan, emergency operations such as warning systems, issuing evacuation recommendations and orders, and flood fighting and relief activities (figure 7.2). In cases where a municipality is so widely and heavily devastated that it cannot carry out many of its primary roles, the prefectural government shall

Figure 7.2 Outline of Japan's disaster management system

Source: Cabinet Office.

Note: NKH = Japan Broadcasting Corporation; NTT = Nippon Telephone and Telegraph.

issue evacuation recommendations and orders instead of the municipality.

A local disaster management plan shall provide for the following:

- Specification of the roles of government organizations, designated public corporations (such as public utilities and the Red Cross), and other relevant public organizations

- Plans by category of activity, including: development or improvement of DRM facilities, investigation and research, education, drills and other preventive measures, collection and dissemination of information, issuing and disseminating of forecasts and warnings, evacuation, firefighting, flood fighting, rescue, hygiene, and other emergency measures and rehabilitation efforts

- Plans for coordination, stockpiling of food and supplies, procurement, distribution, shipping, communication, facilities, equipment, materials, funding, and so on

When a prefectural disaster management council wishes to formulate or revise a local disaster management plan for the prefecture, the council is required to consult the prime minister in advance, who in turn shall consult the Central Disaster Management Council. When the prefectural disaster management council has formulated or revised its local prefectural disaster management plan, the council is required to release and disseminate a summary of the plan or revision.

Following the GEJE, local governments across Japan have started reviewing their disaster management and risk reduction systems to strengthen countermeasures for multihazard, high-impact events.

The Fire and Disaster Management Agency set up a Review Committee on Improvement of Earthquake and Tsunami Countermeasures in Local Disaster Management Plans. This committee aimed to (a) assess countermeasures taken by local governments in the GEJE and (b) support local governments in revising their local disaster management plans, which are the foundation for local disaster management and reduction measures. In particular, emergency measures, including evacuation measures, and emergency training have been emphasized.

The committee made the following key recommendations for revising local disaster management plans:

- Develop action plans with concise descriptions and measurable results by setting quantitative targets.

- Plan the timing of initial actions to be taken in the event of a disaster (manuals and so forth).

- Be sure to specify emergency measures for evacuating local residents (issuing of evacuation orders and other communications with residents).

- Establish procedures in case local disaster management capabilities are lost; for example, prefectures must quickly provide appropriate alternative measures or assistance to municipal governments.

- Clarify the basic principles, including policies and standards, for developing disaster management systems.

- Make full use of emergency disaster management and reduction programs (implemented by individual communities) to further strengthen evacuation measures.

Specific actions to review local disaster management plans

Local governments in the affected areas have started enhancing their disaster management systems. For example, at an informal meeting of municipal mayors in January 2012, the government of Iwate Prefecture proposed amendments to its disaster management plan based on its experience in the GEJE. These amendments aim to improve disaster countermeasures by taking into account scenarios involving the largest possible earthquakes and tsunamis.

The amended plan includes procedures that allow the prefecture to provide support to municipal governments during large-scale disasters before they request it. It also provides for a communications security program for setting up multiple telecommunications systems, including satellite mobile phones, in prefectural and municipal government offices. These amendments were prompted by the experience of damaged or suspended administrative functions after March 11 because of power failures and destruction of offices. According to the amended plan, when contact with the affected municipalities cannot be made, the prefecture will automatically dispatch a survey team. The plan also authorizes the governor to provide support to municipalities in the event of a large-scale disaster.

After the GEJE, many local government organizations across Japan, in addition to the Iwate, Miyagi, and Fukushima prefectures, started reviewing their local disaster management plans. For example, Kawasaki City is in the process of adding tsunami countermeasures, which are hardly mentioned in the current plan, and Saitama Prefecture has decided to review measures for dealing with commuters who can't get home, emergency supply policies, and widespread radiation contamination.

LESSONS

National and local governments in Japan have distinct and complementary roles in DRM planning. The national level is in charge of defining the overall DRM strategy, coordination and legislation, allocation of funds, and deployment of the government budget. In local-level disaster management plans, governments are focusing on coordination of administrative and operational functions; preventive measures, such as education, safety drills, and

issuing and transmitting of information and warnings; evacuation and rescue activities and primary goods supply and distribution in emergency situations; and overall coordination of reconstruction and restoring livelihoods during the recovery phase. The central government provides substantial funding for emergency response and reconstruction.

The lessons learned from the Great Hanshin-Awaji (Kobe) Earthquake in 1995 had already prompted improvements to Japan's DRM legislation and government policies. In recent years, high priority has been given to developing countermeasures for large-scale earthquake disasters. Legislation has been passed on countering large-scale ocean-trench-type earthquakes, plans for large cities where damage is likely to be wide-ranging have been established, and the overall legislative framework on DRM and disaster countermeasures has been improved. After the GEJE, these kinds of countermeasures have been emphasized even more, and the Disaster Countermeasures Basic Act was revised in June 2012.

The main drivers of the latest revision were the need to account for low-probability, high-impact, and multilocation hazards and to strengthen the local government's role in providing training and planning emergency measures and evacuations.

RECOMMENDATIONS FOR DEVELOPING COUNTRIES

Keep plans up to date. Plans at the national and local levels should be revised frequently, based on lessons learned from other disasters in and outside the region. It could be helpful to set up a committee at the national level to coordinate the timing and content of revisions at the national and local levels. Consultations between national and local government representatives could assure complementarities and synergies across roles and activities.

Keep plans local. A local disaster management plan is useful in specifying countermeasures against future natural disasters, as long as clear roles and responsibilities are assigned to each tier of government regarding preventive measures, emergency response and rescue, and recovery and reconstruction activities. It is also useful to identify capacities that may need to be strengthened.

Agreements made during normal times ensure quick postdisaster responses. Agreements could be designed and signed at the local level with key sectors, specifying responsibilities for emergency response measures, rescue operations, and evacuation plans. Private companies, as well as community-based organizations entering into those agreements, could develop services (in coordination with local governments) that can be delivered as soon as a disaster strikes, even without a formal request or authorization from the local government (chapter 20).

NOTE

Prepared by Makoto Ikeda, Asian Disaster Reduction Center.

REFERENCES

Cabinet Office, Japan. 2004. "White Paper on Disaster Management." Tokyo.

———. 2011. *Disaster Management in Japan.* Tokyo.

Fire and Disaster Management Agency. 2011. *Report of Review Committee on Improvement of Earthquake and Tsunami Countermeasures in Local Disaster Management Plans* [in Japanese]. http://www.fdma.go.jp/disaster/chiikibousai_kento/houkokusyo/index.pdf.

Nishikawa, S. 2010. "From Yokohama Strategy to Hyogo Framework: Sharing the Japanese Experience of Disaster Risk Management." *Asian Journal of Environment and Disaster Management* 2 (3): 249–62.

Tanaka, S. 2008. *Local Disaster Management and Hazard Mapping.* Tsukuba, Japan: International Centre for Water Hazard and Risk Management, Public Works Research Institute.

CHAPTER 8

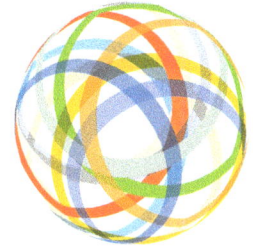

The Education Sector

Although the education sector sustained considerable damage in the Great East Japan Earthquake (GEJE), it also played a key role in protecting lives. Importantly, it provided both civil protection "hardware" and "software": school buildings served as evacuation shelters and transition shelters, and the school curricula ensured that children knew how to prepare for and react in emergencies. The performance of schools in responding to the GEJE provided a number of important lessons about the role of the education sector in disaster risk management.

DAMAGE TO THE EDUCATION SECTOR

The Great East Japan Earthquake (GEJE) caused severe structural damage to schools. In total, 6,284 public schools were damaged. The Ministry of Education, Culture, Sports, Science and Technology classified the schools into three categories according to the level of damage they sustained: 193 schools were completely destroyed (level 1); 747 schools sustained heavy damage and needed structural repairs (level 2); and over 5,000 schools had minor, mostly nonstructural damage (level 3) (figure 8.1).

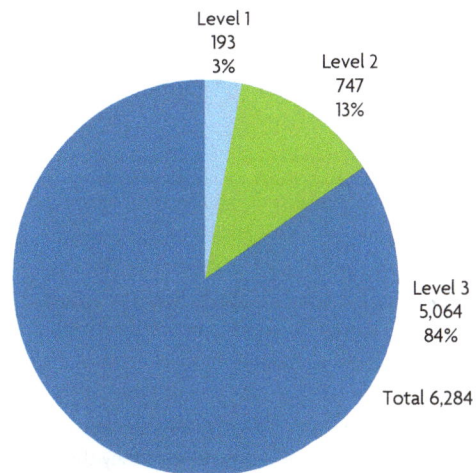

Figure 8.1 Categorization of schools by level of damage

Source: Ministry of Education, Culture, Sports, Science and Technology (MEXT)

Seven hundred and thirty-three students and teachers were killed or are missing. The proximity of the schools to the coastline was a contributing factor. The students and teachers in the Okawa Elementary School building in Ishinomaki City died tragically because they did not evacuate to higher ground (box 11.3). Where students in coastal schools survived, the school structure, disaster risk management (DRM) education, and linkages to community preparedness played critical roles.

THE ROLE OF DRM EDUCATION IN COMMUNITY PREPAREDNESS

DRM education conducted effectively and in cooperation with other local community preparedness efforts saved many lives after the GEJE. For example, in Kamaishi, where the number of casualties reached 1,000, there were 5 deaths reported among 2,900 school children, and not a single child present in school that day was killed. The so-called Kamaishi Miracle is attributed to strong DRM education, including a longstanding local tradition of teaching children the culture of *tendenko,* which means to evacuate to higher ground on their own without searching for relatives or friends (see box 8.1). This practice resulted

from many years of experience with disasters, and is based on a strong mutual understanding and trust that family members will also evacuate to safety.

Kamaishi City has been conducting DRM education programs since 2005 in cooperation with Gunma University. The programs are not mandatory, but are being implemented in selected schools in vulnerable coastal areas. Two such schools are Unosumai Elementary and Kamaishi-Higashi Junior High School (figure 8.2). The program engages the local community in preparing disaster risk maps and holds evacuation drills four times a year—one joint drill with the elementary and junior high school and one annual drill with the local community.

Kesennuma City provides another excellent example of how DRM is integrated into

Figure 8.2 Kamaishi-higashi Junior High School (top) and Unosumai Elementary School (bottom) are located near the sea

Source: © Kyoto University. Used with permission. Further permission required for reuse.

BOX 8.1

Kamaishi Miracle

When the earthquake hit on March 11, students of the Kamaishi-Higashi Junior High School evacuated together with the children of the Unosumai Elementary School. They had conducted joint evacuation drills. They reached the first evacuation point located 700 meters from the school, where they noticed a cliff had collapsed. A resident stated that she had never seen this happen there before and advised the students to move to a safer place. They moved to another point that was 400 meters higher, where they at first felt safe. However, when they heard the tremendous roar of the tsunami 30 minutes after the earthquake, they decided they should go to an even higher evacuation point, a decision that saved their lives.

school curricula, where education for sustainable development (ESD) has taught students for years about local environmental issues and how to value and protect natural resources, the environment, and cultural heritage assets. ESD also includes a strong focus on DRM.

In Kesennuma, students at the Hashikami Junior High School are taught DRM as part of the ESD program. The school served as an evacuation center for more than 1,500 people after the GEJE, which occurred just before graduation. A graduation ceremony took place in the gymnasium, and was attended by the evacuees. During the ceremony, a student gave a speech in which he honored two students who had lost their lives in the tsunami:

"People are talking about Hashikami Junior High School as the 'School of disaster prevention education,' and we are being praised around the world. We trained ourselves thoroughly and have been doing disaster prevention drills regularly. But our power as human beings was dwarfed by nature's violence, and nature deprived us mercilessly of some of our most important things. This disaster was too cruel to simply be called a trial sent from heaven . . . I feel angry and hardened. But our future lies not in blaming God but in helping each other and persevering, as difficult as that may be."

His words reflect the anguish of the community, and at the same time the recognition, gained from the ESD curriculum, that the community's responsibility is not to despair, nor to consider the disaster an "act of God," but to rely on one another for support and to improve their risk management capacity.

THE ROLE OF EDUCATIONAL FACILITIES IN DISASTER RESPONSE AND RECOVERY

As noted earlier, schools played a critical role in the immediate response to the GEJE (as evacuation shelters) and in the recovery process (as transition shelters) as shown in figure 8.3. The

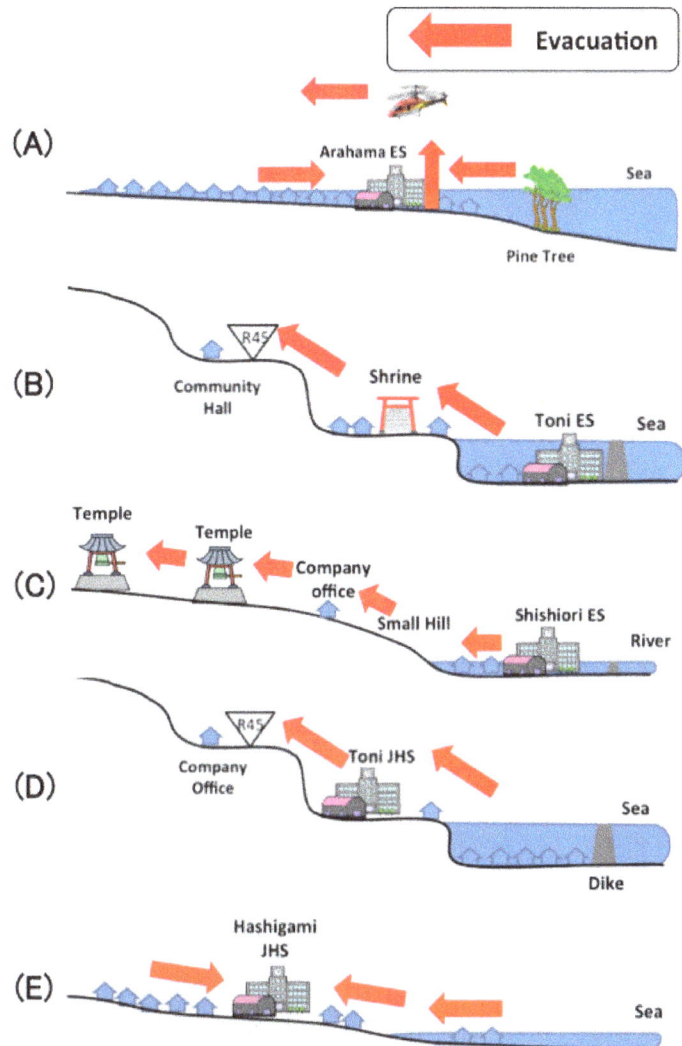

Figure 8.3 Location of schools in various areas

Source: Kyoto University.

Note: ES = elementary school; JHS = junior high school.

arrows represent the evacuation routes that people followed. Balancing the need to provide evacuation centers for communities and the need to reconvene classes for students has been a challenge, particularly where limited availability of suitable land has made housing reconstruction difficult, and the move from transitional shelter to permanent housing has been delayed.

Another challenge relates to the future role of school buildings in civil protection as demographics shift. While schools have traditionally been the most important public facilities in local communities, declining birth rates and a rapidly aging population make it difficult to justify rebuilding them in the same numbers and sizes. The following examples illustrate these issues in more detail:

- In the Arahama area, a school building served as an important evacuation shelter because of the flatness of the surrounding terrain and the building's height. The reconstructed school building should be able to withstand future earthquakes, have a flat rooftop to which people can evacuate, be situated away from the coast, and be kept stocked with emergency supplies. Since the disaster, a large proportion of the local community has relocated elsewhere because of a lack of jobs, adequate housing, and infrastructure. These issues will need to be examined before rebuilding the school (figures 8.3 [a] and 8.4 [a]).

- In the Toni area, both the elementary and junior high schools need to be rebuilt. However, it is difficult to justify the construction of new schools of the same size because of the decrease in the number of school-aged children. Therefore, a single building will be developed, jointly housing the elementary school, the junior high school, and other public community facilities. The security of school children also needs to be ensured given that the school building will be shared with the general public, and anyone can access it (figure 8.3 [b] and [d], and figure 8.4 [b] and [d]).

- Although Shishiori Elementary School is not located on the coast, the tsunami flooded the ground floor as it moved upstream along the river. While it only reached the ground floor, the school was nevertheless evacuated as there was no way of telling whether the upper level would be affected (figures 8.3 [c] and 8.4 [c]).

- Hashikami Junior High School was used as an evacuation shelter following the disaster and is still being used for transition shelters. Since the school gymnasium has not been available for more than a year, the quality of educational services is being affected (figure 8.3 [e]).

The loss of teachers who died in the tsunami has created a shortage of staff in many schools, posing an additional challenge for the continuity of education. Finally, counseling services for school children suffering from posttraumatic stress disorder must also be provided.

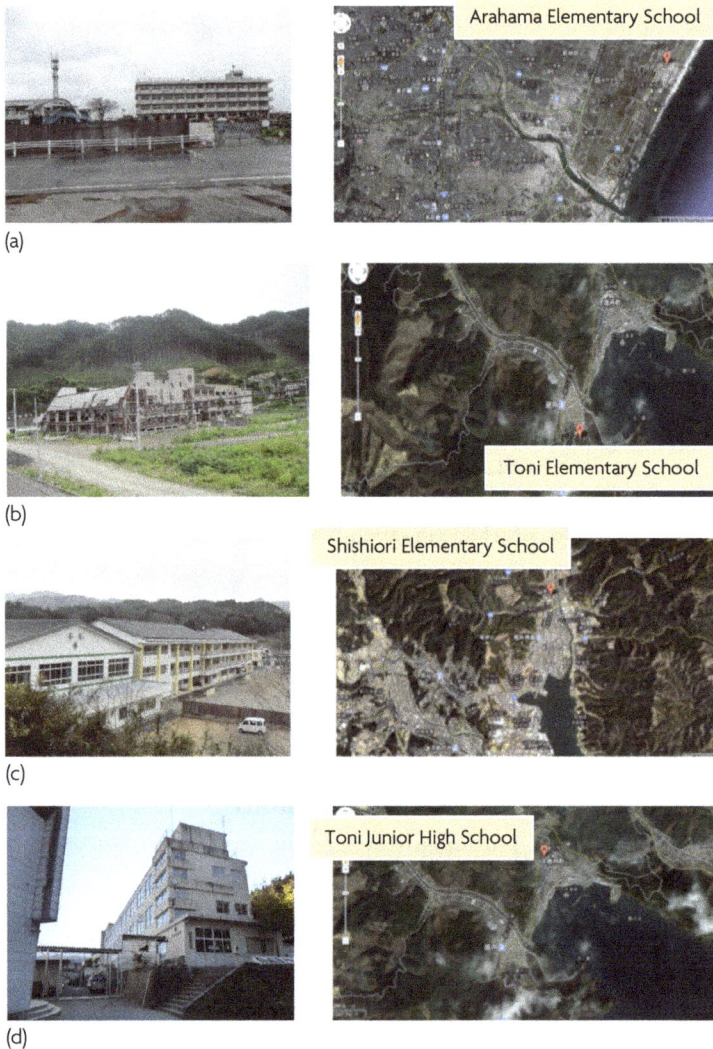

(a)

(b)

(c)

(d)

Arahama Elementary School

Toni Elementary School

Shishiori Elementary School

Toni Junior High School

Figure 8.4 Schools and locations

Source: © Kyoto University. Used with permission. Further permission required for reuse.

Considerable resources are required if the education sector is to recover fully. Funds need to be allocated for temporary schools, followed by site selection and construction of new schools, and repairs to buildings that remain structurally sound. An aging population and declining numbers of students presents a special challenge, as communities balance using school facilities for both education and civil protection purposes.

LESSONS

Key lessons from the GEJE experience for the education sector are as follows:

- *Importance of DRM education.* As exemplified by the "Kamaishi Miracle," DRM education played an important role in determining the students' evacuation behavior. The role of teachers in implementing DRM in schools should be emphasized.
  - *Structure, location, and layout of schools.* The location of school buildings is a crucial issue. In most cases, the buildings were located in close proximity (within 100–200 meters) of the coastline. Newer schools have slanted roofs to avoid water logging and structural decay. However, this prohibited people from taking shelter on rooftops. Also, it has been observed that schools that were parallel to the coast sustained greater damage than those set perpendicular to the coastline.
  - *Function of schools and educational continuity.* While schools were to be used as evacuation centers, in several cases people from local communities remained sheltered in schools for more than six months. This has serious implications for the restoration of educational services and children's educational development. This issue needs to be considered in future school-level contingency planning.

- *Human resources and training.* In the aftermath of the GEJE, schools face a shortage of teachers, which has affected the continuity of education. Students from the education faculties of local universities have tried to fill this gap; however, this also needs to be addressed in planning for educational continuity in postdisaster situations.
  - *New role of schools and multistakeholder dialogue.* Given the changing demographic conditions, schools need to play a bigger role as community facilities. Therefore, a broader range of stakeholders, including community members, needs to be included in reconstruction decision making.

RECOMMENDATIONS FOR DEVELOPING COUNTRIES

The education sector plays an important dual role in the provision of civil protection hardware and software. School buildings serve as evacuation shelters and transitional shelters, and school curricula help instill a culture of DRM and preparedness in the community. The recovery of the education sector is directly linked to the recovery of the entire community.

DRM education saves lives, as the "Kamaishi Miracle" shows. Students save their own lives and the lives of others when they lead evacuations in communities. DRM in the education sector should not be limited to the education curriculum, but should also include related issues such as structural and nonstructural safety measures; legislative measures supporting the integration, implementation, and funding of DRM in the education sector; risk assessments and early warning systems; and DRM training for school staff.

Include community members in planning. The postdisaster reconstruction process offers an opportunity for communities to reconsider their future needs regarding both the education of their children and their community facilities. A multistakeholder dialogue can help determine the optimal arrangements and design reconstruction plans accordingly.

NOTE

Prepared by Rajib Shaw and Yukiko Takeuchi, Kyoto University; Margaret Arnold, World Bank; and Masaru Arakida (box 8.1), Asian Disaster Reduction Center.

BIBLIOGRAPHY

Fernandez, G., R. Shaw, and Y. Takeuchi. 2012. "School Damage in Asian Countries and Its Implications for the Tohoku Recovery." In *East Japan Earthquake and Tsunami,* edited by R. Shaw and Y. Takeuchi. Singapore: Research Publishers.

Gwee, Qi Ru, R. Shaw, and Y. Takeuchi. 2011. "Disaster Education Policy: Current and Future." In *Disaster Education,* edited by R. Shaw, K. Shiwaku, and Y. Takeuchi. Bingley, U.K.: Emerald Group Publishing.

Takeuchi, Y., and R. Shaw. 2012. "Damage to the Education Sector." In *East Japan Earthquake and Tsunami,* edited by R. Shaw and Y. Takeuchi. Singapore: Research Publishers.

CHAPTER 9

Business Continuity Plans

A business continuity plan (BCP) identifies the potential effects of disruptions to an organization's critical operations if a disaster were to occur, and specifies effective response actions and quick recovery measures. In the Great East Japan Earthquake, BCPs served their purpose to some extent, but certain weaknesses were identified. While BCPs helped to keep critical operational functions going, and then to rehabilitate general operations, most small- and medium-size enterprises had, unfortunately, not even prepared BCPs. Since the private sector plays a major role in creating jobs and supporting local economies, it should be required to prepare BCPs, but with support from the government.

INTRODUCTION

Why is private sector preparedness important?

Because social functions and stakeholders in modern developed societies are highly interconnected and interdependent, any disruptive incident can affect an entire region. A single incident can have an extensive impact both domestically and internationally, by undermining supply chains and value chains (chapter 30).

Examples of direct and indirect negative effects include the following:

- Loss of human life and injury

- Damage to physical assets, the environment, and natural resources

- Disruption of public utilities, such as electricity, water, transport, and telecommunications

- Disruption of citizens' daily livelihoods

- Disruption of local government administrative functions

- Reduced supplies of daily goods and services

- Bankruptcy of private companies, lost economic opportunities, and income loss

- Unemployment and economic downturns

The private sector plays a major role in creating employment and supporting the local economy, thereby ensuring regional sustainability (chapter 24). In the event of a disaster, the role of the private sector becomes even more important in this respect. In each phase of disaster risk management (DRM), the private sector:

- Provides evacuation shelters and relief goods

- Ensures employment so that victims can regain their livelihoods quickly

- Provides labor, services, and products essential to the speedy recovery of social functions, roads, transportation, supermarkets, schools, hospitals, and other functions

Accident at a microchip plant

In 2000, lightning struck a Philips microchip plant in New Mexico, in the United States, causing a fire that contaminated millions of mobile phone chips. Nokia and Ericsson, Philips's biggest customers, reacted differently to their supplier's plight. Nokia's supply-chain management strategy allowed it to switch suppliers quickly; it even reengineered some of its phones to accept other types of chips. Ericsson took no action and waited for Philips to resume production. That decision cost Ericsson more than $400 million in annual earnings and, perhaps more significantly, some of its market share. By contrast, Nokia's profits rose by 42 percent that year.

Effective cooperation among disaster-resilient private sector players helps ensure a resilient and sustainable civil society. One lesson learned from past catastrophic events such as the Great Hanshin-Awaji (Kobe) Earthquake, Hurricane Katrina, the Great East Japan Earthquake (GEJE), and the Thailand flood is that when the private sector is well prepared, it plays an important role in reducing national and regional economic damage.

What is a business continuity plan?

A business continuity plan (BCP) identifies the critical operational functions of an organization and the potential impacts of a threat prior to its occurrence. It specifies effective ways of responding and quick recovery measures so that a business can continue to operate at acceptable levels and avoid disruptions for a specified period of time (box 9.1). The process of developing and deploying a BCP strategically within the organization is referred to as business continuity management (BCM).

BCM is a risk management strategy that focuses on maintaining the continuity of critical operations to ensure the supply of goods and services, and thereby the organization's survival. Figure 9.1 shows the concept of business

Figure 9.1 The business continuity plan concept

Source: Cabinet Office.

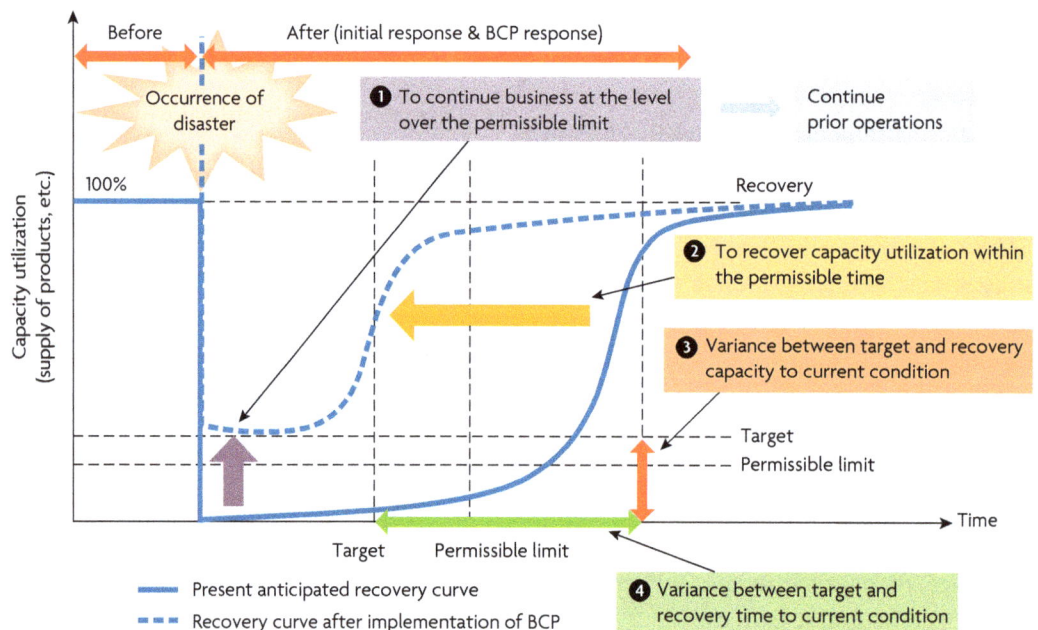

continuity and the recovery curve of an organization's level of service before, during, and after a disaster. Developing a BCP helps an organization identify what preparations must be made before a disaster strikes to secure its employees, assets, information technology (IT) systems, and information, as well as its reputation.

Business continuity plans in the Asia-Pacific region

The Asia-Pacific Economic Cooperation (APEC) region accounts for approximately 40 percent of the world's land area, more than 40 percent of the world's population, and around half of global gross domestic product (GDP). And yet, regrettably, it sustains almost 70 percent of the world's natural disasters. As the APEC region's supply chains are closely intertwined, and a single disaster can affect the economic activities of the entire region, it is essential and urgent that efforts be made to strengthen the private sector's capacity for disaster preparedness and recovery by promoting BCP development among APEC member economies. A survey was conducted in 2011 to better understand the current level of BCP awareness and adoption in the private sector.

Substantial differences were found in the level of BCP development between small and medium enterprises (SMEs) and large companies, listed and unlisted companies, and between companies that have actually experienced disaster-related disruptions and those that have not. The level of BCP development varies greatly by firm size: only 15.9 percent of SME respondents have a written BCP, while 52.0 percent of large company respondents have one. Also, there are considerable differences among APEC economies.

FINDINGS

Business continuity plans in Japan

The Central Disaster Management Council chaired by the prime minister has carried out damage estimates for the Tokyo metropolitan area in the event of a strong inland earthquake. A magnitude 7.3 earthquake with an epicenter in the northern part of Tokyo Bay has been forecasted and one scenario assumes extensive damage, including a death toll of approximately 11,000 people, the total collapse of 850,000 buildings, and a maximum economic loss of ¥112 trillion (more than $1 trillion). After the GEJE, governments are currently revising this damage estimate to verify if even worse figures are possible or probable.

In 2005 the council established the Policy Framework for Tokyo Inland Earthquakes to ensure the continuity of functions in the capital, and to establish countermeasures for reducing the death toll by 50 percent and economic losses by 40 percent. It also set strategic goals that included increasing the earthquake-proof rating of houses and buildings to 90 percent, increasing the fixed furniture rate to 60 percent, and increasing the BCP adoption rate to 100 percent for large companies and 50 percent for medium-size companies within a 10-year period. In addition, it published business continuity guidelines to help companies develop their BCPs. Forty-six percent of large companies and 21 percent of medium-size enterprises had developed BCPs in 2011.

Damage and recovery after the GEJE

The GEJE caused 656 private companies, which employed 10,757 workers, to go bankrupt within one year. But only 79 companies of them, 12 percent, were located in the Tohoku region while the others were located all over Japan. The reason for bankruptcies among the latter group was indirect loss or damage caused by disruptions in their supply chains.

The BCPs functioned to some extent but with some problems. The ratio of companies without a BCP was still high at the time of the GEJE and differed according to company size. Among large companies, 40 percent had prepared BCPs before March 11, while only 12 percent of medium-size enterprises had done so. Approximately 80 to 90 percent of the medium-size and large companies indicated that their BCPs were effective in the response-and-recovery phase after the March 11 disaster. All SMEs indicated that their BCPs were effective to some degree, while the ratio of SMEs that produced BCPs was low. Workers' capacities had been developed by formulating BCPs, so they were able to respond to even unexpected events. SMEs were able to start alternative production by collaborating with companies in other prefectures and were willing and able to collaborate in BCPs, because they do not compete with one another on a national scale.

The main reasons that BCPs did not function are as follows:

- *The damage was much greater than predicted,* because the companies followed government scenarios that underestimated reality.

- *Not enough training was conducted.* Workers who had not seen the BCP documents could not take the necessary actions.

Practices following the GEJE

The case of a large distribution company

Seven & i Holdings Co., Ltd. operates convenience stores, general merchandise stores, department stores, and supermarkets. The company had revised its BCPs seven times since the Kobe earthquake in 1995. A supermarket in Ishinomaki City, one of the most devastated cities, started selling foods and other goods outside its own buildings starting at 6 p.m. on March 11. On the next day, all 10 supermarkets opened in the Tohoku Region. The decision to reopen in times of disaster was delegated to the individual shops, which could assess the situation quickly. Multiple logistics routes were secured and 400 workers were brought from other areas to support the stores in the devastated areas.

The case of an SME

The Suzuki Kogyo Co. is a waste management company with 67 employees in Sendai City, which suffered from the GEJE. The company equipped itself with satellite phones and standby generators, and conducted training and drills based on a BCP formulated in 2008. The emergency center was established at 3:30 p.m., 45 minutes following the earthquake on March 11. Two days later the company resumed the critical operation of treating medical waste from dialysis. Other companies took over the waste management operations.

How payment and settlement systems and financial institutions responded to the GEJE

Financial services are a basic lifeline in a society, supporting many kinds of economic activities. The failure of payment and settlement systems could prevent customers from making deposits, cash withdrawals, and payments, thereby intensifying public anxiety in times of disaster. The financial sector was seriously affected by both the physical damages and the indirect effects of the disaster. Nevertheless, even in the aftermath of the earthquake, the nation's payment and settlement systems and financial institutions, including the Bank of Japan, continued to operate in a stable manner and, on the whole, managed to function normally (box 9.2).

LESSONS

The private sector in Japan has made substantial efforts to adopt BCPs, which proved to be useful when put into action following the GEJE. At the same time, however, some lessons were learned that could make corporate BCPs even stronger and more effective. Until recently there had been an attitude of tolerance toward business disruptions caused by disasters of a certain scale, as they were considered to constitute force majeure. Public opinion has shifted since March 11. Now, even if the scale and intensity of a disaster exceeds

BOX 9.2

How the GEJE affected payment and settlement systems and financial institutions

The Bank of Japan (BOJ) responded to the disaster by

- *Supplying a massive amount of cash to financial institutions.* The cash paid out by BOJ branches and local offices in the Tohoku region of northeastern Japan in the first week after the earthquake amounted to approximately ¥310 billion, about three times the amount paid out over the same period in the previous year.

- *Exchanging damaged banknotes and coins* for clean ones through the Bank's branches in the Tohoku region and a special window in Morioka City, which amounted to ¥2.42 billion starting after the earthquake up through June 21.

- *Ensuring the stable operation of the BOJ-NET,* which is used for funds transfer and services related to Japanese government bonds as well as the BOJ's market operations.

- *The minister for financial services and the BOJ governor jointly requested financial measures,* such as allowing withdrawals of deposits upon the verification of the depositor's identity even in cases where deposit certificates or bank passbooks had been lost.

- *Arranging treasury funds services and government bond services* at its head office and branches, where treasury agents were unable to continue those services.

- *Gathering information,* in cooperation with the Financial Services Agency, on damage to and the actions taken by payment and settlement systems as well as financial institutions, and providing accurate and timely information to domestic and overseas markets on the operational status of the Japanese financial infrastructure.

The private sector responded as follows:

- *To meet the needs of depositors and borrowers, financial institutions opened temporary offices* and then opened windows on Saturday, March 12, and Sunday, March 13, 2011, the weekend immediately after the earthquake. Of the total of about 2,700 offices of the 72 financial institutions headquartered in 1 of the 6 prefectures in the Tohoku region or Ibaraki Prefecture in the Kanto region, some 310 offices (11 percent of the total) were closed as of March 16.

- *The financial institutions worked in close coordination,* such as by delivering cash to other institutions that needed additional cash.

- *Major bill and check clearing houses expanded their areas of coverage,* so that participating financial institutions could bring in bills and checks that normally would be processed by the clearing houses that were not operating.

- *Payment and settlement systems as well as financial institutions across Japan generally continued to operate stably.* There were also procedures and systems in place to address the temporary inability of affected financial institutions to participate in the payment and settlement systems.

- *Marketwide business continuity arrangements* developed in the money market, and the foreign exchange market and the securities market functioned smoothly.

- *The stock market infrastructure was able to provide smooth and uninterrupted processing* with a high level of operational capacity despite the surge in trading volume following the earthquake.

Lessons

- Payment and settlement systems and financial institutions need to review the severity and scope of the scenarios used in designing their business continuity arrangements, to see whether they address potential stress events sufficiently in light of the recent disaster.

- It is crucial to enhance business continuity arrangements in line with the identified scenarios. This includes enhancing backup arrangements for computer systems and headquarters functions, increasing in-house power-generating capabilities against potential long-term constraints on the electricity supply, enhancing arrangements for securing necessary staff in the event of prolonged disruption of public transportation services, and securing system-processing capacity to withstand a surge in trading activity.

- Implementing and enhancing "streetwide exercises," with participation of the overall financial industry, and eventually with the cooperation of nonfinancial firms such as the providers of social infrastructure, ensures the consistency of arrangements across institutions.

For details, see Bank of Japan (2011).

assumptions and predictions, disruptions are deemed to constitute negligence, and top managers are expected to be able to take appropriate measures to ensure the continuity of critical operations. Companies should

- *Ensure BCP effectiveness through regular drills and continuous education.* These drills and training must target specific departments in the company and should address specific capacities and skills; generic training is of no use. The plan should list specific activities and give detailed directions to be followed in emergencies and to facilitate recovery. These should be explained in detail to those officials and employees who are expected to implement them. Drills and training should be regular and ongoing, and some coordination at the sectoral level is recommended.

- *Radically shift from a "disaster-based" to a "consequence-based" approach to strategy development.* Private companies should formulate their BCPs to reflect the results or outcomes they expect from implementation, rather than specific measures to counter specific disasters. They should identify key services, and examine how long the service will be disrupted and how they can shorten the disruption time.

- *Focus more on supply chain disruption risk by knowing more about the situations of stakeholders.* In addition to the company's own operations, BCPs should address supply chain issues that affect other companies and markets. To facilitate this, meetings should be held regularly with companies in the same sector and with supply chain companies, first to assess the potential risks and then to develop concerted measures to ensure business continuity throughout the supply chain.

RECOMMENDATIONS FOR DEVELOPING COUNTRIES

If well prepared for disasters, the private sector can play an important role in reducing local and regional economic damage. BCPs are an effective tool for strengthening the private sector's disaster resilience.

- *Raise public awareness.* Private companies and organizations do not always recognize the importance and usefulness of BCPs. Efforts should be made to raise awareness about BCPs and develop effective BCPs to achieve greater regional resilience. Practices and lessons from disasters should be widely shared with private companies and organizations.

- *Start from a small disaster.* Private companies could begin with a small hazard scenario as the first step in formulating BCPs, and then add greater or different kinds of hazards. For example, in Japan, since earthquakes are a very familiar hazard, most companies start by preparing BCPs for earthquakes, which are considered easier to produce. They then proceed to develop BCPs for more complicated disasters, such as pandemics.

- *Mobilize government support.* Governments may feel that providing support to BCPs for the private sector is not their role. But securing livelihoods and the local economy is certainly a relevant public sector concern. Governments should provide private companies with the necessary information such as risk assessments and guidelines for producing BCPs. Also, governments should collaborate with chambers of commerce and other industrial associations that provide support to these companies.

NOTE

Prepared by Takahiro Ono, Asian Disaster Reduction Center and Mitsubishi Corporation Insurance, and Mikio Ishiwatari, World Bank.

BIBLIOGRAPHY

Asian Disaster Reduction Center. 2011. *BCP Status of the Private Sector in the APEC Region 2011.* http://publications.apec.org/publication-detail .php?pub_id=1234.

BOJ (Bank of Japan). 2011. *Responses to the Great East Japan Earthquake by Payment and Settlement Systems and Financial Institutions in Japan.* BOJ Report and Research Papers, Payment and Settlement Systems Department, Bank of Japan, Tokyo.

Cabinet Office, Cabinet of Japan. 2009. *BCP Guidelines* [in Japanese]. http://www.bousai.go.jp/MinkanTo Shijyou/guideline02.pdf.

——. 2011. *Survey on BCPs* [in Japanese]. http://www .bousai.go.jp/kigyoubousai/jigyou/keizoku12 /kentoukai12_05.pdf.

Financial Times. 2008. "The Fire That Changed an Industry: A Case Study on Thriving in a Networked World." http://www.ftpress.com/articles /article.aspx?p=1244469.

Teikoku Data Bank. 2012. *Higashinihon daishinsai kanren tosan no doukou chousa* [Survey of bankruptcies caused by GEJE]. http://www.tdb.co.jp /report/watching/press/pdf/p120303.pdf.

Tokyo Marine Nichido. 2011. *Survey on GEJE and BCP* [in Japanese]. http://www.tokiorisk.co.jp/cgi-bin /risk_info/page.cgi?no=753.

CHAPTER 10

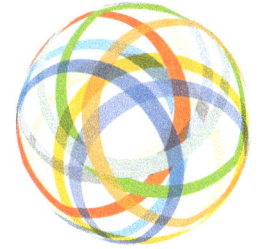

Tsunami and Earthquake Warning Systems

Warning systems can mitigate the damage caused by tsunamis and other natural events, and prevent the loss of human life and properties. Countermeasures, such as evacuation to higher ground and the stopping of trains, depend on getting the right information and disseminating it in a timely manner. Warning systems must also be aligned with community response. While Japan has developed the most sophisticated tsunami-warning system in the world, the system underestimated the tsunami height on March 11, 2011, and may have misled the evacuees and increased human losses.

FINDINGS

Community-based tsunami-warning systems

Before March 11, 2011, Japan had already developed sophisticated high-technology tsunami-warning systems that included satellite communications and hundreds of real-time monitoring stations. But on March 11 the community-level response (and community-based warnings) was the key that saved countless human lives. The volunteer fire corps—which are community-based organizations (CBOs) trained in disaster management (see chapter 6)—used various tools such as handheld loud speakers, fire bells, sirens, and fire engine loudspeakers to warn communities throughout the affected areas. In Katsurashima, Shiogama City, all community members including 30 disabled people were safely evacuated because the fire corps went door to door to every house, helping community members move to higher ground. In Otsuchi Town and Natori City, some members of the corps kept ringing fire bells or giving directions on their loud speakers right up until the tsunami hit—some at the expense of their own lives.

Tsunami warnings on a national scale

Japan Meteorological Agency (JMA) monitors seismic activity throughout Japan around the clock. The agency can quickly calculate the hypocenter and magnitude (M) of an earthquake, and issue a tsunami warning within three minutes after the earthquake. The information is immediately disseminated to the public by disaster management authorities, local governments, and the mass media (figure 10.1 and box 10.1). The JMA has recently invested some ¥2 billion in tsunami and earthquake monitoring and warning systems.

On March 11, 2011, the JMA issued the first tsunami warning at 14:49, three minutes after the earthquake. People started evacuating and organizations concerned started preparing for the tsunami.

Critical problems were found in estimating the tsunami's height and getting information out to the public. Underestimation of the tsunami's height likely contributed to the delay in people's evacuation. The agency at first estimated tsunami heights of 3 to 6 meters in Iwate, Miyagi, and Fukushima prefectures, well below the actual heights. This was because the agency underestimated the earthquake magnitude as Mj (JMA magnitude) 7.9, while the actual magnitude was Mw (moment magnitude) 9.0.[1] The agency could not calculate the Mw within 15 minutes, as with a normal operation, because of the scaling-off of most broadband seismographs. Cable-type offshore hydraulic gauges, which provide useful forecasting information, were not installed to revise tsunami information. Also, the JMA issued information on a 0.2-meter-height tsunami 13 minutes after the earthquake. The agency revised the estimated height to 6 to 10 meters at Iwate, Miyagi, and Fukushima prefectures 30 minutes after the quake, and then to more than 10 meters in

Figure 10.1 Information flow in the tsunami-warning system

Source: Japan Meteorological Agency (JMA).

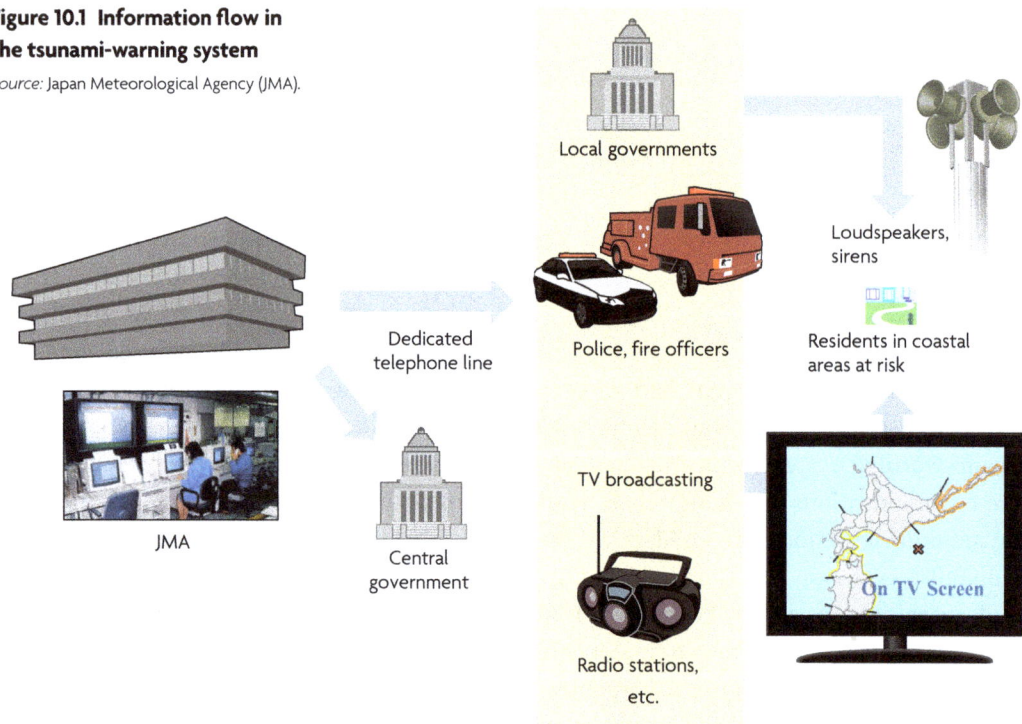

JMA

Dedicated telephone line

Central government

Local governments

Police, fire officers

TV broadcasting

Radio stations, etc.

Loudspeakers, sirens

Residents in coastal areas at risk

On TV Screen

BOX 10.1

Tsunami warnings in Japan

The Japan Meteorological Agency (JMA) conducted tsunami computer simulations for various earthquake scenarios and stored the results (which included tsunami arrival times and heights) in a database. Since the simulation takes some time, the agency cannot promptly issue warnings using real-time simulation following an earthquake. When a large earthquake occurs, the operating system quickly calculates the hypocenter and magnitude, searches the tsunami database for this hypocenter and magnitude, and selects the most appropriate simulation results from the database. Based on the estimated height of the tsunami, the JMA issues a tsunami forecast. Tsunami forecasts fall into two categories: tsunami warnings and tsunami advisories. Warnings are further divided into two classes: tsunami and major tsunami.

The JMA has improved the warning systems following the establishment of a tsunami-warning unit for the Sanriku coast in 1941. The agency expanded it into a nationwide service in 1952, and after the 1960 Chilean earthquake tsunami, the system started covering long-distance tsunamis as well. In the Hokkaido Nansei-Oki Earthquake of 1993, the tsunami arrived before any warning was issued. The JMA improved the system, and started issuing estimated tsunami heights in 1999.

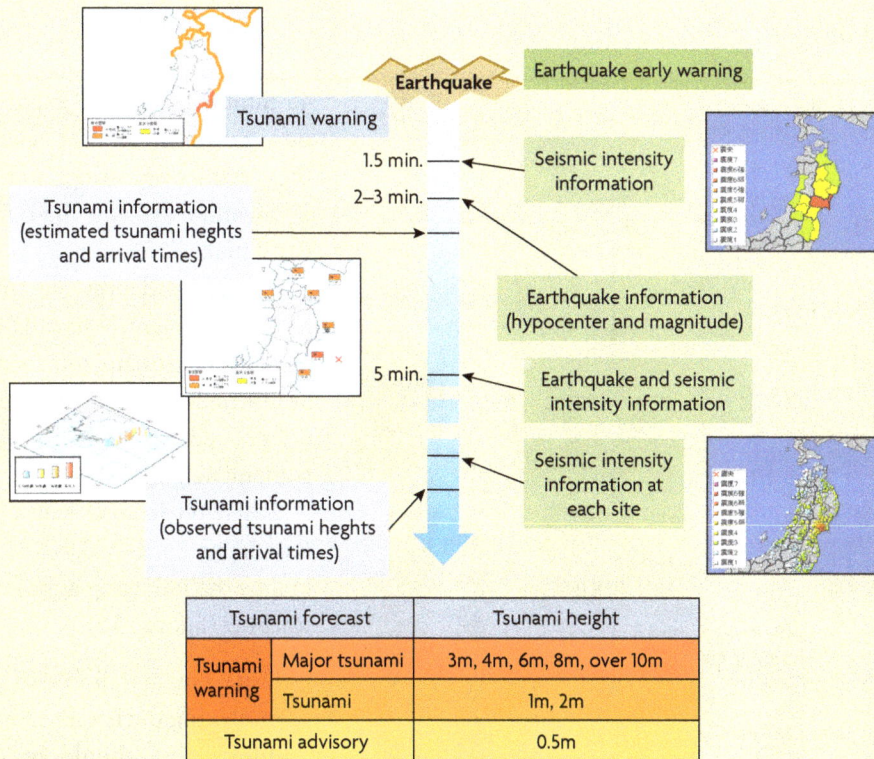

Tsunami forecast		Tsunami height
Tsunami warning	Major tsunami	3m, 4m, 6m, 8m, over 10m
	Tsunami	1m, 2m
Tsunami advisory		0.5m

Source: JMA

14:49 3 minutes following earthquake	15:14 28 minutes following earthquake	15:30 44 minutes following earthquake
• Observed Mj 7.9 • Issued tsunami information: 3 meters in Iwate and Fukushima, and 6 meters in Miyagi.	• Observed rapid rise of offshore tsunami height by global positioning system (GPS) buoys. • Revised information: 6 meters in Iwate and Fukushima, and over 10 meters in Miyagi.	• Tide gauges scaled out. • Revised information: over 10 meters in Iwate, Miyagi, and Fukushima.

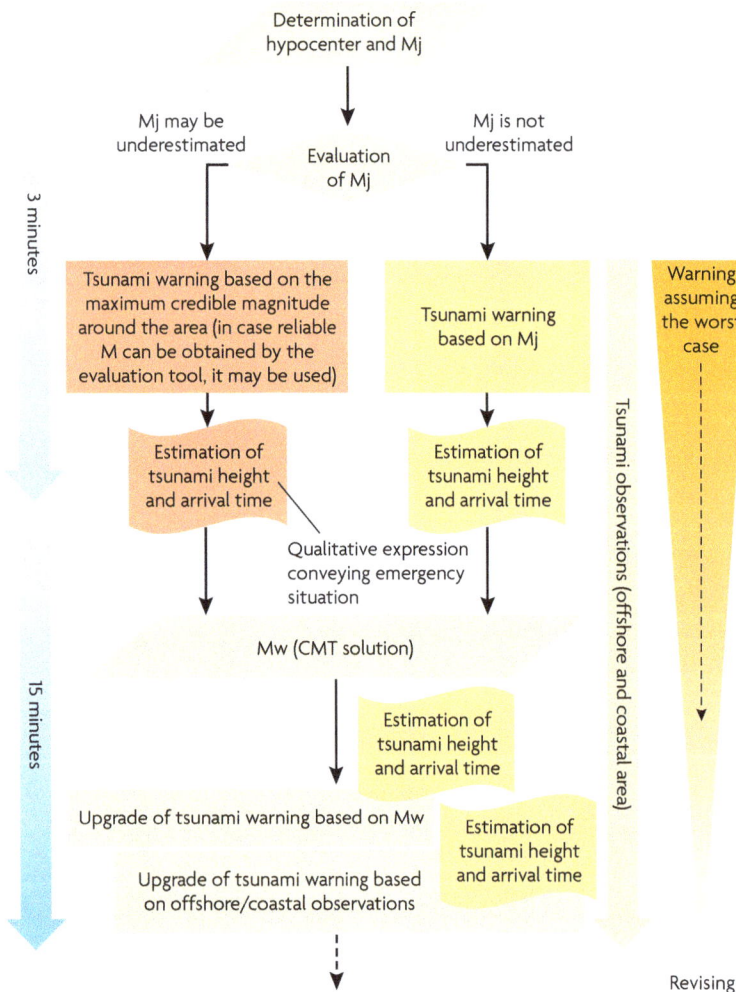

Figure 10.2 New methods for tsunami warnings

Source: JMA.

Note: CMT = centroid-moment-tensor; M = magnitude; Mj = JMA magnitude; Mw = moment magnitude.

45 minutes (map 10.1). The revised information, however, did not reach everyone since they were evacuating, and power and communication systems had failed due to the earthquake.

According to interview surveys by the Japanese government, almost half of the population received no tsunami information or evacuation orders in the affected areas, and 60–70 percent did not receive the revised information about tsunami heights.

Based on lessons learned from the Great East Japan Earthquake (GEJE), the JMA plans to take the following approach while issuing a warning (figure 10.2):

- Issue tsunami information that is useful in making decisions about evacuation; the information should be clear and timely, emphasize crucial messages, and encourage evacuation.

- Issue the first warning quickly, within three minutes following an earthquake, and revise it according to observed data.

- Provide tsunami height predictions qualitatively, instead of numerically in the first warning for possible megatsunamis caused by earthquakes greater than M8, considering the uncertainty of tsunami estimates.

- Raise public awareness of the principle that people should take the initiative on their own to escape from tsunamis when they feel any quakes.

- Improve the accuracy of warnings about frequently occurring tsunamis to better inform people's evacuation decisions by improving confidence.

NHK (Nippon Hoso Kyokai, or Japan Broadcasting Corporation) reviewed programs during the GEJE and found that the tones of the warning announcers on television were rather flat and lacked urgency. The corporation is revising the warning methods issued through television to encourage evacuation by announcements that are persuasive.

The Earthquake Early Warning system

The Earthquake Early Warning (EEW) system aims at mitigating earthquake damage by providing a lead time to slow down trains, stop elevators, and give people time to take protective measures (figure 10.3). The JMA quickly determines the hypocenter and magnitude of an earthquake based on real-time monitoring data. The agency estimates the distribution of strong ground shocks, and issues warnings to government officials and the mass media, such as radio, television, and communication companies before the shocks reach them. For example, gas and railway companies use this warning to control their operations. Also, warnings are issued to the public through SMS (short message service) alerts. The JMA launched this EEW service in 2007.

During the GEJE, the JMA issued the first EEW 8.6 seconds after detecting the first primary wave (P-wave) at the nearest seismic station. There were 15 to 20 seconds of lead time after the warning and before the main shock hit Sendai. At Seisho High School, Kanagawa

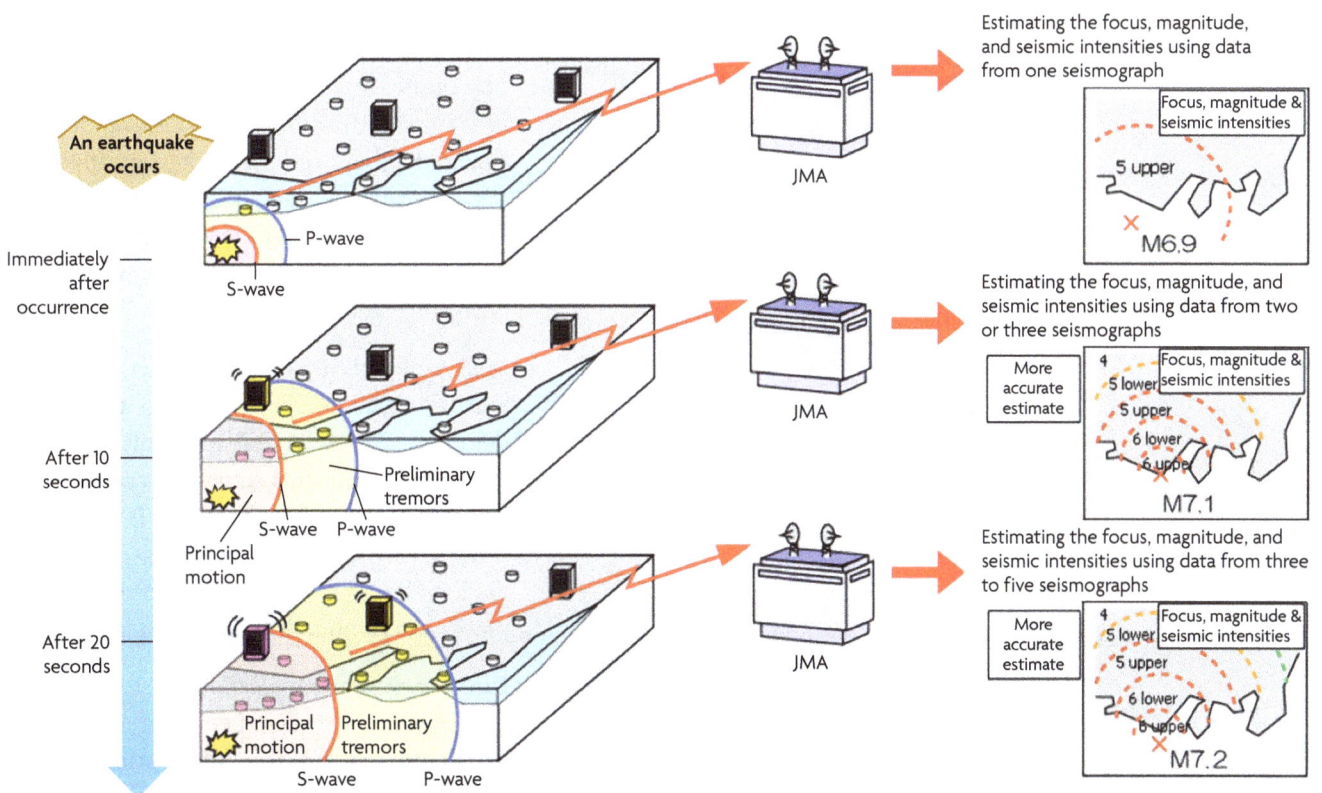

Figure 10.3 Earthquake early warning system

Source: JMA.

Note: JMA = Japan Meteorological Agency; M = magnitude; P-wave = primary wave; S-wave = secondary wave.

Prefecture, students used this time to get under their desks or leave at-risk spots. Also, at a primary school where teachers and students had conducted practice evacuation drills, they calmly began evacuating as soon as they got the warning.

According to a JMA survey, over 80 percent of people believe the EEW information helps them protect themselves. Some 60 percent took action, such as taking shelter under desks, upon receiving the EEW. Although some 40 percent of EEWs have been incorrect and underestimated the actual size of quakes, over 80 percent of respondents want to keep using the system. The JMA is improving the accuracy of the EEW by upgrading prediction models.

Bullet trains' earthquake detection system

On March 11, 19 bullet trains (including two traveling close to the maximum speed of 270 kilometers per hour) were running on the Tohoku Sinkansen Line. All trains were able to stop safely soon after the earthquake occurred without incurring any casualties. The system detected the P-wave and stopped the trains by automatically cutting their electricity supplies (figure 10.4). The railway companies started using the system in 1992, and have improved it since then. During the Chuetsu Earthquake in 2004, a bullet train derailed because it was traveling right above the epicenter, although no casualties were reported. The companies shortened the lead time between detecting a P-wave and issuing the warning, from three seconds to

between one and two seconds. The number of earthquake monitoring and detection stations has also increased to 239 across the country.

LESSONS

The following lessons should help inform the development of warning systems:

- *Japan's earthquake warning systems were able to reduce economic damages and loss of life* by shutting down bullet trains and providing lead time for people to take protective measures. Japan has developed new technologies to improve these systems.

- *Using warning systems to trigger timely community response is the key to disaster management.* No matter how advanced technology becomes, the guiding principle is that people should take the initiative to escape from a tsunami on their own as soon as they feel any quakes.

- *Inaccurate or inappropriate information in a tsunami warning could mislead,* delay evacuation, and increase the loss of lives. Warning information should be issued on the side of safety, considering the possible inaccuracy of estimates and the limitations of the forecasting technology.

- *Multiple methods of information sharing must be secured.* While warnings must be delivered to everyone at risk, only half of the affected residents actually received the information following the GEJE. It was difficult to provide people with revised information during the evacuation because of power and communication system failures.

- *Disaster risk communication must be practiced regularly,* so that people are able to better understand the information, and messages and agencies can better understand the mechanisms that local people use to cope with disasters (chapter 27).

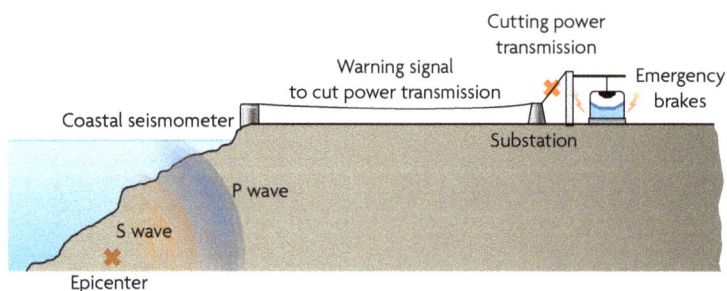

Figure 10.4 Earthquake early detection system

Source: Ministry of Land, Infrastructure, Transport and Tourism (MLIT)..

RECOMMENDATIONS FOR DEVELOPING COUNTRIES

Warning systems can save people's lives and reduce economic damages from natural disasters such as floods, tsunamis, earthquakes, landslides, and other events. People can take countermeasures, such as escaping to higher ground, protecting themselves from falling debris, and stopping trains before they are overtaken by these events.

Start with low-cost systems. Warning systems can start with simple methods. Low-cost equipment, such as fire bells and sirens, were widely utilized as warning tools during the GEJE. Observers in communities once monitored water levels in rivers and sent information to concerned organizations by phone until a decade ago in Japan. Warning systems can evolve by replacing equipment, such as automatic monitoring equipment and telemeter systems, based on these basic systems.

Link with community-based activities. Actions at the community level are crucial as demonstrated by the volunteer fire corps that issued warnings and saved lives on March 11. Warning systems and other measures organized by communities may be particularly relevant in developing countries where government capacity and resources are limited (box 10.2).

Develop technology, and understand its limitations. Although various technologies, such as flood prediction, tsunami simulations, communication systems, and earthquake monitoring are all needed to develop effective warning systems, their limitations must be taken into account. On March 11, underestimating the tsunami height likely caused people to delay their evacuation and led to greater losses.

Conduct interactive risk communication. Communities, governments, and experts should exchange information and ideas about potential risks (chapter 27). Communities should be able to understand the information delivered in the warning, while also being aware of the system's limitations. Also, government staff must

BOX 10.2

Community-based warning in Sri Lanka

In Sri Lanka the Disaster Management Center and National Building Research Organization are promoting community-based landslide warning systems. Simple rain gauges, which are bottles for measuring rainfall or bottles equipped with an automatic buzzer (Type OI) were delivered to at-risk communities. When the water level in the bottle reaches the risk level, a warning is issued to community members.

Source: © Mikio Ishiwatari. Used with permission. Further permission required for reuse.

understand communities' response to disasters to design warning systems.

Understand communities' coping mechanisms. Since warning systems are meant to benefit communities on the ground and to inform their actions, the responsible organizations should understand how local people cope with and respond to disasters. Community members decide on their own when, where, and how to escape. The organizations should tailor the contents of warning messages to the users' needs and points of view. Such messages need to be simple, timely, and encourage evacuation.

Establish end-to-end systems to ensure that warnings reach the communities at risk. Multiple communication channels should be established so that information keeps flowing in case of power and communication failures.

Ensure services are available 24/7. Since natural events can happen at any time, the organizations concerned are required to function around the clock—24 hours a day, 7 days a week. Staff rotation should be arranged in the organizations.

NOTES

Prepared by Mikio Ishiwatari, World Bank.

1. The JMA magnitude has the advantage of being calculated quickly within three minutes, but tends to underestimate the magnitude of earthquakes over M8. The moment magnitude is utilized worldwide but takes around 10 minutes to calculate.

BIBLIOGRAPHY

Goto, M., and N. Aihara. 2012. "Development of Education for Natural Disaster Preparedness and Reduction at School Linking to the Community." In *East Japan Earthquake and Tsunami: Evacuation, Communication, Education and Volunteerism,* edited by R. Shaw and Y. Takeuchi. Singapore: Research Publishing.

Ishiwatari, M. 2012. "Review of Countermeasures in the East Japan Earthquake and Tsunami." In *East Japan Earthquake and Tsunami: Evacuation, Communication, Education and Volunteerism,* edited by R. Shaw and Y. Takeuchi. Singapore: Research Publishing.

JMA (Japan Meteorological Agency). 2002. *Survey on Utilization of EWE* [in Japanese]. Tokyo, Japan. http://www.jma.go.jp/jma/press/1203/22c/23 manzokudo_data.pdf.

———. 2011. *Improvement of Tsunami Warning in Light of Tsunami Disaster Off the Pacific Coast of Tohoku Earthquake* [in Japanese]. Tokyo, Japan. http://www.jma.go.jp/jma/press/1109/12a /torimatome.pdf.

Ministry of Education, Culture, Sports, Science and Technology. 2011. *Interim Report of an Expert Panel on Disaster Education and Disaster Management Based on GEJE* [in Japanese]. http://www .mext.go.jp/b_menu/shingi/chousa/sports/012 /attach/1310995.htm.

Ohara, M., K. Meguro, and A. Tanaka. 2011. "Comprehensive Study on People's Awareness of Earthquake Early Warning before and after the 2011 Earthquake Off the Pacific Coast of Tohoku" [in Japanese]. *Seisan Kenkyu* 63 (6): 811–16.

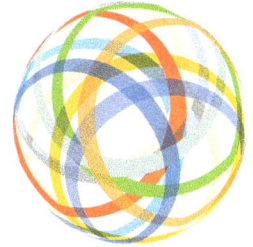

CHAPTER 11

Evacuation

Community evacuation measures should be the centerpiece of disaster risk management systems. Because the Sanriku region has suffered from frequent tsunamis, its local communities have passed their knowledge from generation to generation, mainly by constructing commemorative monuments and by conducting education and drills. Nevertheless, about 20,000 people died or are missing as a result of the catastrophic tsunami on March 11, 2011. Various factors, such as underestimating tsunami heights in warnings and on hazard maps, as well as a lack of awareness, influenced the number of human lives lost. Since neither the local governments nor the electric power company had prepared properly for possible nuclear accidents, evacuation from the accident at the Fukushima Daiichi nuclear power station was chaotic.

FINDINGS

Preparing evacuation measures

Because predictions and other measures to foresee or prevent potential disasters are often unreliable, community evacuation measures should be at the center of disaster risk management (DRM) systems (figure 11.1). Other measures, such as hazard maps, education programs, practice drills, and warning systems all contribute to successful evacuation.

Since the Sanriku region has often sustained severe tsunami damage, its local governments and communities have developed a high level of disaster preparedness. The Meiji Sanriku Tsunami of 1896, with a maximum run-up height

Figure 11.1 The relationship between evacuation and other DRM measures

of 38.2 meters (the highest point that a tsunami reached inland), killed over 22,000 people; the Showa Sanriku Tsunami in 1933 with a maximum run-up height of 23 meters killed approximately 3,000; and a tsunami following the Chilean earthquake in 1960 killed 142. After

Stone monuments transfer local knowledge to the next generations

In Aneyoshi District, Miyako City, Iwate Prefecture, villagers who followed the practices of their ancestors survived and saved their properties from the tsunami. A stone monument, set up after the 1933 Showa Sanriku Tsunami, is 60 meters above sea level—20 meters higher than the level of the 1933 tsunami. The inscription reads as follows:

"Living on higher ground will make the lives of our descendants more peaceful. Remember the catastrophic tsunami. Never build houses below this point. The tsunamis of 1896 and 1933 reached this point, and the villages were completely destroyed, leaving only 2 and 4 survivors. Be careful now, even after many years."

When the tsunami occurred on March 11, villagers working on the coast immediately ran up the winding path toward this monument. A huge, black wave rushed up from the port, stopping 70 meters short of the monument.

Figure 11.2 Evacuation map and information on past tsunamis
Source: © Masaru Arakida. Used with permission. Further permission required for reuse.

Figure 11.3 Evacuation signs
Source: © Masaru Arakida. Used with permission. Further permission required for reuse.

each of these disasters, the local governments revised their DRM plans accordingly, designating shelters, procedures, and other mechanisms to facilitate speedy evacuation (chapter 7).

Communities in the Sanriku region have built 150 monuments to raise public awareness among future generations (box 11.1). Community-based organizations, such as the volunteer fire corps and disaster management organizations, conduct training and regularly schedule practice drills (chapter 6). Schools give classes on local experiences with past disasters and on disaster preparedness (chapter 8). Earthquake and tsunami evacuation drills are also conducted, and local governments designate evacuation routes and shelters at higher elevations based on past tsunami heights. Tsunami hazard maps including the locations of evacuation shelters are displayed on sign boards in town (figure 11.2) and distributed to every household. Past tsunami water levels and the places of evacuation shelters are posted on electricity poles and elsewhere on the roadside (figures 11.2, 11.3). Evacuation routes have been developed to reduce evacuation times, even if only by a few minutes (box 11.2).

Local governments conduct tsunami evacuation drills every year on days commemorating past large-scale tsunamis, and residents learn how to evacuate safely and quickly from their own houses to designated shelters. Volunteer organizations and private companies also participate, demonstrating, for example, how to assist people with disabilities, how to guide evacuees, and how to close tsunami dike gates. In sightseeing areas, tourists are also encouraged to participate in these drills.

Certain issues had been identified in evacuation measures even before the March 11 disaster. Public awareness about the possibility of a tsunami disaster had decreased since large-scale damage had not been sustained in many years. It was mainly the elderly and children who took part in the drills, while other age groups assigned them a lower priority. The number of participants in the drills had also

BOX 11.2

Tsunami evacuation routes for schools

The evacuation bridge. The tsunami nearly reached the roof of the three-story Okirai Elementary School in Ofunato City, Iwate Prefecture, but all students got away safely over the evacuation bridge. The bridge had been built in October 2011, connecting the school building with a nearby road on higher ground. It shortened the evacuation route from 250 meters to 110 meters, and the evacuation time from 6 minutes to 3 minutes.

The evacuation stairway. The Omoto Elementary School in the town of Iwaizumi, in Iwate Prefecture, is located right in front of a cliff more than 10 meters high. To evacuate to safer ground, children had to take a roundabout route, so an evacuation stairway 30 meters long was built in March 2009. The school building and the gymnasium were inundated by the March 11 tsunami.

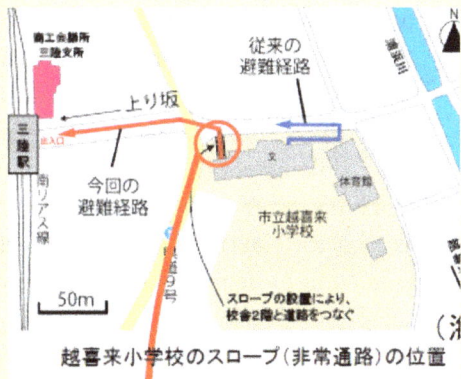

Source: Cabinet Office (CAO) and MLIT.

been decreasing every year. Also, local organizations exhibited varying degrees of evacuation preparedness (box 11.3).

Evacuation scenarios on March 11

Of the approximately 602,000 people in the inundated areas, 582,000 escaped the tsunami, with 20,000 dead or missing. The Japan Meteorological Agency's (JMA) underestimation of the tsunami's height issued three minutes after the earthquake is likely to have delayed the evacuation. Although the agency revised its warnings later through real-time monitoring of the tsunami, all local governments and communities did not receive them because of power and communication failures (see chapter 10). Others, believing that the coastal dikes would protect them, may have delayed evacuations. A survey of evacuees conducted by the government at the evacuation centers revealed the following points.

BOX 11.3

The Okawa tragedy

Seventy-four of the 108 students (70 percent) in the Okawa Elementary School, Ishinomaki City, died or went missing after the tsunami. The school is located about 5 km from the mouth of the Kitakamigawa River. Following the earthquake on March 11, teachers led the children from the school buildings to the playground as they had been trained to do. Since tsunami evacuation sites had not been identified before the disaster, they headed toward an elevated bridge not far away. The tsunami engulfed the students and teachers on the way to the bridge.

Source: © World Bank. Used with permission. Further permission required for reuse.

Not all people evacuated immediately after the earthquake

Fifty-seven percent of the residents evacuated immediately (immediate evacuation), 31 percent evacuated after doing chores, such as clearing debris (delayed evacuation), 11 percent left only when the tsunami was in sight (urgent evacuation), and 1 percent of the residents did not evacuate as they lived on higher ground (figure 11.4).

Early evacuation is the key to staying safe

Most residents who evacuated immediately after the earthquake (immediate evacuation) were safe. But half of residents who did not evacuate immediately (urgent evacuation) had to contend with the tsunami (figure 11.5).

Figure 11.4
Evacuation timing

Source: CAO.

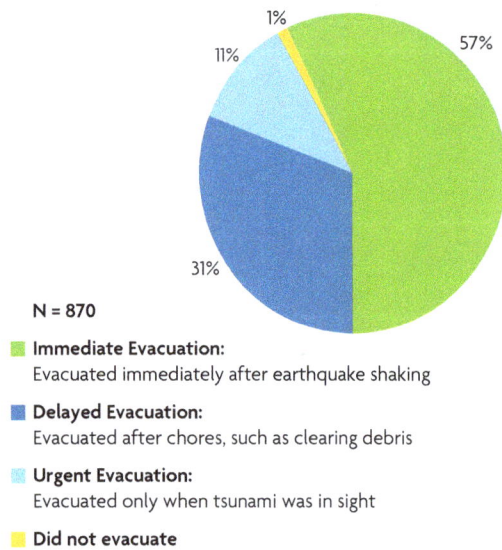

N = 870

■ **Immediate Evacuation:**
Evacuated immediately after earthquake shaking

■ **Delayed Evacuation:**
Evacuated after chores, such as clearing debris

■ **Urgent Evacuation:**
Evacuated only when tsunami was in sight

■ **Did not evacuate**

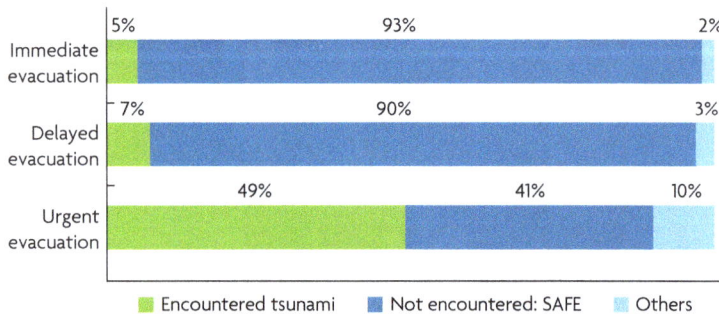

Figure 11.5 Evacuation pattern and encounter with the tsunami

Source: CAO.

Residents with a high level of awareness are likely to evacuate immediately

Half of the residents who evacuated immediately (immediate evacuation) thought that the tsunami would reach them, while 70 percent of urgent evacuees didn't think it would or were not concerned about it.

Over half the residents evacuated by vehicle

Many wanted to leave with their family members, or thought that the tsunami would catch up to them if they left on foot. One-third of them were stuck in traffic jams. The average evacuation distance on foot was 450 meters, while the average distance to evacuate by car was 2 kilometers (km). While evacuation on foot is the general rule, vehicles are also needed to carry the elderly and disabled. Measures for evacuating by vehicle need to be improved.

Some designated evacuation shelters were submerged

Some 40 percent of the evacuees went to shelters that had been designated by the local governments. Among them, some 30 percent of the evacuees were submerged at the shelters by the tsunami.

People's behavior is influenced by group actions—during the Great East Japan Earthquake (GEJE), residents were influenced by their neighbors' decisions

People escaped as a group, though they were encouraged to escape the tsunami independently—*tendenko*. A survey found that some families were saved with their adjacent families, but others were not in Yuriage village in Natori City. In New York City on 9/11, too, people escaped from the World Trade Center with their office colleagues or in groups.

Commuters and school children stranded in Tokyo

On March 11, 5.15 million people in the national capital region, including Tokyo, could not get home from schools, offices, and other venues because of traffic disruptions. In Tokyo, some 94,000 people stayed in about 1,030 facilities, including a city hall building. In Sendai City, 50,000–100,000 people, including tourists, had to stay at evacuation shelters. In November 2011, local governments asked private companies to shelter their employees for three days following future disasters. This promises to facilitate response activities by keeping people off the streets. Companies are required to store emergency food rations, water, and other amenities for a three-day stay.

Safety for tourists and visitors

Tourists and other visitors do not have enough information on tsunami risks and emergency evacuation centers in unfamiliar places. The Japanese government proposed pictographic signs of tsunami disasters to the International Organization for Standardization, based on global and national standards (figure 11.6).

THE ACCIDENT AT THE FUKUSHIMA DAIICHI NUCLEAR POWER STATION

The first stage

As the Government Investigation Committee on the Accident at the Fukushima Nuclear Power Station (2011) explains,

> [e]vacuation instructions from the central government did not reach all the relevant local governments in a timely manner; and there was a great deal of confusion during the evacuation. Moreover, the instructions were not specific or detailed enough. With insufficient information the local governments had to make decisions about whether to evacuate and evacuation procedures, locate evacuation sites, and so forth.

Fifty patients evacuated from the Futaba Hospital died by March 31. One of the main reasons for the confusion was that neither the central government nor the electric power companies had prepared well enough.

The governments issued six different evacuation directives within 24 hours: four revisions for the Daiichi Station and two for the Daini Station, as follows:

The Daiichi Station

MARCH 11

20:50   Fukushima's governor gives an order to evacuate the area within a 2-km radius of the station.

21:52   The chief cabinet secretary gives another order at a press conference to evacuate the area within 3 km, and in-house evacuation within 10 km.

MARCH 12

09:35   The chief cabinet secretary gives an order at a press conference to evacuate the area within 10 km.

20:32   The prime minister gives another order at a press conference to evacuate from within 20 km.

The Daini Station

MARCH 12

07:45   Evacuation order within 3 km, and in-house evacuation within 10 km.

17:39   Evacuation order within 10 km.

In addition, at a press conference at 11:00 hours on March 15, the prime minister issued an in-house evacuation order within 30 km.

Figure 11.6 Pictographic signs: Safe place from tsunamis, tsunami evacuation shelter, and tsunami risk area

Source: Ministry of Economy, Trade and Industry (METI).

Long-term evacuation

On April 22, 2011, the government defined the following evacuation zones (map 11.1):

- *Restricted area.* The area within a 20-km radius where some 78,000 people live.

- *Deliberate evacuation area.* The area where the cumulative dose of radiation might reach 20 mm Sievert within one year. Some 10,000 residents were requested to evacuate within a month.

Map 11.1
Evacuation areas

Source: METI.

Note: mSv = millisievert (radiation).

Restricted area, deliberate evacuation area, evacuation-prepared area in case of emergency, and regions including specific spots recommended for evacuation (as of August 3, 2011)

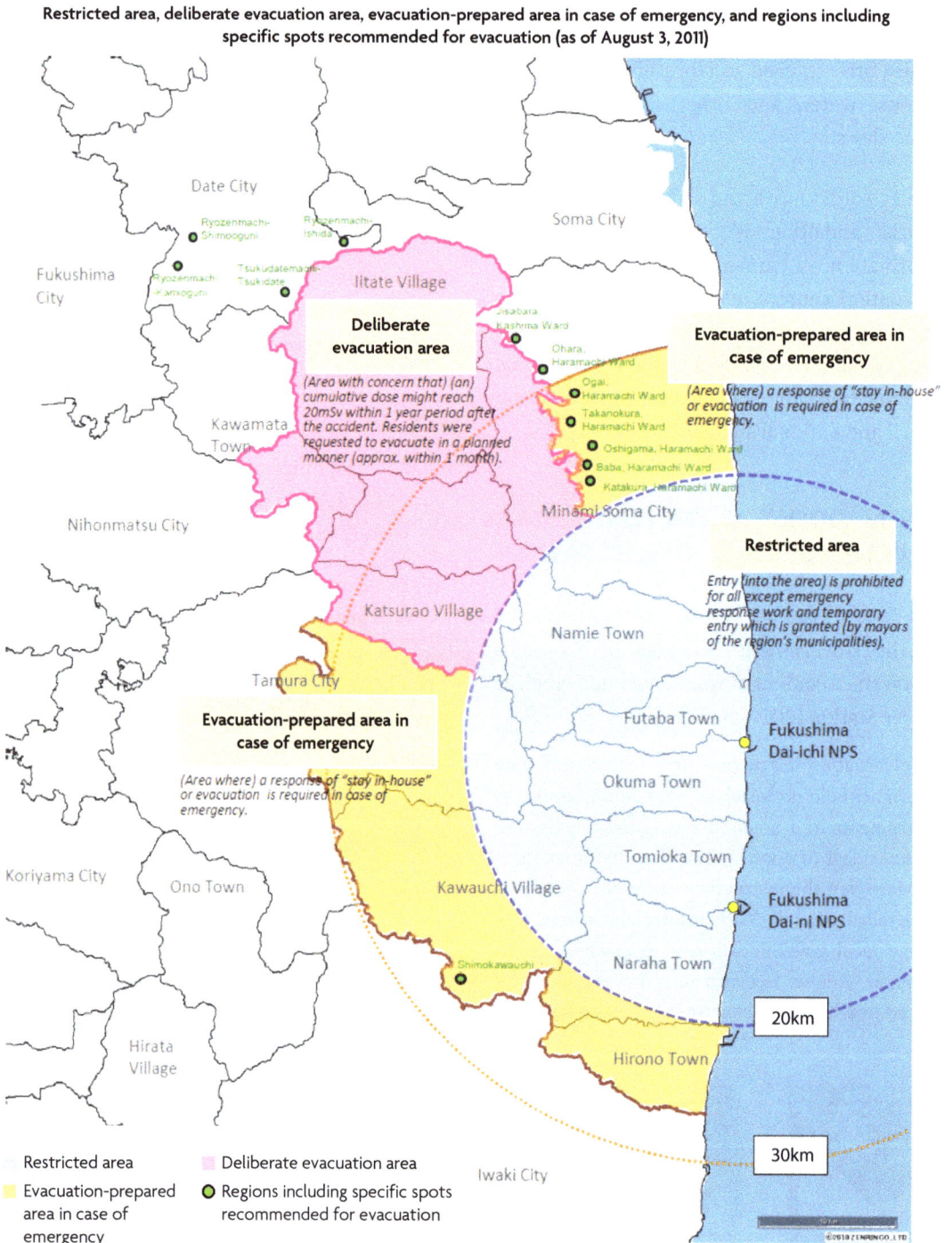

Date City

Soma City

Fukushima City

Ryozenmachi-Shimooguni

Ryozenmachi-Ishida

Ryozenmachi-Kamoguchi

Tsukudatemachi-Tsukidate

Iitate Village

Deliberate evacuation area

(Area with concern that) (an) cumulative dose might reach 20mSv within 1 year period after the accident. Residents were requested to evacuate in a planned manner (approx. within 1 month).

Jisabara, Kashima Ward

Ohara, Haramachi Ward

Ogai, Haramachi Ward

Takanokura, Haramachi Ward

Oshigama, Haramachi Ward

Baba, Haramachi Ward

Katakura, Haramachi Ward

Evacuation-prepared area in case of emergency

(Area where) a response of "stay in-house" or evacuation is required in case of emergency.

Kawamata Town

Nihonmatsu City

Minami Soma City

Restricted area

Entry (into the area) is prohibited for all except emergency response work and temporary entry which is granted (by mayors of the region's municipalities).

Katsurao Village

Namie Town

Tamura City

Evacuation-prepared area in case of emergency

(Area where) a response of "stay in-house" or evacuation is required in case of emergency.

Futaba Town

Fukushima Dai-ichi NPS

Okuma Town

Koriyama City

Ono Town

Kawauchi Village

Tomioka Town

Fukushima Dai-ni NPS

Shimokawauchi

Naraha Town

Hirata Village

Hirono Town

20km

30km

Iwaki City

☐ Restricted area	▨ Deliberate evacuation area
▨ Evacuation-prepared area in case of emergency	⬤ Regions including specific spots recommended for evacuation

©2010 ZENRIN CO.,LTD

- *Evacuation-prepared area in case of an emergency.* The area where a directive of either "stay in-house" or an evacuation might be required in case of an emergency, affecting some 58,500 people. This was lifted on September 30, 2011.

People in the affected areas experienced all kinds of difficulties during the evacuation. Updated information is recorded in chapter 36. They were forced to change shelters as the government expanded the evacuation zone. Some 82 percent of the evacuees changed shelters more than three times, and one-third of them changed more than five times. The death toll among the elderly who were evacuated from long-term care facilities increased substantially in 2011. It was also reported that dementia worsened among the elderly.

People in the Fukushima Prefecture had continued to be evacuated for the first year (figure 11.7). More than 150,000 people were evacuated, of whom over 60,000 were located in other prefectures across the country as of the end of 2011. Also, nine city governments moved to other locations. This evacuation scenario is expected to continue, since detailed plans for relocation back to hometowns have not been formulated. According to an interview survey, one-fourth of the evacuees say they are "unwilling to return" to their towns of origin, and another one-fourth say that they would "return only after others have returned." Younger people show less willingness to go back.

LESSONS

Japan has experienced many tsunamis and has made ongoing efforts for over a century to strengthen evacuation measures and mitigate damages. Japan has already started modifying its DRM plans and developing new systems to prepare for the next tsunami by incorporating the following lessons:

Public awareness programs must be supported by action. Although most residents had

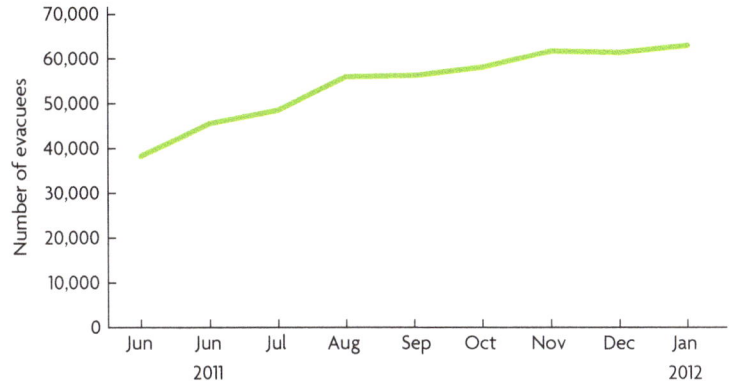

Figure 11.7 Number of evacuees moved to other prefectures, June 2011– January 2012

Source: Fukushima Prefecture.

enough knowledge about earthquakes and tsunamis, some failed to survive because they waited too long to evacuate. Public awareness programs must be designed to encourage evacuation. Without practice drills and trainings during normal times, people fail to evacuate properly and in a timely manner.

Public awareness programs should include practical knowledge. The programs should include the following messages:

- *Don't stick to past experiences.* No one knows how big a tsunami can be, and every tsunami is a new event. If someone says, "It is safe here because no tsunami has ever in my lifetime come this far up," this only reflects a few decades of experience.

- *Don't wait for your family to return.* Some people went to meet their children or waited for family members to get home. These people lost valuable evacuation time.

- *Don't wait for others to decide.* Some people couldn't decide whether to evacuate. They waited and watched what their neighbors were doing.

- *Don't stay in your car.* Some people evacuated in vehicles and got stuck in heavy traffic jams; they didn't leave their vehicles until the tsunami caught up with them.

- *Keep up to date with tsunami evacuation information.* Designated tsunami evacuation sites are sometimes changed based on recent scientific tsunami simulations or new developments in cities. Participating in evacuation drills in your own community, school, or company is an important way of keeping up to date with new information

- *Don't try to figure out for yourself what will or will not happen next,* as tsunami waves come repeatedly.

- *Never go back home to pack an evacuation bag before leaving.* Some people returned to their houses to retrieve valuables and other household items, and the second tsunami came and swept them away.

- *Update information after evacuation.* After the quake, blackouts occurred in most of the affected areas and telephone lines were congested. Portable radios are useful for staying abreast of the latest information and local news.

The limitations of various technologies must be understood. People who believed that tide walls and seawalls would hold off the tsunami delayed their evacuation. Also, some people felt secure because they believed that the tsunami levels estimated by the JMA would be lower than the walls, but at many points the tsunami exceeded the estimated heights (chapter 10). People who lived in areas that were indicated as being safe on hazard maps also delayed leaving (chapter 27). Others evacuated to shelters that appeared on the hazard maps, which had been officially designated as safe by the government, but were nevertheless engulfed by the tsunami.

Evacuation by vehicle should be considered only if needed. The elderly cannot walk for long distances, and in flat areas, it is difficult to walk several kilometers. Measures for evacuation by vehicle should be improved.

Procedures for evacuation from nuclear accidents should be prepared. The Government Investigation Committee on the Fukushima Accident (2011) stressed that "organizations concerned had not prepared because of the myth that all nuclear power stations are perfectly safe, and they therefore ignored the risks." The committee recommended the following:

- *Activities to raise public awareness are needed to provide residents* with a basic knowledge of how radioactive substances are released during a major nuclear accident, how they are dispersed by wind and other agents, and how they fall back to earth; also, the harmful health effects of radiation exposure should be made known.

- *Local governments need to prepare evacuation plans* that take into account the exceptionally serious nature of a nuclear accident, to conduct evacuation drills periodically under realistic circumstances, and to encourage residents to participate in those drills (chapter 36).

- *During normal times, there is a need to make preparations,* such as drafting detailed plans for choosing and arranging of transportation, establishing of evacuation sites in outlying areas, and ensuring water and food supplies at evacuation shelters, considering that evacuees may number in the thousands or tens of thousands. It is especially important to develop measures for the evacuation of the disadvantaged, such as the seriously ill or disabled, including those in medical institutions, homes for the aged, and social welfare facilities.

- *The types of measures listed above also need to actively involve prefectural and national governments* to draw up and administer evacuation and disaster management plans, in the event that a nuclear emergency were to affect a large area. These precautions should not be left up to local municipal governments alone.

RECOMMENDATIONS FOR DEVELOPING COUNTRIES

Promote evacuation measures as the heart of DRM. Evacuation, along with other non-structural measures, is relevant to any other country, while the more sophisticated communication systems are costly and need many years to develop. Other measures, such as education and warnings, should be developed as support to the evacuation measures.

Support the community. Governments should support communities to prepare evacuation measures by providing hazard maps and warnings, mobilizing drills, constructing shelters and evacuation routes, and conducting education programs at school as explained in figure 11.1 (chapter 6). Also, governments should formulate DRM plans by incorporating these measures (chapter 7).

Transfer memory to next generation. Memories and experiences of dealing with disasters should be passed from generation to generation. In Japan local communities constructed stone monuments recording tsunami disasters. Simeulue Island, northwest of Indonesia's Sumatra Island, had less damage than other areas after the Indian Ocean tsunami in 2004. The local residents evacuated as soon as they felt the earthquake, because they knew that after a quake, sea water would come rushing in. They have passed along their tsunami experiences to the next generation through children's songs. Also, they had already relocated their towns from the coast to higher ground after the 1907 tsunami.

Raise public awareness. DRM education in schools, including evacuation drills, is essential to ensure successful tsunami evacuation at the community level. Children will bring back and share their knowledge with their families, which will help educate the whole neighborhood (chapter 8).

NOTE

Prepared by Masaru Arakida, Asian Disaster Reduction Center, and Mikio Ishiwatari, World Bank.

BIBLIOGRAPHY

Arakida, M., A. Koresawa, and Y. Kawawaki. 2011. "Damage from the Great East Japan Earthquake and the Contributions of Space Technology." In *32nd Asian Conference on Remote Sensing 2011 (ACRS 2011)* (3-7 October 2011). Tapei, Taiwan: Asian Association on Remote Sensing.

Cabinet Office (CAO). 2011. *Report on the Evacuation from the Great East Japan Earthquake* [in Japanese]. http://www.bousai.go.jp/jishin/chubou/higashinihon/7/1.pdf.

Committee for Technical Investigation on Preservation of Lessons Learned from Disasters. 2005. *Report on 1896 Meiji Sanriku Earthquake Tsunami* [in Japanese]. http://www.bousai.go.jp/jishin/chubou/kyoukun/rep/1896-meiji-sanriku JISHINTSUNAMI/index.html.

———. 2010. *Report on 1960 Chile Earthquake Tsunami* [in Japanese]. http://www.bousai.go.jp/jishin/chubou/kyoukun/rep/1960-chile%20 JISHINTSUNAMI/.

Fukushima Prefecture. 2011. *Preliminary Draft of 6th Fukushima Prefecture Elderly Welfare Plan and 5th Fukushima Prefecture Nursing Care Insurance Support Plan* [in Japanese]. Fukushima, Japan.

Fukushima University Reconstruction Institute. 2012. *Survey on Recovery Situation of Eight Futaba Towns* [in Japanese]. Fukushima, Japan.

Hirose, H. 2012. "Perceptional Behavioral Biases of Evacuees in Emergencies" [in Japanese]. *Civil Engineering* 97 (6).

Investigation Committee on the Accident at Fukushima Nuclear Power Stations of Tokyo Electric Power Company. 2011. *Interim Report.* http://icanps.go.jp/eng/interim-report.html.

Mizushima, T. 2011. *Review of the Behavior of People during the Great East Japan Earthquake.* The Second Expert Group Meeting on GEJE—Learning from the Mega-Tsunami Disaster. http://www.bousai.go.jp/kyoryoku/Session/Session1/01.pdf.

Mizutani, T. 2012. "Emergency Evacuation and Human Losses from the 2011 Earthquake and Tsunami off the Pacific Coast of Tohoku." *Natural Disaster Research Report of NIED* 48: 91–104.

Nakahara, S. 2011. "Lessons Learned from the Recent Tsunami in Japan: The Need for Epidemiological Evidence to Strengthen Community-based Preparedness and Emergency Response Plans." *Injury Prevention* 17: 361–64.

Nakajima, N., and A. Tanaka. 2011. "Past Tsunami Disasters and Reconstruction Planning in the Sanriku Region." *Urban Planning* 291: 45–48.

Sotooka, H. 2012. *3.11 Complex Disaster* [in Japanese]. Iwanami Shinsyo, Tokyo.

UNESCAP (UN Economic and Social Commission for Asia and the Pacific). 2011. *Thematic Session Report for Session 1*. The Second Expert Group Meeting on the Great East Japan Earthquake—Learning from the Mega-Tsunami Disaster. http://www.unescap.org/idd/events/2011-December-Japan-Earthquake/Report-of-the-Thematic-Session-1.pdf.

Yun, N., and M. Hamada. 2012. "A Comparative Study on Human Impacts Caused by the 2011 Great East Japan Earthquake and Disaster Mitigation." In *Proceedings of the International Symposium on Engineering Lessons Learned from the 2011 Great East Japan Earthquake* (1666–79), Tokyo, Japan.

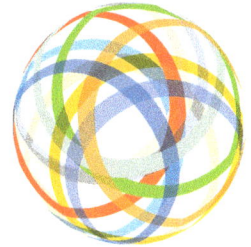

CHAPTER 12

Urban Planning, Land Use Regulations, and Relocation

Reconstruction should include a range of measures to enhance safety: disaster prevention facilities, relocation of communities to higher ground, and evacuation facilities. A community should not, however, rely too heavily on any one of these as being sufficient, because the next tsunami could be even larger than the last. Communities also need to rebuild their industries and create jobs to keep their residents from moving away. The challenge is to find sufficiently large relocation sites on high ground, and also to regulate land use in lowland areas.

FINDINGS

Reconstruction after the March 2011 disaster

Reconstruction after the Great East Japan Earthquake (GEJE) has been slow compared to the Great Hanshin-Awaji (Kobe) Earthquake that hit the city of Kobe and killed 6,400 people in 1995. The seismic shocks experienced during the GEJE affected a much broader area. A number of characteristics of the GEJE made reconstruction more difficult and lengthy.

First, since tsunamis tend to hit the same areas repeatedly over several decades or even several hundred years, some affected people wanted to reconstruct their houses at suitable new locations instead of the damaged sites. Although the rubble has been removed, full-scale reconstruction has not yet begun. Planning and local consensus-building for relocating communities to high ground has been attempted. It takes time to find places to live and to reach agreement as a community to move together to a safer place. Since it takes several years to rebuild completely, it is unclear whether local employment and population levels can be sustained (see chapter 33).

Second, the radiation contamination from the accident at the Fukushima Daiichi Nuclear

Power Station will last a long time and prevent the local people from returning to their homes. Reconstruction projects may be delayed since it is still unclear when or if people will be able to return to their places of residence. There is also a concern that many people, especially younger families, may choose not to go back to their hometowns (chapter 36).

While the nuclear accident in Fukushima was a bit less serious than in Chernobyl, it was ranked the same on the International Nuclear Event Scale. No major emissions of radioactive material from the collapsed nuclear power plants have been observed since April 2011. A ban or restriction on land use will be introduced to prevent exposure to high levels of radiation. The government plans to reclassify the Warning Zone and Planned Evacuation Zone into three new categories: *long-term habitation difficult zone* (more than 50 millisieverts [mSv] of annual radiation exposure), *prioritized decontamination zone,* and *decontamination and possible to return zone.* All the nuclear power plants in Japan were shut down in May 2012 for maintenance and evaluation, and two units started operation in June 2012.

There are two tiers of local government in Japan, prefectures and local municipalities, which are responsible for disaster response and reconstruction. Municipal governments play the most important role because they are closest to the victims and the stricken areas. The prefectural governments are grappling with broad reconstruction issues. For example, they have supported municipal governments in debris management by coordinating solid waste management facilities in the prefectures (chapter 23).

Managing disaster risk: A three-pronged approach

All reconstruction plans aim at rebuilding towns and communities that are resilient to major disasters. The most important lesson from the GEJE is that there are many disasters we cannot prevent; all we can do is reduce

the damages. Sometimes we cannot predict, or even imagine, the severity of future natural hazards and so we will be unprepared. Although many breakwaters and tsunami dikes were built in the stricken areas, the tsunami nevertheless destroyed or overtopped most of them, and poured into the towns and villages behind them. Reducing damages means first and foremost preventing the loss of human lives; however, property damage to houses, infrastructure, and various manmade facilities may be unavoidable.

In regards to building relocation and reconstruction, disaster risk management (DRM) consists of three components: disaster prevention facilities, community relocation to safer ground, and evacuation facilities. This approach was reflected in the government's basic policy on reconstruction, after the GEJE Reconstruction Council's report recommended a shift in DRM from prevention to risk reduction.

Disaster prevention facilities included tsunami breakwaters or dikes. It is important to recognize both their usefulness and their limitations (as explained in chapter 1); damages would have been even worse without them. At the same time, the facilities could not prevent the huge tsunami from destroying areas behind them. Most of the breakwaters and dikes will be rebuilt to be even stronger and larger, but these facilities can only resist tsunamis of limited size.

Community relocation and redesign are also important ways of reducing damage. Clearly, when communities are located on sufficiently high ground, the tsunami can't reach them. This was well known in areas that had been repeatedly hit. After the Showa Sanriku Tsunami in 1933, which killed about 3,000 people, the government promoted reconstruction on higher ground; but this policy could not be fully implemented since it was difficult to find suitable locations.

Evacuation facilities consist of escape routes and shelters. Escape routes should be easy to follow and clear of debris. Although evacuation drills and instructions discourage the use of vehicles, escape routes must nevertheless accommodate both pedestrians and cars (chapter 11). Evacuation shelters should be multilevel structures to accommodate evacuees safely as water levels rise.

All three components must be used together in a holistic system. Using only one or two elements is not enough. While disaster prevention facilities and the location of communities are based on forecasts and estimates, the actual hazard may be larger; life-saving evacuation facilities will also be required.

Although these strategies are being applied in the reconstruction of tsunami-stricken areas, experience has shown that relocating communities to higher ground has been difficult to implement. And while relocation of communities and construction of evacuation facilities may be possible in newly reconstructed areas, people are also worried about areas that are under threat of being hit by tsunamis in the near future. In these areas, construction of disaster prevention facilities takes longer and the relocation of communities to higher ground is more difficult than in those areas destroyed by the GEJE: compensation has to be paid for the existing buildings, and consensus has to be established among affected residents.

Learning from past tsunami reconstruction

The following three examples illustrate the challenges of reconstruction. Dikes alone cannot protect communities, so locating communities at higher elevations is key. But it is difficult to find suitable locations and to sustain people's livelihoods.

Building on higher ground saves lives and property

The Yoshihama fishing and farming village in Ofunato City, in the Iwate Prefecture, was successfully relocated to available land close to the original residential area, with financial assistance from the government. The village began moving to higher land following the Meiji Sanriku Tsunami in 1896, which washed away almost the entire village. The residents found and developed the relocation site themselves, and the relocation was completed with government financial support after the Showa Sanriku Tsunami in 1933. Fortunately, there was a hill above the old village that sloped gently to the beach. The villagers moved all their houses to the hill and turned the lowlands, where they had lived, into farmland. A 3-meter high tsunami dike was built in the 1970s. On March 11, the tsunami hit the village, flooding most of the farmland but not the residential zone. Only a couple of houses, located lower down, were washed away, and one person was killed.

Half-measures do not suffice

Another example is the Touni-hongo village in Kamaishi City, Iwate Prefecture. This is a well-known village that relocated after the Showa Sanriku Tsunami in 1933 to a newly developed site on hilly ground nearby. One of the community leaders, who owned the land, donated it to the community. The Iwate prefectural government developed the relocation site with financial support from the central government. One hundred houses were moved to the new site and the old location was turned into farmland.

The GEJE tsunami flooded and washed away all 50 houses located on lower ground, but it didn't reach the houses relocated to higher ground.

The houses on the lower level were built after the 10-meter-high tsunami dike had been constructed. The dike was expected to protect the hinterland. But the tsunami broke into the village at a point beyond the dike, and another tsunami wave came in through a tunnel behind the village that connects it with the neighboring village. One of the reasons for building houses on lower ground is to make daily life easier for the elderly, who have a hard time on steep slopes. In many similar cases, communities were partially damaged on low ground. Constructing large dikes may even have encouraged building on lower ground.

A low-lying community destroyed

The final example is Taro, Miyako City, Iwate Prefecture. Taro was once known around the world for its long tsunami dikes (chapter 1). Taro was hit by the 1896 Meiji Sanriku Tsunami, losing 1,867 people—about 83 percent of its population of 2,248. It was then hit again in 1933 and lost 911 residents, or 32 percent of the population. After the Showa Sanriku Tsunami in 1933, Taro considered following the central government's recommendation and relocating the entire community to higher ground. They could not, however, find a suitable site where the people could see the fishing port or build their houses facing south, among other important conditions. Because Taro was a large village, the residents finally gave up looking for a new site; they decided to build a dike around the residential area, and paid for it themselves.

After the first year of construction, the central and prefectural governments approved the project as a disaster prevention public work and provided the rest of the funding. A second dike with almost the same dimensions as the first one was built after the 1960 Chilean earthquake tsunami, to prepare for larger tsunamis.

But even with these two dikes, Taro, this time, was utterly destroyed. An estimated 200 out of 4,400 residents perished. The newer dike closest to the beach was destroyed and the other was overtopped. There were several cases in Sanriku where previously stricken communities had not moved but had simply added landfill. All of these incurred severe damage.

Recovering industries and jobs

Another serious problem came up while planning for reconstruction: out-migration. A survey showed that the population had decreased by 46,000 between 2005 and 2010 before the disaster in the coastal municipalities ranging from the Iwate to the Fukushima prefectures. According to residential statistics, the same area lost 57,000 people between March and November 2011, including about 15,000 people who were taken by the disaster. If people are not strongly induced to stay in these areas through economic incentives such as industrial recovery and job creation, even more residents may leave in spite of physical reconstruction (chapter 24).

The urgent need for development requires that part of the huge national reconstruction budget be used to develop new job-creating industries and to attract entrepreneurs from outside the region.

The first step is to rebuild existing enterprises, especially in the fishing and marine-product-processing industries, including shipbuilding, freezing, and warehousing. But these cannot be relied on alone, since they have been gradually losing jobs to heavy international competition.

A second important initiative is setting up new industries that may increase future employment. All local government reconstruction plans include activities such as tourism, renewable energy production, and manufacturing of products that respond to local demand.

In Fukushima the outlook is worse. The government announced that certain parts of Fukushima will not be habitable for a long time because of high radiation levels. The government must therefore help people relocate.

Toward building communities resilient to disaster

Local governments did not effectively regulate land use in the affected areas. Lowlands had been developed for residential, commercial, and industrial purposes. Meanwhile, economic development, urbanization, and population growth increased residents' vulnerability to tsunami damage along the coast.

The population in the coastal areas of Iwate Prefecture tripled over the past century: from about 76,000 at the time of the Meiji Sanriku Tsunami in 1896 to some 274,000 in 2011.

The Japanese government is reinforcing DRM systems by introducing land-use regulations based on lessons learned from the GEJE. The Act on Building Communities Resilient to Tsunami was legislated in December 2011 to prepare for low-probability, high-impact tsunamis (figure 12.1). The goal of the act is to protect human lives at all costs. The following approaches have been adopted:

- Multiple lines of defense, combining structural and nonstructural measures (see part I, and chapters 6, 7, and 13)

- Shifting from a "single line of defense" based on tsunami dikes to a "zone defense" using roads and other structures such as secondary dikes, and land-use regulation

- Instituting practical measures for quick and safe evacuation (chapter 11)

- Assessing tsunami risks based on local conditions, such as industry, commercial activities, history, and culture (chapter 25)

The Ministry of Land, Infrastructure, Transport and Tourism has formulated basic guidelines on tsunami countermeasures for prefectures and municipal governments. The guidelines specify that prefectural governors should categorize risk areas as "yellow zone," "orange zone," and "red zone." In municipalities, mayors formulate countermeasure action plans. The governors and mayors designate structures such as highways as disaster management facilities.

In yellow zones, where residents are likely to lose their lives, evacuation measures such as evacuation shelters, drills, and hazard maps, are required. In the orange zones, where

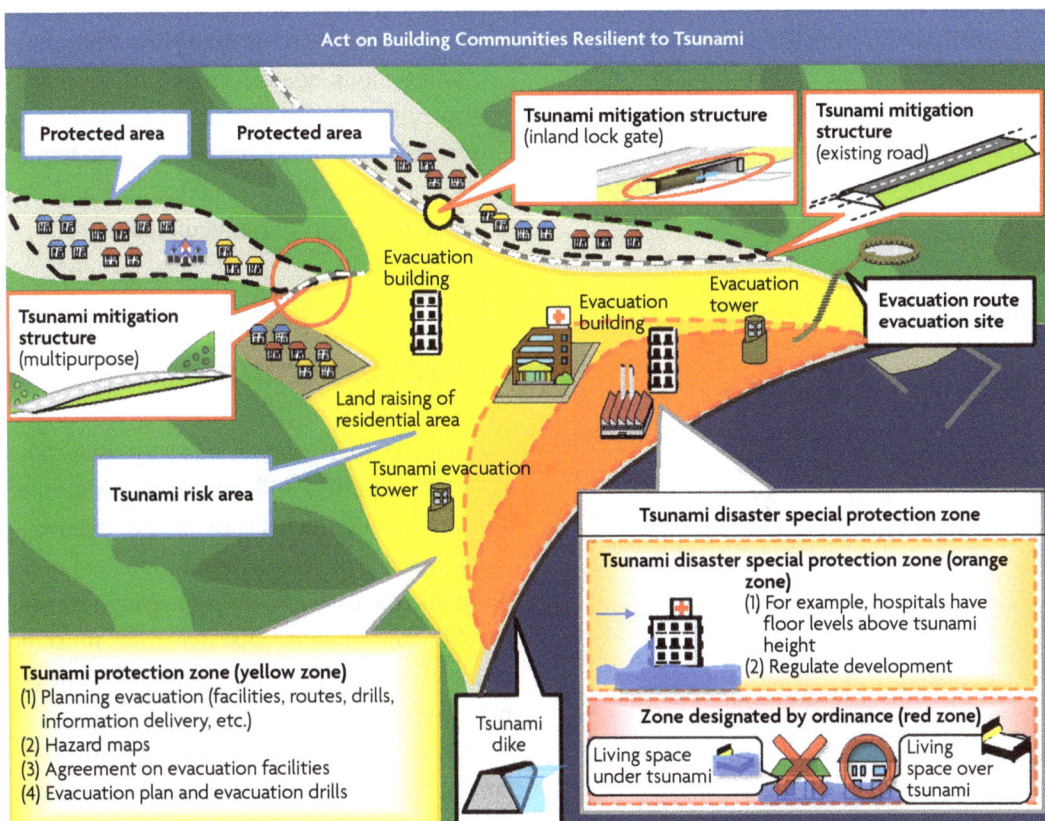

Figure 12.1 Building communities resilient to tsunamis

Source: Ministry of Land, Infrastructure, Transport and Tourism.

residents are highly likely to lose their lives, key facilities such as hospitals are to be set up in tsunami-resilient structures. In red zones, where residents cannot escape a tsunami, all buildings including residences must be tsunami resilient, such as having multiple stories that rise high enough to evade the tsunami waters.

Cost sharing and various incentives are used in implementing these measures. Local governments may provide the private sector with incentives to secure evacuation facilities. Additional floor-space ratios for evacuation spaces on high floors are given as bonuses. They may also be exempted from paying 50 percent of the building tax on evacuation space. Participating organizations share the costs of multipurpose structures. For example, DRM organizations will share the additional construction costs for roads used as secondary dikes.

The central government and local governments provide financial assistance for developing safe relocation sites on high ground. Community members must reach a consensus on relocation before it begins. The community bears the cost of building new houses, while local governments are responsible for developing the infrastructure associated with the relocation sites.

LESSONS

- *Tsunami-prone areas must be ready for recurring disasters.* Reconstruction must include three key safety measures: disaster prevention facilities, relocation of communities to higher ground out of reach of tsunamis, and evacuation facilities. The community must not rely too heavily on any one of these, since the next tsunami may be much larger than the last and require a broader range of precautions.

- *Industrial recovery is indispensable for economic sustainability.* In the absence of

businesses and job opportunities, people will leave their disaster-stricken communities. Simply rebuilding houses will not induce people to stay; industrial recovery policies must also be strengthened (chapter 24).

- *Public-private partnerships are crucial.* Enormous sums of public money are being spent on reconstruction projects and to stimulate the local economy. But this will end in several years. It is important to create as many business activities as possible to promote economic growth and opportunities in the long term.

- *Relocation effectively mitigates damage and loss of life, but implementation is a challenge.* Three examples from past tsunamis illustrate that although relocation measures are effective, they are not easy to implement. In the village of Yoshihama, houses that had already been relocated following a tsunami did not suffer from the GEJE. But finding suitable relocation sites around the mountainous coastal village of Taro was difficult, and in the village of Touni-hongo, where houses had been relocated to higher ground following a tsunami, lowland use could not be properly regulated.

RECOMMENDATIONS FOR DEVELOPING COUNTRIES

Understand and manage disaster risk. The Japanese experience illustrates that improper land-use regulation leads to increased damage from disasters. Urbanization in lowland areas has made the eastern coast of Japan more vulnerable to tsunamis. Disaster risks must be properly understood and managed in urban planning.

Develop facilities, live in safe places, and prepare for evacuation. The approach of integrating three elements—setting up proper facilities, settling in safe areas, and properly planning evacuations—can be used to manage disaster risk in developing countries. Since every country has its own geographic, socioeconomic, and budgetary characteristics, and also faces hazards of different dimensions, practical approaches will differ from country to country. Since most developing countries have limited resources for constructing facilities, people should focus on living in safe areas and putting rigorous evacuation measures in place.

Protect by zone and multiline. "Zone defense" and "multiline" approaches can be effective against tsunamis, as well as other disasters such as floods, landslides, and mud flows. Infrastructure, such as highways and railways, help mitigate disaster risks in both rural and urban areas. In the Philippines, a megadike constructed to protect against *lahars* (volcanic mud flows) from Mount Pinatubo is also used as a highway. Disaster management organizations and infrastructure organizations should coordinate in planning and sharing the costs of multipurpose infrastructure (chapter 4).

Promote relocation where feasible, acknowledging difficulties. As Japan's experiences with tsunami disaster recovery illustrates, relocation to safer sites and land use regulations in risk-prone areas are effective but challenging to implement (chapter 33). Even though people may be ready to relocate to higher ground right after a disaster, they may also change their minds, preferring to live in the lowlands because it is more convenient for daily life. After the Indian Ocean tsunami in 2004, the Indonesian and Sri Lankan government tried to introduce similar regulatory approaches, but they did not succeed because of opposition from the communities and limited enforcement mechanisms.

NOTE

Prepared by Takashi Onishi, University of Tokyo, and Mikio Ishiwatari, World Bank.

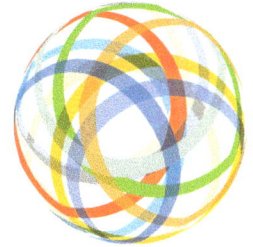

CHAPTER 13

Green Belts and Coastal Risk Management

For more than four centuries Japan has been developing forested green belts to mitigate coastal hazards such as sandstorms, salty winds, high tides, and tsunamis. Although Japan's green belts were severely damaged by the March 11 tsunami, they did reduce the impact of waves and protected houses by capturing floating debris. Local governments are planning to reconstruct the green belts as a countermeasure against tsunamis. While local communities have traditionally taken charge of maintaining green belts, their role has been weakened because of changes in society brought about by economic development and urbanization.

FINDINGS

Japan is surrounded by the sea; its coastline measures approximately 34,000 kilometers (km), with 1,640 km² of a forested green belt distributed along its sandy coast. For more than four centuries Japan has been developing this green belt. Composed mainly of Japanese black pine (box 13.1), it serves various functions. It reduces the impact of coastal hazards such as blown sand, salty winds, high tides, and tsunamis. Japan's Forest Law stipulates that disaster risk management (DRM) forests should be planted in coastal areas to prevent damages from wind, airborne sand, and tsunamis.

Another benefit of the green belt is that it is a scenic landscape called *hakusa-seisyou* in Japanese, which means "beautiful coast with pine trees and sandy beach." Its role as a tourist attraction has become increasingly important as Japanese society has become more affluent.

In the Sendai Plain, a 200- to 400-meter-wide swath of pine forests along Sendai Bay, has for the past four centuries mitigated disasters and provided beautiful scenery consisting of green forests, white sands, and blue ocean. Masamune Date, a distinguished feudal lord, started to plant Japanese black pines along the Teizan Channel on the Sendai Plain in 1600. The people who lived on the dunes along

Takatamatsubara and the hope of recovery

In the disaster-affected areas of Tohoku, there were several famous coastal forests. Takatamatsubara of Rikuzentakata City was a 21-hectare coastal forest, 2 kilometers long and 200 meters wide, consisting of some 70,000 pine trees. In the 17th century, a wealthy merchant started planting pine trees in the barren coastal areas to protect agricultural lands from heavy winds and salt water. Another merchant began planting in the 18th century. The local communities developed and maintained the forests for some 350 years, conducting annual festivals to commemorate the two merchants. These coastal forests had also been a tourist attraction where a million or so people came to bathe or enjoy nature every year. After the GEJE disaster only a single pine tree remained—a meager symbol of hope of recovery.

Sources: (left) Ministry of Environment. (right) © Mikio Ishiwatari. Used with permission. Further permission required for reuse.

the coast had suffered from sandstorms and tidal disasters that damaged their agricultural products, and the pine forests protected their fields. Masamune allowed the people to sell wood from branches that were trimmed or had fallen to cover the expense of maintaining the green belt.

In the late 19th century, the Japanese government designated Reserved Forests, maintaining their DRM function. In 1933 the green belt mitigated damages from the Showa Sanriku tsunami. In 1935 the government started an afforestation program to mitigate tsunami damage and again promoted afforestation following the Chilean earthquake tsunami in 1960.

The green belt became less important after the rapid economic growth of the 1970s, as other more effective DRM measures were developed, and electricity and gas replaced

wood as energy sources for people. The community's role in managing the green belt diminished, and governments took over its maintenance.

Damage to the green belt

In the Great East Japan Earthquake (GEJE) of 2011, 3,660 hectares (ha) of the green belt were damaged by the tsunami, at a cost of ¥55 billion (over $540 million). In the four affected prefectures, 2,825 ha of the green belt were flooded, and 1,069 ha of the green belt were damaged more than 75 percent (map 13.1). The green belt of the Miyagi Prefecture was severely damaged—trees were uprooted or bent, or their trunks were broken.

The green belt reduced the impact of the tsunami, delayed its arrival time, and protected houses by capturing drifting debris. Several ways in which the green belt reduced damages

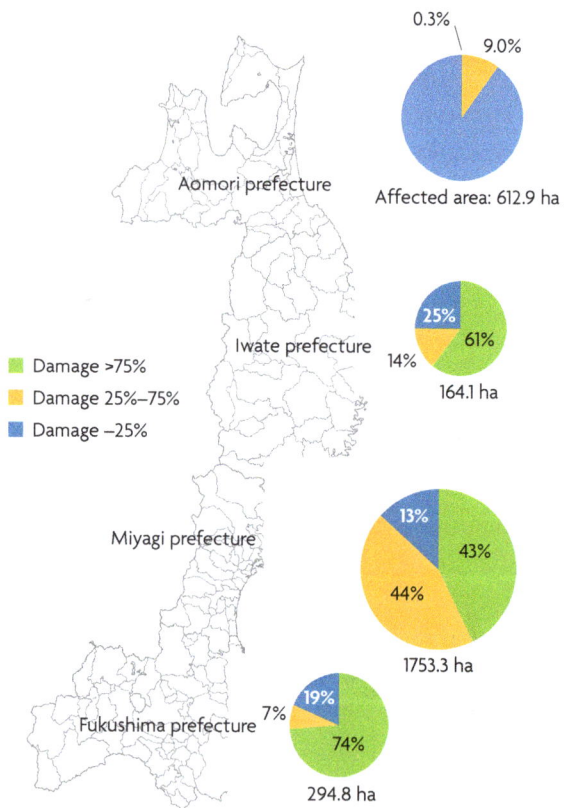

Map 13.1 GEJE tsunami damage to the green belt in four prefectures

Source: Forest Agency.

Figure 13.1 The forest captures a floating ship

Source: Forest Agency.

have been reported. In Hachinohe City, Aomori Prefecture, a forest caught 20 ships washed inland by a 6-meter tsunami, thereby protecting the houses located behind the trees (figure 13.1). Although these houses were inundated by over 3 meters of water, they were not washed away. In past tsunami disasters, the following benefits have been confirmed:

- The energy and speeds of the tsunamis decreased.

- Floating wreckage was blocked.

- People washed away by the tsunami were able to save their lives by clinging to trees.

- The trees helped preserve sand dunes, which in turn mitigated the force of the tsunami.

Natori City was hit by a tsunami of 8.5 meters. Almost all of the green belt was flooded and 106 ha (more than 80 percent) was damaged. Figure 13.2 shows the condition of the green belt in Natori City before and after the tsunami. The extent of the damage differed by location depending on the geographic conditions on the ocean side. In the northern part, which had sand embankments from port construction, the green belt was preserved; in the middle portion, which had no barrier, the green belt was washed away or knocked down; and in the southern part, the presence of tidal dikes preserved the green belt.

Figure 13.2 Condition of the green belt before and after the tsunami in Natori City

Source: © Kyoto University. Used with permission. Further permission required for reuse.

Local governments are planning to restore DRM coastal forests as one of their structural countermeasures, along with dikes and mounds. The Forest Agency suggests that the forests should be at least 50 meters wide, and preferably 200 meters, for effective DRM in coastal areas. DRM effects can be increased with building mounds, and debris, which is a serious obstacle to rehabilitation, can be used for building mounds.

The Miyagi prefectural government recommended the following actions to help the recovery of DRM forests:

- Coordinating with other rehabilitation works, such as coastal dikes and debris management

- Selecting tree species that conform to local conditions and support biodiversity

- Collaborating with nonprofit organizations, volunteers, and the private sector

Maintenance

Community action is essential to maintaining the coastal green belt. Local communities had historically developed and maintained the green belt to protect their houses and agricultural lands from coastal hazards. Proper maintenance is required to preserve the forest's DRM function: trees should be planted with moderate density, and frequent thinning is required otherwise the trees will not develop to their full size.

Since the late 1960s, the community's role in managing the green belt diminished as Japan experienced rapid economic growth; as previously noted, governments took over their management (figure 13.3). Growth led to the development of infrastructure such as dikes and new energy installations, while the fishing and agriculture industries lagged behind. Dikes replaced the green belt in coastal hazard prevention, and communities started using coal instead of pine trees as a fuel source. Community-based organizations that had managed the green belt broke up as communities lost interest and the government was unable to manage and maintain such vast forested areas. Moreover, damage caused by the pine weevil became a serious problem from the 1990s.

LESSONS

- *Green belts can be effective against small tsunamis, sea winds, or sands,* but not against a huge tsunami like that of March 11. Combining green belts with dikes and embankments can strengthen their effectiveness (chapter 12).

- *Green belts reduce tsunami damage by reducing wave energy,* delaying water arrival time, and protecting houses by capturing floating debris.

- *Coastal zone protection.* Green belts also provide other important benefits recognized by communities, such as protection from coastal storms, salt damage, and sand and provide spaces for recreation and wildlife. Forests may also provide psychological safety and augment well-being.

- *Green belts require several decades to develop properly.* Japan has had over four centuries of experience in their development.

- *Local communities can play important roles in green belt maintenance.* Maintenance mechanisms should be modified as society changes. In Japan the government expanded its roles as the economy grew.

RECOMMENDATIONS FOR DEVELOPING COUNTRIES

Forest projects can be effective countermeasures against tsunamis, floods, and other water-related disasters. Forested green belts can decrease disaster risks by reducing the force of natural hazards. Not only in Japan, but also during the Indian Ocean tsunami in 2004, mangroves and other coastal green belts mitigated potential damages due to the disaster.

Understand the DRM function of the green belt. Public awareness of the DRM function of the green belt should be raised. Also,

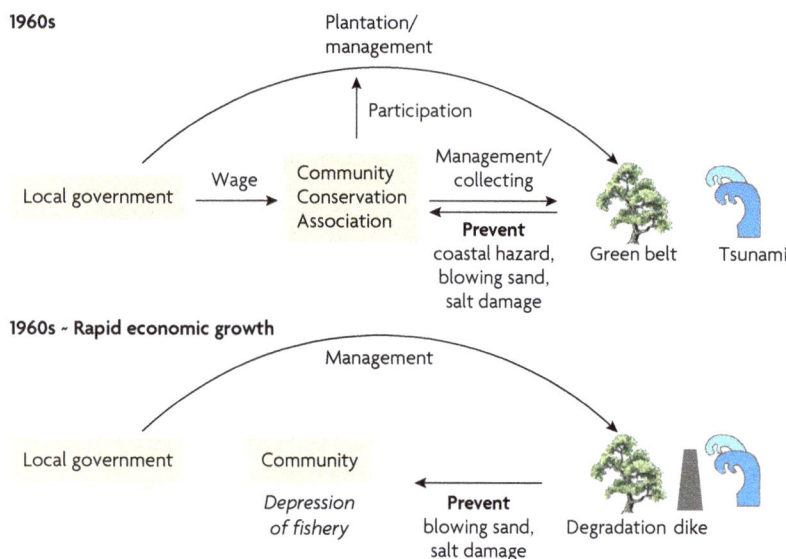

Figure 13.3 Changing approaches to managing the green belt

Source: Kyoto University.

information should be shared with decision makers to promote green belts.

Utilize the forest as a means of livelihood. In Japan forests have been used along rivers to mitigate floods, and farmers use bamboo from the green belts to produce handicrafts that provide them with additional income. Farmers can also earn from fuel woods and nontimber products, such as fruits, flowers, and medicinal plants.

Foster participatory maintenance. Restoring the green belts includes two major activities: cultivation and sustainable management, which should involve several stakeholders. Plantations can be jointly implemented by the government and civil society, including the community. Community participation in cultivation leads to a sense of ownership. Communities can continue using the green belt as a space to learn skills and as a way of maintaining relationships with external organizations.

Support community. Local governments and civil society organizations play an important role in increasing awareness and engaging the local community. DRM education in schools will also raise awareness and encourage participation.

NOTE

Prepared by Rajib Shaw and Yusuke Noguchi, Kyoto University, and Mikio Ishiwatari, World Bank.

BIBLIOGRAPHY

Haraguchi, T., and A. Iwamatsu. 2011. *Detailed Maps of the Impacts of the 2011 Japan Tsunami* [in Japanese]. Tokyo: Koko Shoin Publishers.

Investigative Committee on Revitalization of Coastal Forests Associated with Great East Japan Earthquake. 2012. *Kongoniokeru kaigan bousairinno saiseinitsuite* [Revitalization of coastal forest]. Forest Agency.

Miyagi Prefecture. 2012. *Kaigan bousairin ni tekishita shokusaijyu shu ni kansuru chousa houkokusho* [Report on the survey on appropriate species for plantation of coastal forest]. Miyagi Prefecture, Japan.

Natori City. 2011. *Report on Tsunami Damage in Natori City.* Natori City.

Noguchi, Y., R. Shaw, and Y. Takeuchi. 2012. "Green Belt and Its Implication for Coastal Risk Reduction: The Case of Yuriage." In *East Japan Earthquake and Tsunami: Evacuation, Communication, Education and Volunteerism,* edited by R. Shaw and Y. Takeuchi. Singapore: Research Publishing.

Yuriage-kyoudoshikenkyukai. 1977. *Records of the Culture and Geography of Yuriage (Yuriage Fudoki)* [in Japanese]. Natori: Syouhei Ono.

PART III

Emergency Response

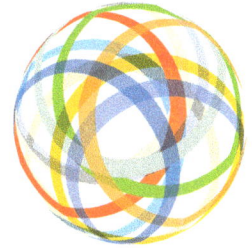

Mobilizing and Coordinating Expert Teams, Nongovernmental Organizations, Nonprofit Organizations, and Volunteers

In response to the Great East Japan Earthquake (GEJE), domestic and international assistance initiatives were launched by a large number of public and private sector organizations; meanwhile, various emergency teams were mobilized through national and international networks. The GEJE served as a reminder that civil society organizations play an indispensible role in disaster management. These organizations have the advantage of flexibility and speed in reaching and caring for affected communities. But the GEJE also revealed that, without prearranged coordination mechanisms, even the best-prepared teams cannot function properly on the ground. Because of the complexity of disaster response operations and the large numbers of actors involved, coordination mechanisms must be established well in advance of any disaster.

FINDINGS

Mobilizing the government's expert teams

Municipality and prefecture governments play a leading role in disaster response in Japan. But because of the magnitude of the March 11 earthquake and tsunami, local governments were unable to respond alone. National agencies as well as prefectures and municipalities outside the affected region were quickly deployed (chapter 17). Organizations concerned had formed a variety of expert teams (see table 14.1) in light of the lessons learned from past disasters, in particular the Great Hanshin-Awaji Earthquake (Kobe earthquake) in 1995. The national government took action immediately by setting up a response office four minutes after the earthquake; and an Emergency Disaster Response Headquarters, headed by the prime minister, within 30 minutes. Its mandate was to oversee and coordinate all response activities.

Table 14.1 Expert teams organized by the government

MINISTRY/AGENCY	EXPERT TEAMS
Ministry of Defense	Self-defense forces
Ministry of Health, Labour and Welfare	Disaster medical assistance team
Ministry of Land, Infrastructure, Transport and Tourism	Technical emergency control force, coast guard
Fire and Disaster Management Agency and prefectural fire departments	Emergency fire response teams
National Police Agency and prefectural police agencies	Interprefectural emergency rescue units

Japan Self-Defense Forces

The total number of Japan Self-Defense Forces (JSDF) personnel in operation reached some 107,000 people with about 540 aircraft and nearly 60 vessels. The JSDF rescued approximately 19,000 disaster victims, or nearly 70 percent of those rescued in the Great East Japan Earthquake (GEJE) event. They also provided transportation assistance to medical teams, patients, and rescue units dispatched from various countries, and livelihood assistance to disaster victims by providing water, food, and other necessities. The JSDF also responded to the nuclear accident, engaging mainly in pumping water for cooling used fuel pools, decontaminating personnel and vehicles, and monitoring amounts of airborne radiation (figure 14.1).

Emergency fire response teams

Following its experience with the Kobe earthquake, the Fire and Disaster Management Agency created fire response teams to mobilize firefighting departments across Japan. After the GEJE, the emergency teams dispatched more than 30,000 firefighters from 712 fire departments in 44 prefectures nationwide over a period of 88 days ending on June 6, 2011. In cooperation with local fire departments, the emergency teams rescued 5,064 people as of June 30, 2011. Most fire departments in devastated areas had lost their radio equipment or base of communications. In light of this experience, the Fire and Disaster Management Agency decided to provide the teams with additional mobile communications equipment and a larger supply of fuel so that they could operate effectively over wider areas and for longer periods of time.

Interprefectural emergency police rescue units

Interprefectural emergency police rescue units are police units that have been set up in prefectures nationwide, based on the experience with the Kobe earthquake. In response to the GEJE, these rescue units conducted such activities as search and rescue and the securing of emergency transportation routes. A total of 750,000 person-days were spent working on-site, with as many as 4,800 personnel working per day (figure 14.2). A review of their operations during the GEJE revealed that the scale was so large that some units could not manage

Figure 14.1 The Japan Self-Defense Forces in action

Source: Ministry of Defense.

Figure 14.2 An interprefectural emergency rescue unit in action

Source: National Police Agency.

their operations on their own, while others had difficulty securing enough personnel. The National Police Agency plans to enhance its response capacity by setting up emergency quick-response teams and long-term response teams numbering 10,000 personnel.

Crimes such as theft were a major concern since many houses had been left vacant after residents fled to evacuation centers. According to the National Police Agency, the number of crimes committed in the disaster-affected areas in the year after the disaster decreased significantly compared to the previous year, while the number of burglaries rose slightly (bold in table 14.2). Many ATM machines were also destroyed. Police teams were deployed to ensure safety in the disaster-affected areas.

Table 14.2 Crime in the disaster-affected areas

	MARCH 2011– FEBRUARY 2012	MARCH 2010– FEBRUARY 2011	CHANGE (%)
Total crimes	42,102	51,305	–18
Felonious	187	245	–24
Violent	1,804	2,008	–10
Larceny	31,894	38,484	–17
Burglary	**5,729**	**5,690**	**0.7**
Vehicle	9,992	12,440	–20
Nonburglary	16,173	20,354	–21
Intellectual, "white collar"	1,150	1,905	–40
Moral, sexual	375	404	–7
Others	6,692	8,259	–19

Source: National Police Agency.

The Disaster Medical Assistance Team

The Disaster Medical Assistance Team (DMAT) is a specialized team of medical doctors, nurses, and operational coordinators trained to conduct emergency operations during the critical period, normally within 48 hours, after a large-scale disaster or accident. The DMAT was established in 1995 after the Kobe earthquake, when it was learned that 500 more people could have been saved if medical support had been provided more promptly. In response to the GEJE, the DMAT sent about 380 teams consisting of 1,800 staff from 47 prefectures for 12 days to provide support to hospitals and to rescue and transport patients. Because the tsunami damage was so extensive and local medical centers had been washed out by the tsunami, the DMAT also had to provide care for people with chronic illnesses. Although the DMAT's operations usually take place within 48 hours after a disaster, they had to operate for a much longer time.

The Technical Emergency Control Force

The Ministry of Land, Infrastructure, Transport and Tourism (MLIT) established the Technical Emergency Control Force (TEC-FORCE) in 2008. The TEC-FORCE is a specialized group made up of ministry staff that helps disaster-affected municipalities to quickly assess damages, identify measures to prevent additional damage, and provide technical assistance for rehabilitation and emergency response activities. In response to the GEJE, more than 18,000 person-days were dispatched, together with disaster management equipment and machinery (figures 14.3

Survey of disaster-affected areas

Supporting affected municipalities (technical assistance)

Information & communication team (satellite communication vehicle)

Survey of disaster-affected rivers

Local needs survey

Disaster emergency response (emergency flood removal)

Figure 14.3 TEC-FORCE activities in response to the GEJE
Source: Ministry of Land, Infrastructure, Transport and Tourism (MLIT).

Figure 14.4 TEC-FORCE equipment

Source: MLIT.

and 14.4). The TEC-FORCE provided satellite communication vehicles, enabling them to connect to public lines and establish communications with other organizations concerned.

The Japanese Red Cross Society

Japanese Red Cross Society (JRCS) is designated as a public relief organization under disaster response law and is the biggest humanitarian organization in Japan. It mobilized relief resources to the affected areas from the onset of the disaster—within 24 hours, 55 medical teams (of which 22 teams were from the DMAT) were dispatched. Subsequently, 935 teams (or 6,700 personnel) were deployed for six months; they treated 87,445 persons and provided psychosocial support to the affected population.

Mobilization of Japanese nongovernmental and nonprofit organizations

Domestic nongovernmental organizations (NGOs) and nonprofit organizations (NPOs) have played a significant role in carrying out disaster management activities. As of January 20, 2012, there were 712 organizations participating in the Japan Civil Network for Disaster Relief in East Japan. There is no limit either on the budget size of the organization that can join this network or its type (such as nonprofit, public-interest, or religious).

In a disaster, the role of NGOs and NPOs is to complement government actions. Since in Japan the government is indeed the primary agent obligated to initiate action in response to a natural disaster, NGOs and NPOs are responsible for filling in where governmental support is lacking. But this by no means implies that NGOs and NPOs are government subcontractors; they have broad autonomy in deciding their activities and are not subordinate to the government. Their roles and responsibilities are far reaching, and they engage in a broad range of activities from awareness raising to fundraising, while also engaging directly in relief activities at disaster sites.

The early responders can be categorized into two groups: Japan-based (mainly Tokyo based) NGOs specializing in international relief operations even before the GEJE, and Japanese NGOs and NPOs based in different parts of Japan that address domestic needs. The Japan Platform, a platform for international emergency humanitarian aid, mobilized funding for relief operations within three hours of the earthquake. Seven registered organizations carried out initial needs assessments with ¥15 million in funding, 5 organizations provided support to education with ¥450 million, 2 organizations provided health-care and hygiene promotion with ¥210 million, 8 organizations engaged in rehabilitation work, and 12 organizations provided food and nonfood support with ¥3.12 billion. These organizations, experienced in providing emergency humanitarian aid overseas, were able to leverage international standards and expertise. They played a pivotal role in mobilizing experts in specialized fields.

The Japanese NGOs and NPOs had been mainly involved in domestic emergency-relief activities. Organizations based and operating in the disaster-affected areas made long-term commitments to sustaining activities such as assessing people-centered needs and facilitating a seamless transition from emergency to recovery support.

The JRCS had pulled together ¥307 billion in donations as of January 19, 2012, and its counterpart, the Central Community Chest of Japan, Red Feather Campaign, garnered

¥38.8 billion in donations as of October 2011. A Central Grant Disbursement Committee was set up to ensure a fair allocation of the funds collected by the JRCS and other designated fundraising organizations, to the affected prefectures. Each prefecture has established a prefectural-level grant disbursement committee that sets criteria for eligible recipients as well as for the amounts to be distributed by the municipal authorities responsible for identifying individual beneficiaries and distributing the cash.

The Japan Platform received ¥6.7 billion from private companies as of July 2011, the Japan Foundation received ¥2.4 billion. The line separating fundraising organizations from private companies has narrowed as private companies actively collect funds and work in parallel with emerging NGOs such as Just Giving Japan, which uses the Internet to solicit donations.

Another important responsibility of NGOs and NPOs is the coordination of relief efforts. A designated agency, in most cases a UN agency, would function as the cluster lead for international relief operations, but no central agency was assigned for overall coordination in Japan. The prefectural offices or the disaster response headquarters at the prefecture levels were the first bodies to be assigned to disaster response, but they did not function as a coordinating body for all NGO and NPO relief operations. The newly established prefectural cooperation recovery centers functioned as networking hubs and grew into a spontaneous coalition for coordination. The Tokyo-based NGO—the Japan NGO Center for International Cooperation (JANIC)—which had already created a network of NGOs, functioned as a provider of pooled information.

The third role of NGOs and NPOs in disaster response is enrollment and management of volunteers. The Ministry of Health, Labour and Welfare named the Japan National Council of Social Welfare, Tasukeai Japan, the 3.11 Reconstruction Aid Information Portal in cooperation with the Reconstruction Agency and Japan Civil Network, as the main contact points for people to apply for volunteering. Over 280,000 people joined as volunteers in the disaster response in the two months after the earthquake.

Support in Fukushima

Apart from the national budget, Fukushima Prefecture received ¥7.2 billion in donations, which were used for activities such as school reconstruction, support for children, and improvement of temporary shelters. Of this, ¥1.3 billion was received and used to provide for disaster orphans. In collaboration with governmental funds, the Japan Platform supported eight projects in Fukushima, funding five organizations with ¥1.8 billion. Apart from the Japan Platform there were several other organizations working separately on relief activities, though the number of NGOs working in Fukushima was much smaller than in the Miyagi and Iwate prefectures. According to the JANIC, between March and June 2011, the number of NGOs working in the Fukushima Prefecture was 17, whereas in Miyagi it was 40 and in Iwate it was 33. The contrast is made clearer by the number of projects provided by NGOs: in Miyagi Prefecture there were 292 projects, 179 in Iwate, and 60 in Fukushima. In the early stages, these concentrated on delivering emergency kits, including food and nonfood items. Following emergency activities, these organizations faced difficulties in supporting rehabilitation programs, which was a completely new and unknown operation for them. The experiences and lessons learned in Fukushima should be passed on and shared with the broader international aid community. To this end, it is advisable that the Japanese NGO community conduct timely and objective evaluations and studies of its March 11 operations.

Volunteers

The Japan National Council of Social Welfare set up volunteer centers in the affected municipalities. The social welfare councils in municipalities nationwide sent more than 30,000 person-days of staff to operate the volunteer centers.

As of January 2012 more than 900,000 person-days had been used in volunteer work through volunteer centers in the three prefectures of Tohoku (figure 14.5). Considering that more than 1 million volunteers were mobilized in the first month after the Kobe earthquake in 1995, the number of volunteers mobilized during the GEJE was relatively small. This was primarily because the affected areas were far from large cities and were rural coastal communities dispersed over a wide area, making it difficult for the volunteers to gain access.

International assistance

As of November 1, 2011, 163 countries and regions and 43 international organizations had offered aid and relief. Emergency assistance squads, medical teams, and reconstruction teams had been dispatched from 24 countries and regions along with expert teams from five international organizations. In regards to

material and monetary support, the Japanese government accepted relief supplies and donations totaling over ¥17.5 billion from 126 countries and regions. By May 17, 2011, 43 overseas NGOs from 16 countries had arrived in Japan. The scale of assistance was larger than for the Kobe earthquake in 1995, when 67 countries and regions provided aid and relief, and the United Kingdom, Switzerland, and France dispatched emergency teams.

The JRCS received financial support from 95 sister Red Cross and Red Crescent national societies from all over the world, which amounted to around $700 million, plus an additional $400 million from Kuwait and €10 million from the European Commission. According to a survey conducted by the Brookings Institution, Japan received $720 million from other countries, which accounted for almost half of the global humanitarian disaster funding in 2011 and some 0.4 percent of the planned reconstruction budget of the Japanese government.

The United States dispatched approximately 16,000 military personnel under Operation Tomodachi ("friends"). It provided various types of assistance, including search-and-rescue efforts, transport of supplies and people, and recovery and reconstruction of the devastated areas. At the peak of the action, approximately 140 aircraft and 15 vessels took part in the operation along with the JSDF.

Coordination

There was no functional coordinating mechanism among the various government organizations, civil society, and the private sector, to help avoid duplication and confusion in relief-and-response activities. Coordination was required at all levels and all phases. On the ground, these organizations needed to coordinate with community-based organizations, and with one another, to assess victims' needs and to carry out activities smoothly and effectively. The JSDF and NGOs did coordinate

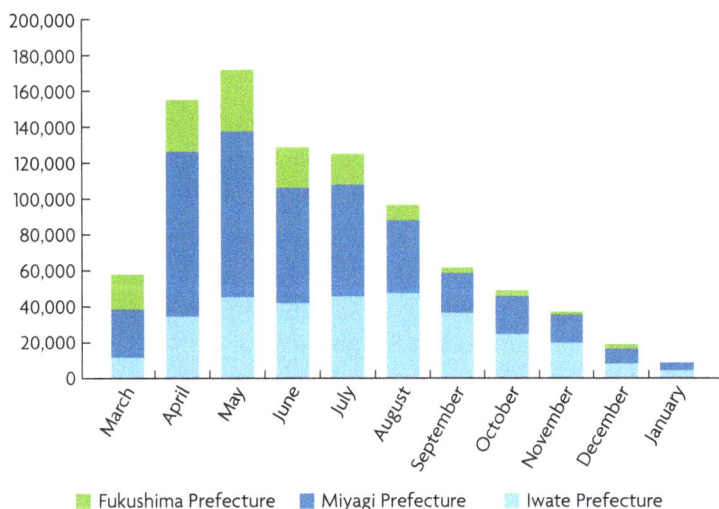

Figure 14.5 Volunteer effort in Tohoku through January 2012

Source: Japanese government data.

emergency food distribution to the evacuation shelters.

Coordination with municipal governments is crucial, since the municipalities have the primary responsibility for disaster management. Since the municipal governments have quite limited experience in working with civil society organizations (CSOs), linkages between the municipalities and CSOs could not be easily established. Municipalities can provide support to evacuees in transition shelters, but not in their homes. This function was instead carried out by CSOs. Coordination was also lacking between the private sector and local governments outside the affected areas, and the overall coordination of international assistance was a challenge.

Coordination is required at all phases of recovery since victims' needs change as recovery progresses. While water and food delivery are key at the emergency phase, needs become more diverse, including sustaining livelihoods, education, and improving the living conditions at evacuation shelters or in transitional housing.

Good practices could be found at specific sectors at some sites. Ishinomaki Red Cross Hospital coordinated all medical teams from the JRCS and other agencies at the 330 evacuation centers throughout Ishinomaki City. The hospital organized survey teams over a month to assess medical and nonmedical conditions, including water and sanitation. These formed the basis for planning and implementing response activities by various organizations and local governments.

LESSONS

- *National networks should be used to mobilize experts,* including search-and-rescue teams, medical teams, and engineers. Organizations should prepare these teams during normal times, by compiling rosters and conducting training.

- *Communication and transport equipment, fuel, food, and water should be stocked.* The teams coming from outside need to independently engage in activities in the disaster field without support, often over a long term.

- *Long-term commitment from experts is expected.* During megadisasters such as the GEJE, expert teams are expected to engage in activities for longer or unpredictable periods. Since an enormous number of public facilities are damaged, expert teams must have the capacity to work for one month or more.

- *Coordination mechanisms are essential,* since enormous numbers of different types of organizations are involved in disaster management. There was no functional coordination mechanism during the GEJE. Without such a mechanism in place, megadisasters overstretch the capacities of local governments, and government staff and facilities in devastated areas suffer. In developing countries, UN cluster systems serve as coordinating mechanisms. Considering the difficulties faced by local governments during the GEJE, similar mechanisms should be established in the central government or under some umbrella organization of CSOs.

RECOMMENDATIONS FOR DEVELOPING COUNTRIES

Prepare response teams. Specialized agencies, such as the police, fire departments, public works, and hospitals should prepare during normal times for the mobilization of response teams. The following activities are required:

- Clarify the chain of command.

- Designate a secretariat function.

- Prepare a roster of emergency team members.

- Conduct emergency drills.

- Keep the necessary equipment in stock.

Develop capacity. Expert teams are required to develop capacities for working independently over the long term. Standby or rotating teams, communication, and transportation should be arranged.

Establish coordinating mechanism. Various types of organizations from inside and outside the country engaged in response-and-recovery activities. Government agencies often have problems coordinating the enormous numbers of organizations carrying out a broad range of activities. Once disasters happen specific teams should come from outside the devastated areas and start coordination among all organizations. The following actions are required:

- *Preparedness.* Establishing face-to-face relationships during normal times facilitates coordination in times of disaster.

- *Networking.* Information, experts, and private sector personnel should be networked to share information, to effectively collaborate with one another, and to mobilize diversified resources.

- *Consideration of vulnerable groups:* Special care is required for vulnerable groups, such as the disabled, the elderly, and children. These groups are easily marginalized (chapter 19).

NOTE

Prepared by Yukie Osa, Association for Aid and Relief; Junko Sagara, CTI Engineering; and Mikio Ishiwatari, World Bank.

BIBLIOGRAPHY

Disaster Medical Assistance Team HP. http://www.dmat.jp/DMAT.html.

Ferris, E., and D. Pets. 2012. "The Year that Shook the Rich: A Review of Natural Disasters in 2011." Project on Internal Displacement, Brookings Institution, London School of Economics.

Fire and Disaster Management Agency. *Emergency Fire Response Teams* [in Japanese]. http://www.fdma.go.jp/neuter/topics/kinkyu/kinshoutai.pdf.

———. 2011. *White Paper* [in Japanese]. http://www.fdma.go.jp/html/hakusho/h23/1-3.pdf.

Japan Civil Network HP. http://www.jpn-civil.net/about_us/group/.

Japan Platform HP. http://www.japanplatform.org/area_works/tohoku/action/ngojyosei20120111.pdf.

Japanese Red Cross. 2011. *Six Months Report on the Great East Japan Earthquake Response.* http://reliefweb.int/sites/reliefweb.int/files/resources/August%20Report_FINAL_rev.pdf.

National Police Agency. 2011a. *White Paper* [in Japanese]. http://www.npa.go.jp/hakusyo/h23/honbun/index.html.

———. 2011b. *Evaluation of Police Operation in GEJE* [in Japanese].

Syakaifukusikyogikai HP. http://www.shakyo.or.jp/saigai/torikumi_01.html.

Technical Panel on Emergency Medical Activity. 2011. *Report on Emergency Medical Activity* [in Japanese].

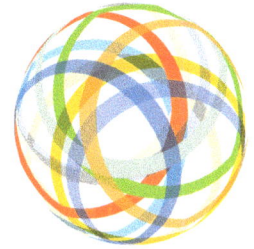

CHAPTER 15

Emergency Communication

The Great East Japan Earthquake caused immense damage to and congestion of telephone infrastructure, including 1.9 million fixed-line services and 29,000 mobile-phone base stations. Government radio communication infrastructure was also seriously damaged. Voice messages were widely used to confirm whether family members and relatives were safe, and satellite phones played a crucial role in emergency communication during the response stage. Social media was extensively used for search and rescue, as well as for fundraising.

FINDINGS

Communication infrastructure is indispensable in securing government functions and protecting lives and property during disasters. Communication systems are used to disseminate warnings to the public, to enable search-and-rescue organizations to communicate among themselves, and to confirm the safety of family members and relatives. Social media is extensively used for search and rescue, as well as fundraising. Community radios can provide local information such as times and locations where emergency water and food supplies or relief goods will be delivered. Social media is most effective in reaching the younger generation and, community radio, the older generation.

Telephone

Damage and subsequent restoration of fixed-line, mobile, and broadband services

The Great East Japan Earthquake (GEJE) caused immense damage to both fixed-line and mobile-phone infrastructure, including flooding of exchange facilities, damage to underground cables and conduits, destruction of telephone poles and overhead cables, destruction and loss of mobile-phone base stations, and draining of backup batteries during the long power outages. In the Tohoku and Kanto regions, an estimated 1.9 million fixed-line services from Nippon Telegraph and Telephone (NTT) East, KDDI, and SoftBank Telecom were rendered inoperable, including subscriber lines, Integrated Services Digital Network (ISDN), and fiber to the home (FTTH), while 29,000 mobile-phone and personal handyphone system (PHS) base stations also stopped functioning.

Telecommunications carriers initially deployed mobile power supply vehicles and mobile base stations to those areas with no commercial power supplies, and set about rebuilding damaged facilities as quickly as possible. The rapid response effort saw full services restored to almost all affected areas, with some exceptions, by the end of April 2011 (figures 15.1, 15.2, and map 15.1).

Figure 15.1 Number of affected fixed lines

Source: Ministry of Internal Affairs and Communications (MIC).

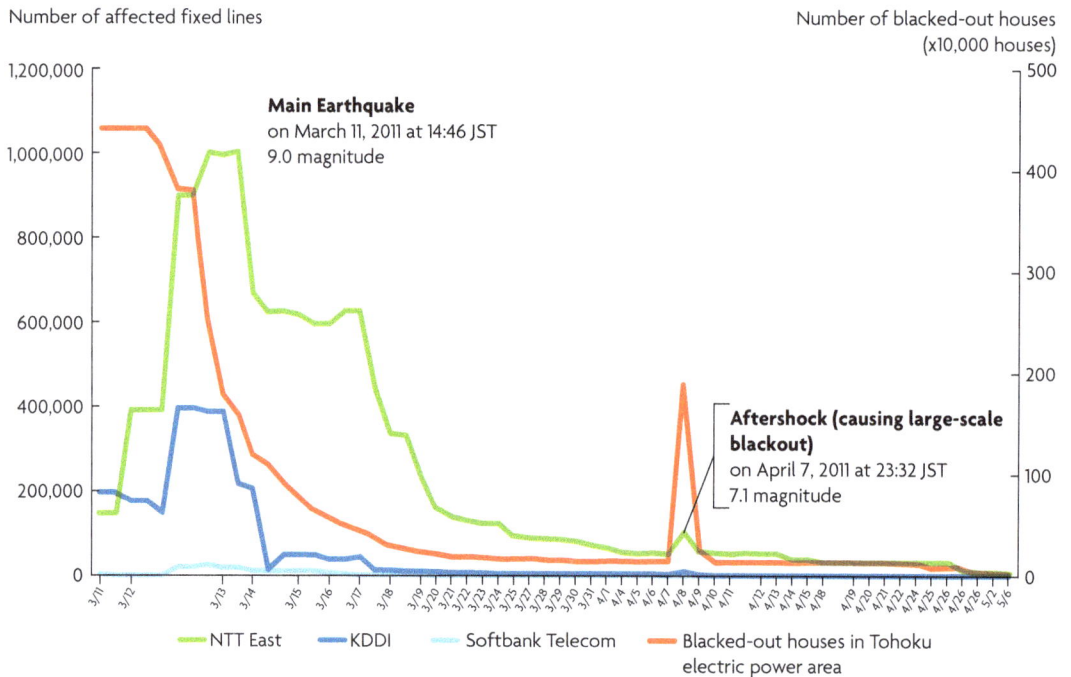

Figure 15.2 Number of affected mobile base stations

Source: MIC.

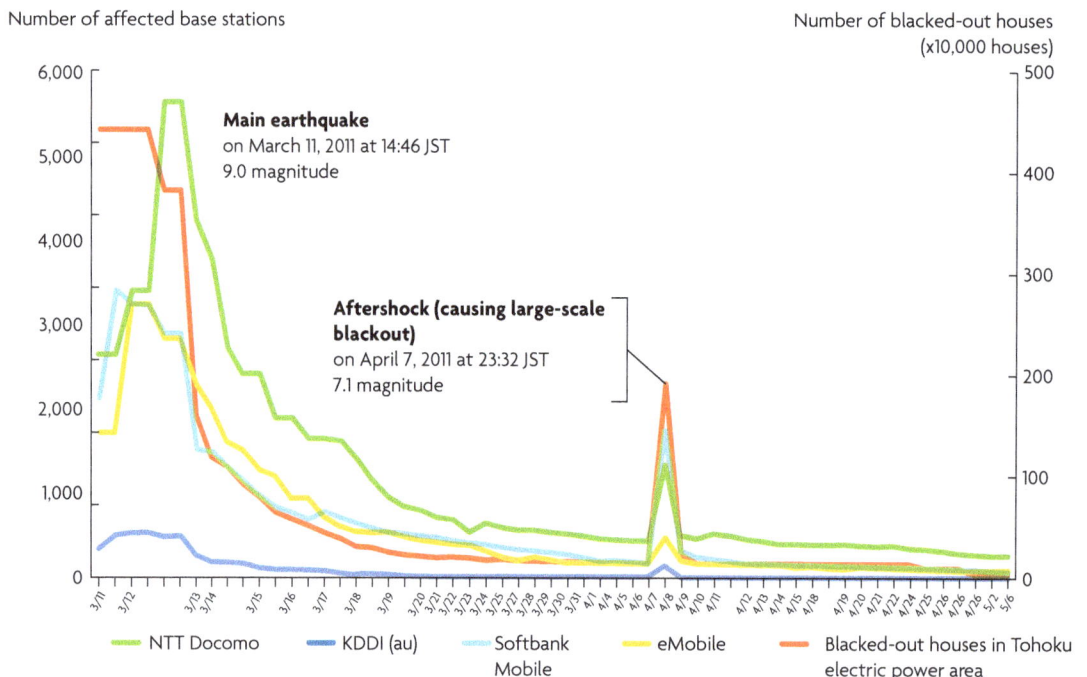

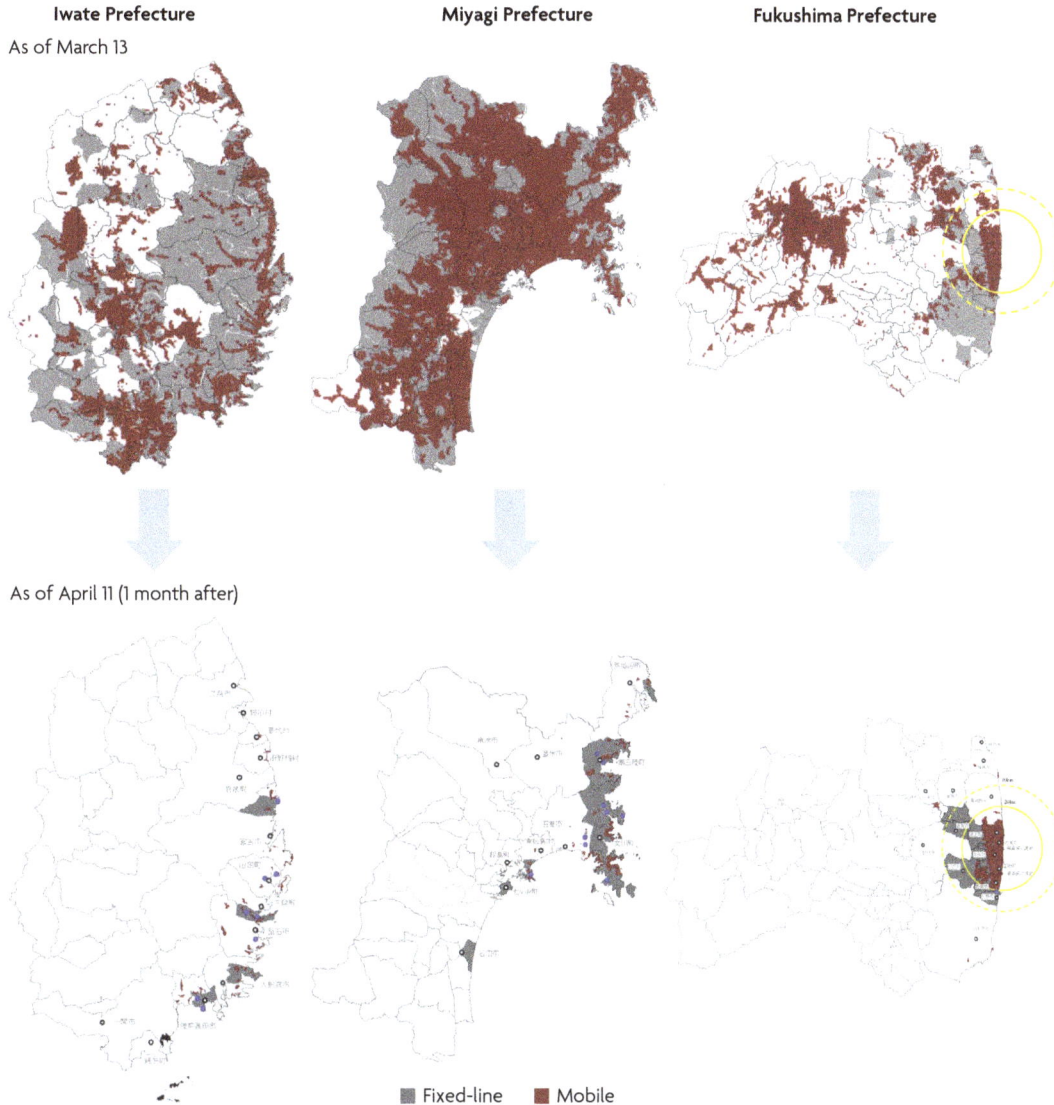

Iwate Prefecture

Miyagi Prefecture

Fukushima Prefecture

As of March 13

As of April 11 (1 month after)

■ Fixed-line ■ Mobile

Map 15.1 Damage to NTT East and NTT Docomo

Source: MIC.

Voice messaging and other services

The sharp increase in voice-call traffic immediately after the earthquake caused significant congestion. Carriers restricted fixed-line traffic by as much as 80–90 percent and mobile services by as much as 70–95 percent to allow emergency calls and other critical communications to go through. Mobile-phone packet communication services such as e-mail were generally not restricted.[1] Even when carriers did impose restrictions, they were generally no more than about 30 percent and were only temporary. Thus, packet communications provided considerably easier access than voice services.

Telecommunications carriers set up emergency messaging services so that people could check on the safety and whereabouts of their families, relatives, and other relevant people (figure 15.3). These services were used some 14 million times following the GEJE. Because of these emergency messaging services, traffic congestion was cleared up on the same day the earthquake struck, in contrast to the Great Hanshin-Awaji Earthquake (Kobe earthquake) in 1995, when congestion continued for five days.

Some mobile-phone carriers introduced an emergency messaging service whereby

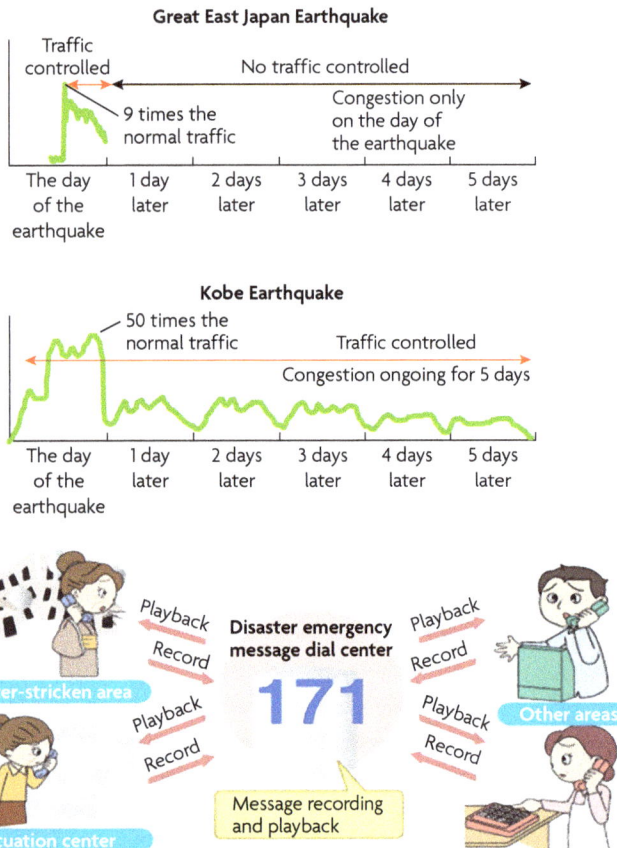

Figure 15.3 **Disaster emergency message traffic during GEJE and Kobe earthquakes**

Source: NTT East.

the terminal device converted voice recordings into voice files that could then be sent via packet transmission. Other mobile-phone carriers are planning to follow suit.

Disaster management radio communications

The disaster management radio communications networks of national and local governments are generally considered to be more robust and resilient than public fixed networks. In the GEJE, however, many towns and villages, particularly those located along the Pacific coastline, suffered various levels of damage to their radio communications systems, including both community announcement systems with loud speakers and mobile systems on emergency vehicles. The main causes were damage to or loss of radio transmission equipment from the earthquake and

tsunami, as well as loss of electric power during sustained blackouts.

In the aftermath of a megadisaster such as the GEJE, a key issue is how to deliver relevant information such as public warnings and evacuation instructions across wide areas in a timely and reliable manner. Local governments are looking at advancing and multiplying ways to deliver emergency information to residents, and improving their disaster resilience.

Satellite communications

Compared with terrestrial communication infrastructure, satellite phones and satellite communication systems are less vulnerable. These systems have the advantage of being available for quick deployment in any region including regions with no land-based communication infrastructure, as well as in marine areas. Satellite phones, in particular, played a vital role after the GEJE in emergency communication among local governments and rescue organizations.

Satellite mobile phones

This system provided voice and Internet communication capabilities for disaster management organizations, evacuation shelters, and staff working on infrastructure rehabilitation (among others), as well as local governments and communities isolated by typhoons and heavy snowfall. In preparing for disasters, batteries and equipment should be stored for rapid deployment.

Very small aperture terminals

Very small aperture terminals (VSATs) provide voice and Internet communication capability by enabling access from multiple mobile terminals via wireless local area network (LAN) technology. They are also used to provide connection through portable and truck-mounted mobile-phone base stations for rapid restoration of the communication infrastructure, and to provide a temporary communication network for disaster relief organizations.

Portable and truck-mounted satellite earth stations

These were used by disaster relief organizations and media entities to transmit video images from disaster sites. The Heli-Sat system, which enables video transmission through satellite, will be introduced in the future.

Marine earth stations

This provided communication for rescue-and-recovery activities by seagoing vessels in cases where land routes were disrupted.

Disaster information broadcasting

After the earthquake occurred, broadcasting companies including National Broadcasting Corporation (NHK; Japan's public broadcasting corporation) and local operators interrupted regular programming to provide disaster-related information. For example, NHK delivered emergency earthquake warnings, followed by news reports, on a continuous basis starting two minutes after the earthquake occurred. These were carried on the company's eight channels, including its general programming channel, educational channel, and radio channels. The general programming channel continued to provide news reports and programs related to the earthquake and tsunami for 12 days up until March 22, and the total time devoted to disaster-related news and reports was about 254 hours. People were able to watch many of those programs on their mobile phones in areas where electricity supply had failed. The programs were delivered by one-segment broadcasting.[2]

As many as 120 television relay stations stopped functioning because of the loss of commercial electricity during the initial period of the disaster, and as many as 4 radio relay stations shut down. Master stations continued broadcasting by generating their own power. All the stations within the area were restored by the end of May 2011, except for one radio station within the evacuation zone around Fukushima Daiichi Nuclear Power Station.

This station was restored by March 2012. After the March 11 events, the Ministry of Internal Affairs and Communications (MIC) requested the NHK, the National Association of Commercial Broadcasters (NAB) in Japan, and the radio stations in the affected areas to increase broadcasting disaster information, and on April 1, 2011, the MIC requested that NHK and NAB provide accurate and detailed information as quickly as possible to the public.

Social media

Social media are a set of applications and services that use the Internet to connect people. They combine dynamic, collaborative Internet-based tools, social networks, computers, and, increasingly, mobile devices. Social media consist of social networks such as Twitter and Facebook that connect users, as well as websites and computer applications that enable users to collaborate and create content, such as Wikipedia and YouTube.

Social media were used extensively during the GEJE for various purposes, such as search, rescue, and fundraising. Table 15.1 summarizes how they were used to meet different types of information-sharing needs during the disaster. A questionnaire survey was carried out to learn about the uses of social media by 250 different types of responders: information senders, volunteers, managers of media groups, and so on (figure 15.4).

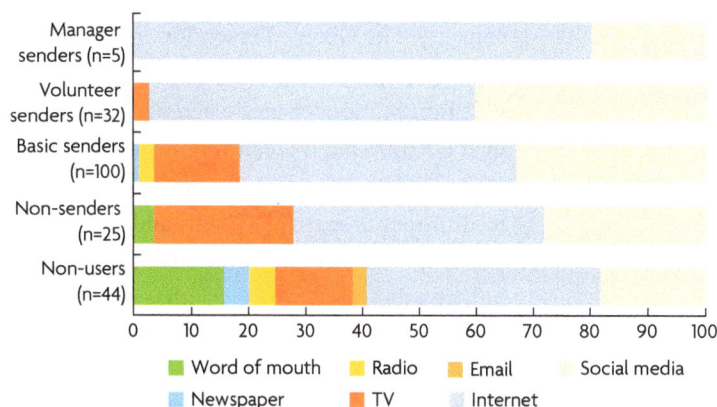

Figure 15.4 Most viable source of information as perceived by message sender group in the GEJE

Source: Kyoto University

Table 15.1 Dominant types of information and how they were shared

	TWITTER	FACEBOOK	MIXI	SMS	E-MAIL	WIKIS	WEB PAGES OR BLOGS	SMARTPHONE APPLICATIONS	MAPS
General disaster information	O	O	O	O	O	O	O	O	O
Safety conformation	O	O	O	O	O			O	
Fundraising	O	O	O	O			O		
Infrastructure status notification/ regional facility status	O		O			O			O
Housing provision		O							O
Goods provision	O						O		O
Moral support	O	O	O	O			O		
Resource saving	O	O	O				O		
Volunteer recruitment		O	O		O		O		O
Special needs support	O	O							

Source: Kyoto University.

Social media and the Internet were found to be highly reliable regardless of the users' role, location, or the extent to which they were affected by the disaster. Users found social media to be extremely beneficial to an overwhelming degree. For directly affected individuals and people in the affected areas, the strongest reasons for using social media were convenience and their mass dissemination capacity. The Google Person Finder let people enter an inquiry about a missing person or provide information for interested parties. In total over 600,000 names were registered.

Support for government use of social media in disasters is extremely high and during the GEJE it was highest among directly affected individuals, individuals in disaster-stricken areas, and those involved in disseminating information to groups.

A general note about social media is that the information is not always trustworthy, especially about infrastructure. The higher the level of participation in sharing information through social media, the more likely an individual is to receive and share large amounts of information, and believe that the information comes from credible sources.

Emergency FM radio

Emergency FM radio also played a crucial role in providing information to local residents. In the Tohoku area, 25 emergency broadcasting stations specializing in disaster information were set up. Immediately after the disaster, the communication systems developed by local governments did not work because of power failures and a lack of emergency backup power supply. The MIC distributed 10,000 portable radio receivers to evacuation shelters and requested equipment manufacturers, such as Panasonic and Sony, to distribute over 40,000 portable radio receivers.

FM radio provided locally customized information, such as information about aftershocks or the availability of local services and activities related to people's everyday needs. This kind of information was beneficial immediately after the disaster, while different information was required as reconstruction progressed. Some entertainment programs were presented six to nine months after the disaster (box 15.1).

Several problems were identified. Ensuring sufficient human resources is a key issue. Immediately after the disaster, a significant number of volunteers provided the radio stations with various kinds of help, but over time the number decreased. A sustainable funding source is needed to continue radio broadcasting. FM radio users in Natori City are keen on having local residents continue to participate in broadcasting activities, and on gradually changing over to community FM with funding from the community and subsidies from local governments.

LESSONS

- *To reduce telephone network congestion, packet communications and emergency message services should be expanded.* The MIC is raising public awareness about using these services in times of disaster.

- *Backup systems are needed.* The GEJE reminded us that resilient and redundant communication systems should be established. Batteries and generators with enough fuel should be acquired and stored in higher locations to avoid flooding.

- *Social media and FM radio have played a crucial role* in providing information to local communities; they reached two distinct age groups: while the former is used more by the younger generation, the latter is used mainly by the older generation.

- *Information through social media changes over time.* The way in which social media and FM radio are used changes over time

Ringo ("apple") radio of Yamamoto Town, Miyagi Prefecture

A temporary emergency radio station was set up inside the Yamamoto Town Hall with the help of FM Nagaoka of Nagaoka City, Niigata Prefecture. Ringo FM started broadcasting on March 21, from 7:00 a.m. to 7:00 p.m. At first, it only announced information such as bathing times and food rationing information for those living in the town. Later the content became less about daily life and more about supporting and comforting the residents. According to the coordinator, "We will never be able to completely eliminate the sadness of the victims, but we would like to provide them with encouragement from the bottom of our heart."

Source: © Kyoto University. Used with permission. Further permission required for reuse.

in the aftermath of a disaster—from sharing information about the safety of family and friends to disseminating information about relief goods and services and, gradually, to livelihood-related information.

- *City and local governments should use social media in emergencies* for regular communication relating to city news and events, in order to enhance their effectiveness. In Japan, the prime minister's office launched a new Twitter feed after the disaster.

- *For FM radio, sustainability is a key issue.* Off-air activities, in which communities participate in producing radio programs, should be strengthened so that communities will be invested in supporting the continuation of FM radio.

RECOMMENDATIONS FOR DEVELOPING COUNTRIES

When disasters strike, communications infrastructure should be used to disseminate warnings to the public, to enable communication among search-and-rescue organizations, and to confirm the safety of family members and relatives. Immediately after the disaster, however, communication systems often break down because of power failures, damage to infrastructure, and congestion.

Improve the reliability of communication networks. The following actions are required:

- *Reducing damage* by developing backup systems, such as batteries, generators, and backup trunk lines

- *Mitigating congestion* by increasing the capacity of facilities, such as switching equipment

- *Restoring services* by deploying emergency facilities, such as portable switching equipment and portable satellite stations

Utilize social media. The increasingly higher levels of mobile-phone penetration in developing countries can allow for the effective use of social media during disasters, provided they set a precedent for use during normal times. Social media can also provide information to communities outside the disaster-stricken areas, and facilitate the acquisition and appropriate allocation of aid and assistance. Starting with the Haiti earthquake of 2010, the use of social media during disasters has significantly increased in other countries. There is a strong potential for cultivating the use of social media among different groups and for developing a social-media-based platform designed for emergency situations.

Improve accessibility. Local accessibility is a key issue in many developing countries. Using mobile networks and social media can help in collecting and disseminating local information before and during disasters.

Enhance reliability of social media. The trustworthiness of information is extremely important for users to trust social media. Local government or relevant national government agencies should consider using social media in their public relations activities during normal times. When disasters occur, those social media channels can be used to share disaster-related information with the public.

Utilize radio to share information in communities. FM radio is commonly used in developing countries to share information in communities. Community radio is a rather low-cost and effective means of reaching small groups that are usually not served by the national and international media. Radios can provide information such as times and locations for provision of emergency water and food supplies or distribution of relief goods in the immediate aftermath of a disaster, and then gradually shift to providing different information for daily living or to help lift the spirits of people in the local communities. Radio is also appreciated by the elderly who may not have access to Internet-based information.

Enlist community participation to ensure sustainability. For FM radio to be effective, there needs to be a balance between on-air and off-air activities. Community participation is the key to the long-term survival of FM radio, and therefore, off-air community activities, such as workshops, are very important. These activities can also be linked to local schools and educational systems for greater sustainability.

NOTES

Prepared by Rajib Shaw, Brett Peary, Ai Ideta, and Yukiko Takeuchi, Kyoto University; and Japan's Ministry of Internal Affairs and Communication.

1. A data stream is divided into packets, or units, that are separately routed to a destination where the original message is then reconstituted.
2. A mobile terrestrial digital audio video and data broadcasting service in Japan. People can watch television programs by mobile phone.

BIBLIOGRAPHY

Ideta, A., R. Shaw, and Y. Takeuchi. 2012. "Post Disaster Communication and Role of FM Radio: Case of Natori." In *East Japan Earthquake and Tsunami: Evacuation, Communication, Education and Volunteerism,* edited by R. Shaw and Y. Takeuchi. Singapore: Research Publishing.

Peary, B. D. M., R. Shaw, and Y. Takeuchi. 2012. "Role of Social Media in Japan Earthquake and Tsunami." In *East Japan Earthquake and Tsunami: Evacuation, Communication, Education and Volunteerism,* edited by R. Shaw and Y. Takeuchi. Singapore: Research Publishing.

CHAPTER 16

Management of Logistics Chain for Emergency Supplies

In response to the Great East Japan Earthquake disaster, relief goods were distributed and delivered through prefectural- and municipal-level depots. This delivery system faced several problems including fuel shortages, interruption of telecommunication services, and supply and demand mismatches, resulting in stockpiling of the goods in depots and delayed delivery to the people in need. Several measures can be taken to address these issues, including prior surveys of depot facilities, advance estimates of the quantities of emergency goods that will be required, the enlisted support of professional logistics specialists, and the promotion of logistics information management in unaffected areas, among others.

FINDINGS

The damage from the earthquake and tsunami of March 11, 2011, was enormous; over 120,000 houses were totally damaged, and more than 470,000 people had to leave their home and evacuate to over 2,400 shelters.

Delivery of relief goods was planned to be executed through depots at two levels—prefectural and municipal. Especially in the first two weeks, fuel shortages made downstream deliveries from prefectural depots very difficult. Also, manpower shortages and the inconvenient building specifications of depots were the main causes of unnecessary stockpiling in depots. Telecommunications disruptions furthered mismatches between real needs and supplies. But the professional support of logistics specialists was effective in relieving the bottlenecks in depots.

The relief goods delivery system in Japan

In Japan delivery of relief goods is the responsibility of the prefectural governor, who responds to requests from the municipalities. According to the postdisaster plan, delivery of relief goods was to be executed using depots at two levels: prefectural and municipal, as

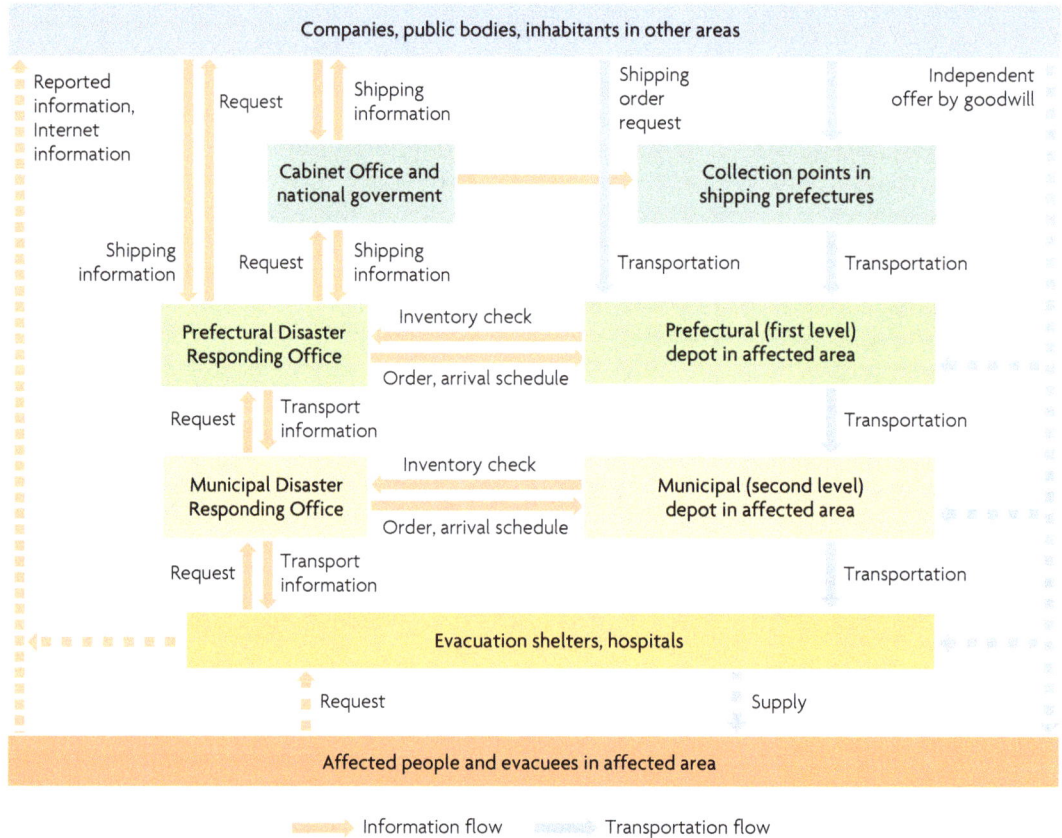

Figure 16.1 Information and transportation flows in the official relief goods delivery system

shown in figure 16.1. As illustrated in green in the figure, the national government (Cabinet Office) was also included in the plan to facilitate nationwide distribution. By April 20, the national goods distribution component had mobilized 26 million meals, 8 million bottles of beverages, and 410,000 blankets using 1,900 trucks, 150 aircraft, 5 helicopters, and 8 ships.

Delivering several kinds of goods—such as food, drinking water, clothing, and bedding—either to people's homes or to more than 2,000 shelters, was a challenge, especially in the first several weeks when fuel was in short supply. This was especially true for the smaller local transport companies that did not have their own storage facilities. By the end of June, 1,800, 1,400, and 2,400 trucks were dedicated to transporting goods from prefectural depots to municipal depots in Iwate, Miyagi, and Fukushima, respectively. Fuel shortages combined with power outages and telecommunications

failures hampered local government efforts to meet emergency needs.

Although many believe that transportation problems were the critical factor, several other forces were at play. The workload spiked at the same time that many staff were being lost to the disaster. Moreover, while the disaster countermeasure manuals state that the economic or industrial support branch of the local government is responsible for the delivery system, workers in that section did not have enough knowledge or experience with logistics and supply chain management. They simply stored the goods in public buildings, with no logistics management plan, so the space was quickly filled (as shown in figure 16.2).

The building specifications and design of the depots was also a contributing factor. The depots require large storage and handling capacity as well as easy access to expressways, especially prefectural depots. Privately owned warehouses would have been ideal if they had not been damaged. The space under viewing stands in athletic fields, race courses, and indoor gymnasiums also served well as depots (figure 16.3). In Miyagi Prefecture, large warehouses located near Sendai Port were severely damaged by the tsunami.

Neither Yume Messe Miyagi, the convention complex at Sendai Port, nor the Miyagi Prefectural Sports Park could be used as depots since they had already been designated as mortuaries.

Telecommunications disruptions and information bottlenecks

The disaster disrupted business operations such as information aggregation; meanwhile, the failure of some communications systems hampered the evacuation of people to safe areas. Very little of the real-time information that was needed to ensure timely and accurate procurement of goods was available, including the location of the shelters, the correct addresses of the recipients of goods, or information about the type and amount of assistance that communities needed. Information about whether relief goods had actually been received could not be easily communicated among depots for several weeks after the earthquake.

LESSONS

- *Plan for logistics and design ahead.* Suitably designed depots with cargo-handling equipment such as forklifts are needed, along with the support of logistics professionals.

- *Information on arrival times at each depot is crucial for planning* storage and location management.

- *Estimate needs ahead of time, based on demographics.* Prior quantitative estimates of urgently needed goods should be carried out based on regional demographic statistics. This helps arrange "push delivery," supply-driven deliveries, in the first few days after the disaster.

Figure 16.2 Badly organized inventory in an initially assigned depot (Iwaki Civic Hall, March 23, 2012)

Source: © Makoto Okumura. Used with permission. Further permission required for reuse.

Figure 16.3 Well-organized inventory in a municipal depot (Taira bicycle race track at Iwaki City, April 6, 2012)

Source: © Makoto Okumura. Used with permission. Further permission required for reuse.

- *Get back to normal soon.* Emergency delivery systems should be closed down as soon as feasible to allow normal commercial distribution systems to take over. They are capable of serving a variety of consumers and are more flexible and demand driven.

- *Logistics need to be managed locally.* At the intermediate stage, logistics management is best delegated to designated municipal authorities in unaffected areas.

The need for specialized support

As stated earlier, local government officials without sufficient knowledge, training, or experience in logistics management performed the specialized functions of receiving, sorting, and dispatching emergency supplies at distribution depots. This resulted in confusion and massive congestion of the delivery networks.

In large-scale disasters, local government staff are called upon to discharge a variety of functions related to emergency management. The government should enlist business logistics professionals and draw on the capacity of the private sector as much as possible, to ensure properly integrated management of the distribution depots. Many local public bodies hesitated to hire private companies for relief goods distribution and management because they were not sure that they would be able to pay them under the Disaster Assistance Law. In the future, a case can be made for putting in place agreements and contracts with the private sector for specialized logistics management services.

Getting information from upstream

For distribution depots to operate smoothly, local decision makers need to have real-time information about the kinds of goods being transported and the timing of shipments. This information enables them to arrange for the personnel and space needed to accommodate consignments. In normal times, this information can be obtained from, for example, point of sales (POS) systems.

In the aftermath of the disaster, this kind of information about the emergency goods ordered by the national government was not available to prefectures and municipalities in time. In addition, relief goods often arrived unexpectedly from various private companies, nonprofit organizations, and individuals with no prior information, which seriously reduced the processing capacity of distribution depots (box 16.1).

Preparing a "push" logistics plan

Since it is impossible immediately after a disaster to collect information about affected populations and the extent of damages and loss, it is helpful to design simulations of different scenarios to generate data on the expected number of victims, including data on vulnerable groups such as the elderly, disabled, women, children, and so on. Based on these simulations, contingent emergency stocks of basic goods—packages of water, food, household goods (such as tableware, kitchen wrap, tissues, towels, toothbrushes, masks, and blankets) and emergency medicines for the first three days following the disaster should be stored locally, typically at community-level schools and centers.

Since the initial disaster response is invariably carried out rapidly without geographical or population information from the affected areas, data need to be gathered or forecast in advance and stored in databases to implement "push delivery" of first-response aid.

BOX 16.1

The negative effect of goods sent with goodwill

The demand for different kinds of emergency supplies continued to change over time. There were many instances where in a certain area emergency goods were in high demand one day, and no longer needed after a few days.

Relief goods resulting from a spontaneous outpouring of goodwill but sent without making any prior arrangements with the recipient municipal bodies and without clearly marked declarations of contents did not meet people's needs and further burdened an already strained distribution network with dead stock and inventory.

Unpacking and sorting the emergency supplies sent by goodwill alone was an enormous amount of work. As these kinds of donations mounted, they clogged and undermined the efficiency of the distribution depots.

Many such goods arrived in Onagawa City, in Miyagi Prefecture. Used clothing was sent to the temporary shelters; however, 80 percent of the clothes, or 200 cartons, were returned to the gymnasium of the junior high school, which was the distribution center. About 7.7 tonnes of donated goods were not used and had to be recycled.

Switching back to commercial systems

National and local governments should use supply chain and logistics management as they respond to victims' changing needs. As many victims move from shelters into temporary housing, and as normal distributors such as shops, supermarkets, and convenience stores gradually recover, national and local governments should facilitate the return to normal commercial supply.

More specifically, the early restoration of commercial demand and supply chains, the rapid restoration of market dynamics, and the speedy distribution of donations to increase local purchasing power and liquidity should be a priority for municipal and local authorities. Job creation and conditional or unconditional cash transfers are highly effective short-term postdisaster measures, and are often more important than continuing the supply and distribution of relief goods by public agencies.

The speed and manner of the transition from public to private supply logistics should be determined by how dependent the affected population is on relief supplies, and on the robustness and speed with which the private sector networks can restore commercial operations. In the case of the Great East Japan Earthquake (GEJE), delivery of relief goods lasted for 40–50 days after the disaster. Commercial businesses reappeared in about a month.

RECOMMENDATIONS FOR DEVELOPING COUNTRIES

- *Public facilities, such as gymnasiums and community halls, can be used as logistics depots* as they are well designed with strong-enough floors, wide-enough entrances and exits, and good accessibility for cargo handling.

- *Prior agreements can be put in place between the government and logistics companies* specifying the terms and conditions and payment methods for hiring logistics professionals, machinery, and depot facilities.

- *Engage local government officials in normal times.* There should be prior identification and training of local government staff who will be tasked with responding to large-scale disasters.

- *Prepare lists and define delivery modes in normal times.* There should be prior formulation of a list of goods and a standard format for shipments and orders for smooth and seamless activation of the disaster response.

Planning public facilities

Building specifications for new public facilities, such as gymnasiums and meeting halls, should take into account their possible use as relief goods distribution depots. Floor strength, entry and exit widths, accessibility for cargo handling, as well their geographical locations, should be assessed. If private sector warehouses already exist in the region, agreements for diverting their use in case of disaster, as well as for the provision of labor and for allocating costs, should be signed in advance.

Building a resilient information system

Information on the needs of affected populations must guide procurement agents in purchasing the right goods and quantities to be delivered to distribution depots. In the wake of a disaster, communication must be maintained between municipal offices, prefectural offices, and the national government. Communication networks can be made more resilient by using satellite communication systems and on-site power generation equipment (chapter 15). Communication networks also need to support two-way connectivity between distribution depots and those facilities that can be used as evacuation shelters.

With respect to reliable road transportation, road status information gathered by probe cars linked to a global positioning system (GPS) is very helpful in determining delivery routes. To provide real-time information for emergency administrative and service-truck drivers, a system should be designed to integrate road status information from probe vehicles, road opening status from each road management authority, and traffic regulations from the police.

Multiple execution systems and paired administrations

In the aftermath of the GEJE, the national government formed a special team to take charge of the logistics of relief supplies. Ideally, every disaster response unit—at the national, prefectural, and municipal levels—should do the same.

Since the affected regions cannot be expected to provide sufficient information after large-scale disasters, municipalities outside the disaster area should initiate the information management functions for relief logistics. When municipalities are matched up in predetermined pairs based on their disaster profiles and spatial distribution, there are more chances of success.

The need for information sharing and coordination

Information about goods, such as the volume, size, and weight of unit packages; number of individual items packed in a unit package; and the need for temperature control is indispensable for logistics managers to calculate the type and number of trucks required and to determine where and how to store the cargo

in the distribution depots. Thus, it is important to create a mechanism for responsible parties to properly collect and share this essential information.

There is an equal case to be made for adopting universal definitions of various items and ensuring accurate and smooth information exchange about logistics by determining corresponding units among national and local government agencies, logistics operators, providers of goods, and so forth. As the first step, standard order forms, transportation request forms, and cargo transportation certifications should be prepared and adopted across the board.

In each region, the division of roles, cost-sharing arrangements among the related organizations, as well as appropriate workflow, should be discussed in an interdepartmental council. In addition, training in logistics management should be conducted regularly to make sure that the workflow is smoothly implemented in the wake of disaster.

NOTE

Prepared by Makoto Okumura, Tohoku University.

BIBLIOGRAPHY

Caunhyea, A. M., X. Niea, and S. Pokharelb. 2012. "Optimization Models in Emergency Logistics: A Literature Review." *Socio-Economic Planning Sciences* 46 (1): 4–13.

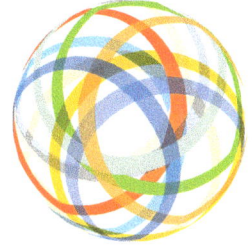

CHAPTER 17

Supporting and Empowering Municipal Functions and Staff

A megadisaster can destroy government offices and kill public officials. In the Great East Japan Earthquake, many municipalities in Tohoku suffered serious damage to their office buildings and incurred considerable staff losses, which hampered their disaster response timing and effectiveness. To compensate for this, many kinds of partnership arrangements were formed between localities in the affected areas and their counterparts in unaffected areas. Formalizing these partnership arrangements and building local government capacities to deal with emergency situations are key success factors for mitigating the effects of disasters in developed and developing countries alike.

FINDINGS

Office damages and staff losses

A disaster can destroy government offices and undermine government functions. Local governments are expected to play a critical leading role in disaster response and relief activities. In the case of the Great East Japan Earthquake (GEJE), many affected municipalities suffered serious damage to their offices and lost many of their public officials, which initially prevented them from undertaking relief activities in a timely manner.

A total of 62 municipalities in six prefectures (Aomori, Iwate, Miyagi, Fukushima,

Ibaraki, and Chiba) in northeastern Japan were affected by the GEJE tsunami. Among them, 28 municipalities in the three worst-affected prefectures (Iwate, Miyagi, and Fukushima) suffered at least partial damage to their office facilities. Sixteen of them had to relocate their administrative functions to other buildings or temporary offices. Furthermore, computer servers in some of these municipalities were seriously damaged or destroyed, resulting in a loss of information about residents and other data critical to providing municipal services.

Fukushima's case was slightly different. Nine municipalities near the crippled

Fukushima Daiichi Nuclear Power Station had to relocate their offices relatively far from the station (mostly within the same prefecture) because of concerns about radiation levels in their jurisdictions, even in cases where the physical damages from the earthquake and the tsunami were very limited.

To make matters worse, many municipalities in the hardest-hit areas lost their public officials: a total of 221 officials were reported dead or missing from 17 municipalities in the three hardest-hit prefectures. In particular, the town of Otsuchi in Iwate Prefecture lost its mayor and 32 officials including seven managers, out of a total of 139 staff (figure 17.1). The town was left without a mayor for five months. Rikuzentakata City, also in Iwate Prefecture, lost 68 officials out of a total staff of 295, while the town of Minamisanriku in Miyagi Prefecture lost 39 out of 240 officials.

Evolving partnerships among localities

One of the most interesting developments to occur after March 11, 2011, was the evolution of various partnership arrangements among those local governments affected by the disaster and those unaffected. Many prefectures and municipalities outside Tohoku took the initiative to quickly send a large number of their own public officials to the disaster-affected areas to help them with postdisaster relief activities and other emergency operations.

According to Japan's Ministry of Internal Affairs and Communications, some 79,000 local government officials were dispatched to the affected prefectures and municipalities from all over Japan between March 11, 2011, and January 4, 2012. After a year, many had been still serving there in every possible field—from civil engineering and urban planning to social work and finance. In fiscal year (FY) 2012, at least 1,200 officials from local governments around Japan spent a significant period working in the three hardest-hit prefectures as part of the reconstruction effort.

Most of the local governments outside Tohoku did this out of altruism, but they also considered it an opportunity for their officials to gain experience in dealing with postdisaster situations. So, it is a win-win arrangement. Various kinds of partnership arrangements are described as follows.

Rikuzentakata City, adopted by Nagoya City

Rikuzentakata City lost about one-fourth of its officials in the disaster, which was a huge loss. Soon after March 11, Nagoya City, one of the biggest cities in central Japan, started exploring how it could best help the disaster-affected areas of Tohoku, and decided to adopt one of the most affected cities, Rikuzentakata.

Nagoya City had sent 144 officials to Rikuzentakata, for a maximum term of one year before March 2012, and 13 officials from Nagoya were still working there as of March 2014. Nagoya sent a variety of experts such as urban planners, public health specialists, and statisticians. Rikuzentakata has gradually recruited more staff and become self-sufficient.

Figure 17.1 The municipal office in Otsuchi Town was damaged by the tsunami

Nagoya will continue to help and send officials to Rikuzentakata, but on a decreasing basis.

Tono as a hub for tsunami relief

The inland city of Tono, in Iwate Prefecture, is located within 50 kilometers of many of the hardest-hit coastal cities and towns in Iwate, such as Miyako City, Yamada Town, Otsuchi Town, Kamaishi City, Ofunato City, and Rikuzentakata City. Tono is about an hour by car from any one of these, and only 15 minutes by helicopter. Taking advantage of its strategic location, Tono established itself rapidly and effectively as a hub for tsunami relief by making the city's 144 facilities (schools, community centers, public parks, and so forth) available for logistics supply and other relief activities. As a result, 3,500 emergency relief workers from the Japan Self-Defense Forces (JSDF), and police and fire departments based themselves in Tono within 10 days of the disaster and started their relief operations from there. Furthermore, about 250 organizations and agencies used Tono as a base for their relief activities, coordinated and supported by Tono City. Tono's initiative was possible because the city had been discussing this kind of support mechanism with the tsunami-prone coastal cities since 2007, and Tono's officials were trained and well prepared for disasters.

Disaster relief agreements

During the past couple of decades, more and more local governments in Japan have signed disaster relief agreements with one another. A typical agreement involves two localities, located far enough apart so that both are not affected by the same disaster; the understanding is that if either party is affected by a disaster, the other is supposed to help. As of April 2010, 1,571 municipalities (or 89.8 percent of all) had signed such an agreement, of which 820 were signed with a municipality outside their own prefecture. Various kinds of support were provided to the municipalities affected by the GEJE based on these agreements.

The Union of Kansai Governments

In the wake of a megadisaster such as the GEJE, mutual support among local governments within the same region may not be available if the entire region is severely affected, and therefore local governments in unaffected regions may need to play a bigger role.

A coalition of prefectural governments in western Japan called the Union of Kansai Governments (UKG) quickly stepped in after the GEJE to help the three most affected prefectures in Tohoku in an organized fashion. To distribute the UKG's support equitably, each UKG member prefecture was assigned to assist only one of the hardest-hit prefectures (table 17.1). After being assigned a prefecture to support, the UKG prefecture dispatched its personnel to gather information, identify needs, and coordinate relief activities.

This is a Japanese version of the "twinning arrangement" that was used in China during the recovery following the Sichuan Earthquake of 2008. This type of partnership is efficient and effective because it is facilitated by local governments, which have a better grasp of the needs of their disaster-affected counterparts.

Among other advantages, the twinning arrangement avoids an overlap of support; clarifies responsibilities; and achieves efficiency, speed, continuity, and accountability.

Table 17.1 Beneficiary and supporting prefectures

BENEFICIARY PREFECTURES	SUPPORTING PREFECTURES
Iwate	Osaka, Wakayama
Miyagi	Hyogo, Tottori, Tokushima
Fukushima	Shiga, Kyoto

Under this arrangement by the UKG, the Hyogo Prefecture was assigned to assist the Miyagi Prefecture. The Hyogo Prefecture extended the following support:

- *Provision of relief supplies* (clothes, food, water, and so on).

- *Dispatch of its own officials* (54,589, as of December 1, 2011).

- *Acceptance of evacuees.*

Recognizing that continuing support is needed in the affected areas, Hyogo Prefecture is now developing a mid- to long-term support plan. This plan includes assigning technical officials such as urban development specialists, as well as those who can share lessons from the experience of the Great Hanshin-Awaji (Kobe) Earthquake of 1995.

Fukushima's problem

While municipalities in the Iwate and Miyagi prefectures mainly receive as many officials as they ask for from the unaffected areas, municipalities in Fukushima Prefecture had difficulty filling their staffing needs because of concerns about radiation risks. According to the Fukushima Prefectural Government, the number of additional staff requested by its 21 disaster-hit municipalities was 178 for FY 2012, but only about 40 percent of that demand was met.

Municipal data protection

In addition to office damages and staff losses, some Tohoku municipalities lost residential information and other critical data because their computer servers were damaged. One of these municipalities, the town of Otsuchi, which lost its on-site computer server, considered adopting cloud backup solutions for storing vital information and other key data. Cloud server backup solutions allow data to be transferred to an off-site location for secure storage, reducing the risk of losing data in times of disaster.

LESSONS

- *City halls and municipal offices should be focal points for disaster response initiatives;* they also play a critical leading role in relief activities. Therefore, they must be located in safer areas, or built or retrofitted to be disaster resistant.

- *Define partnerships in normal times.* Japan's experience shows that partnership arrangements between localities in disaster-affected areas and their counterparts in unaffected areas were effective. Some of these arrangements were based on formal agreements, but others on goodwill. It is advisable to formalize these mechanisms among local governments before disasters strike in order to obtain the necessary legal backing and clarify cost-sharing arrangements. Right after the GEJE, the Japanese central government decided to shoulder the cost of dispatching local officials to disaster-affected areas, which was believed to be instrumental in promoting the emerging partnerships among localities.

- *Geography matters.* When it comes to disaster relief agreements, it is essential that partnering prefectures and municipalities are geographically distant or in different regions. Agreements within the same region may not be effective in a large-scale disaster such as the GEJE, which damaged virtually the entire region.

- *Fair and equitable allocations.* In a large-scale disaster, it is important to allocate the support fairly and equitably across affected areas. The UKG's initiative to assign its member prefectures to support various individual localities was exemplary in this regard.

- *ITC and databases protection.* Disaster preparedness by local governments should include a plan to minimize the damage to their information systems and protect critical databases so that they can continue to function and provide emergency services to disaster victims and residents.

RECOMMENDATIONS FOR DEVELOPING COUNTRIES

Enhance coordination across government levels. The roles that local governments play in the aftermath of a disaster are critical. Clear roles and responsibilities must first be assigned to each tier of government (specifying what needs to be done and by which level in case of a disaster) and its capacities strengthened accordingly.

Review the location of government offices. In disaster-prone developing countries, the locations of municipal offices should be reviewed along with their vulnerability to disasters. Relocating or retrofitting them should be considered if necessary, so that municipalities can continue to perform their roles in the wake of a disaster.

Coordinate support across locales. Partnering with other localities to conduct emergency relief activities could work in many developing countries, particularly in relatively large countries. Such partnerships are, however, unlikely to work effectively if carried out in an ad hoc manner. Formalizing these agreements and building the emergency response capacities of local officials are the keys to successful partnerships. Cost sharing under the partnerships also needs to be clarified up-front.

Be sure to back up data. Municipalities in developing countries should be aware of the risk of losing their digital information and databases in a disaster, and need to come up with a cost-effective solution to minimize that risk.

NOTE

Prepared by Toshiaki Keicho, World Bank, and the International Recovery Platform (Union of Kansai Governments).

BIBLIOGRAPHY

Asahi Shimbun. 2012. "Accelerating Development of Cloud in Local Government" [in Japanese]. February 12.

Fire and Disaster Management Agency. 2011. *Enhancement and Strengthening of Earthquake and Tsunami Countermeasures in the Regional Disaster Management Plan* [in Japanese].

Imai, T., T. Kakimi, and S. Tateishi. 2011. *Study on Transferring Government Functions by Nuclear Accident* [in Japanese]. http://gakkei.net .fukushima-u.ac.jp/files/shinsai11.pdf.

Kahokunippo. 2011. "Focus/14 Minutes Following the Earthquake Support Preparation as Center Hub in Tono" [in Japanese]. August 12. http://www .kahoku.co.jp/spe/spe_sys1071/20110816_01.htm.

———. 2012. "Dispatch Staff by Avoiding Fukushima" [in Japanese]. March 15. http://www.kahoku.co .jp/news/2012/03/20120315t61005.htm.

Mainichi Newspaper. 2012. "Writer's Eye: One Year from GEJE: Whole Support by Nagoya to Rikuzentakata" [in Japanese]. March 8. http://mainichi.jp/select/opinion/eye/news /20120308ddm004070002000c.html.

Takenaka, H., and Y. Funabashi. 2012. *Lessons from Japanese Mega-disasters* [in Japanese]. Tokyo: Toyokeizai.

Technical Committee on Earthquake Disaster Management in Rural Towns. 2011. *Case of GEJE* [in Japanese]. http://www.bousai.go.jp/jishin /chubou/toshibu_jishin/7/sub2.pdf.

Technical Committee on Emergency Response. 2011. *Interim Report* [in Japanese]. http://www.bousai .go.jp/3oukyutaisaku/higashinihon_kentoukai /cyukan_torimatome.pdf.

Union of Kansai Governments. 2011. *The Emergency Proposal for the Great East Japan Earthquake* [in Japanese]. http://www.kouiki-kansai.jp /data_upload/1315378856.pdf.

CHAPTER 18

Evacuation Center Management

A megadisaster necessarily results in an enormous number of evacuees staying in evacuation centers for a significant time period. This note describes how Japan managed its evacuation centers after the Great East Japan Earthquake. It highlights important management issues including shortages of essential supplies and services, successful self-management practices initiated by the affected people themselves, good management practices by local governments, and the sensitivity required to accommodate diverse groups of evacuees with special needs.

FINDINGS

After the Great East Japan Earthquake (GEJE), nearly 2,500 evacuation centers were established in the disaster-affected Tohoku region; additional centers were also located outside of Tohoku. At peak occupancy, more than 470,000 people were staying at these centers (see figure 18.1). Most facilities, such as schools and community centers, were publicly owned and had been designated as evacuation centers even before the GEJE. Right after the GEJE, a number of private facilities such as hotels and temples were also enlisted as the need for centers far exceeded expectations (figure 18.2); also, a number of evacuees stayed with their relatives

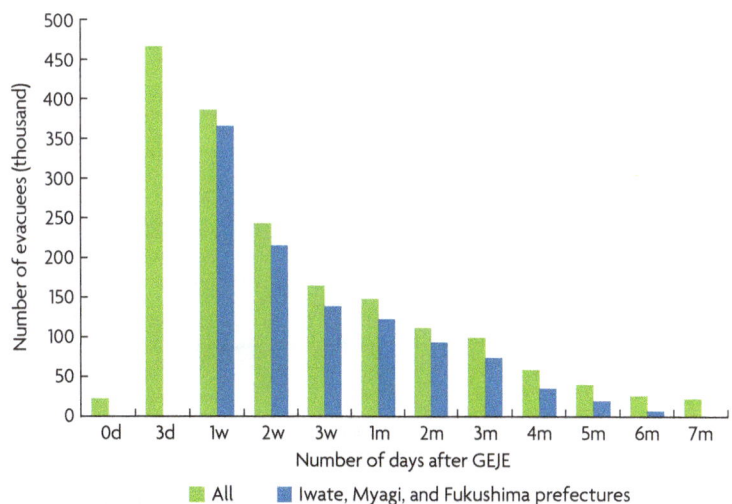

Figure 18.1 Number of evacuees after the GEJE

Source: Cabinet Office (CAO).

Note: d, w, m = day, week, month, respectively.

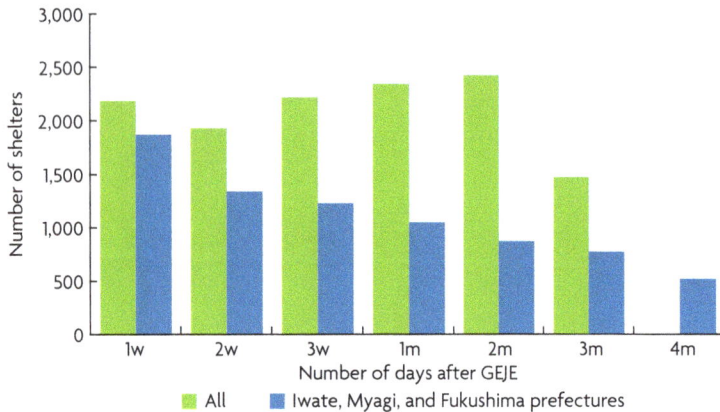

Figure 18.2 Number of evacuation centers

Source: CAO.

or friends. Evacuees gradually moved out of the centers as the construction of transition shelters progressed. Within four months after the disaster, about 75 percent of evacuation centers were closed, although some centers in Tohoku stayed open as long as nine months.

The evacuation pattern in Fukushima, where the nuclear accident occurred after the GEJE, was very different from other disaster-affected areas in Tohoku. In Fukushima many people had to relocate from one center to another, moving further from the crippled nuclear power plant as information became available on the risk of radiation exposure. More than 10,000 people had to change evacuation centers three or more times, with some people moving as many as 10 times (figure 18.3 and chapter 11).

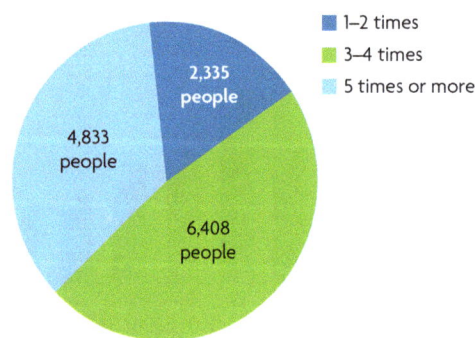

Figure 18.3 Number of times people in Fukushima^a had to evacuate

Source: Fukushima University

a. 8 towns and villages in Futaba region, Fukushima Prefecture.

This knowledge note will focus mainly on the management of publicly owned centers, since collecting information on private centers has been difficult.

Not enough supplies

Given the magnitude of the disaster and the number of evacuees, most evacuation facilities lacked sufficient supplies of food, water, clothes, and blankets. In the first days and weeks following the disaster, transporting these essentials to the centers was hampered by damaged roads and a shortage of vehicles and fuel (chapter 16). This problem was exacerbated by the fact that the many temporary facilities were not formally designated centers and therefore had not been stocked with essentials.

Lack of water and power

Furthermore, water and power supply systems were damaged in most of the disaster-affected areas, and in some places were not restored even after one month (chapter 20). These problems made life miserable for the evacuees. For example, they had difficulty using toilets without water for flushing. The cold weather in northeastern Japan and lack of electric heating in the facilities made many evacuees vulnerable to illness, especially the elderly. As the evacuation period became prolonged, the inability to bathe was also a serious issue.

People could not stay in their high-rise apartments in Sendai City because of water and power failures. Since they could not continue to carry water and food upstairs to the higher stories, they moved to evacuation centers until public services were restored.

Self-management by evacuees

Although managing evacuation centers is a municipal responsibility, most municipalities in the disaster-affected areas suffered badly from a loss of staff, seriously weakening their capacity to cope with the emergency. At the beginning, most facilities were supported by local teachers, volunteers, and other civil society groups. As the evacuation period became extended, evacuees themselves started taking a number of initiatives. At many shelters, a self-governing body emerged, with leaders and members of various committees selected by the evacuees themselves.

For example, evacuees at the Ofunato Junior High School in Iwate Prefecture (figure 18.4) organized themselves into several groups for nursing, sanitation, food, facilities, supplies, and heating. At one school in Minamisanriku in Miyagi Prefecture, evacuees divided themselves into 20 groups, based on the communities they came from before the disaster, and assigned themselves roles and responsibilities for day-to-day activities.

An event hall called the Big Palette in Koriyama, Fukushima Prefecture, admitted more than 2,000 evacuees mainly from Tomioka Town and Kawauchi Village, both affected by the nuclear disaster. These evacuees established a volunteer center at the hall, where volunteers and the evacuees themselves helped organize activities such as opening three cafes, starting an FM radio station, organizing a gardening event, and undertaking a summer festival. The volunteer center provided opportunities for the evacuees to help themselves and engage in productive activities, thus improving their daily well-being.

Gender sensitivity

One of the problems cited at many of the centers was a lack of gender sensitivity (chapter 19). There simply wasn't enough privacy for anyone, particularly not for female evacuees—many did not have private spaces where they could change their clothes or breast-feed their babies. Many centers eventually installed partitions, but it was often too late. It has also been reported that relief goods delivered to these centers were biased in favor of male evacuees. This was mainly because it was mostly men who were managing the centers, whether they were run by municipalities or by evacuees themselves.

Figure 18.4 Evacuation center at the Ofunato Junior High School

Source: © Inabe City. Used with permission. Further permission required for reuse.

Welfare shelters for those with special needs

Many experts have pointed out that evacuees tend to suffer from tremendous stress, especially children, and therefore need special mental health care and counseling services as evacuation periods extend. But the availability of such services varied from center to center.

Taking care of the elderly and others who needed special attention was another big challenge. At many centers, all the special needs groups had to share the facilities with the other evacuees. But Sendai City in Miyagi Prefecture had about 30 special centers called "welfare shelters" that provided nursing and other care for the elderly, the disabled, and other groups. About 250 people and their families were transferred to these from other centers.

Managing with a human face

A close relationship should be established early on between evacuees and local officials who are responsible for managing the centers. A good practice in this regard came from Hachinohe City in Aomori Prefecture. Right after the GEJE, there were about 120 families at eight evacuation centers in Hachinohe. The city government assigned two officials to

Information is both critical and comforting

Keeping evacuees informed is not only critical to their well-being but also comforting. In Rikuzentakata, in Iwate Prefecture, one of the city government's public relations officers continued to publish a special edition of the city's newsletter on a daily basis between March 18 and May 7, 2011, except for one day when a power cut prevented him from printing it. About 2,400 copies were printed every day and distributed to evacuees in more than 70 evacuation centers in the city. He continued publishing the newsletter five times a week for a few more months after May 8, 2011.

The newsletter initially contained information that evacuees really needed, such as procedures to get a disaster victim certificate or be able to receive donations, the locations of temporary public offices and medical facilities, schedules of school events, new public transportation routes and timetables, and so on. The type of information in the newsletter changed over time to meet the evacuees' changing needs. Reading the newsletter became a routine at evacuation centers in Rikuzentakata, and evacuees looked forward to it every day.

every seven or eight evacuated families with whom they could consult on any issue. For example, they had questions about subsidies for future housing and livelihood recovery. The relationship established with the officials at the evacuation centers continued even after the evacuees had resettled in private or public rental houses. Although this arrangement was possible because of the relatively small number of evacuees in a relatively big city with more than 2,000 officials, the city should nevertheless be commended for its initiative.

Disaster relief agreement

In 2006 two cities in Fukushima Prefecture entered into a Disaster Relief Agreement: Naraha City, which was affected by the nuclear disaster, and Aizu-Misato City (located relatively far from the crippled plant), which was not. When the nuclear disaster happened, most evacuees from Naraha City went to evacuation centers in Aizu-Misato City that were managed by local officials. This was a rare example of successful cooperation between two municipalities, strengthened by their long-standing friendly relationship. In Fukushima most

evacuees had to go beyond the prefecture's jurisdictional boundaries because of radiation risks. In most cases, however, the evacuation centers were managed by the evacuees' municipalities rather than by the hosts'.

LESSONS

- *Designate evacuation centers in safe locations.* While it may not be possible to be perfectly prepared for a megadisaster like the GEJE, it is nonetheless essential to designate evacuation centers in safe locations and equip them with as many emergency supplies as possible. Many prefectures and municipalities all over Japan are conducting ex post evaluations to assess the location and number of evacuation centers and the adequacy of supplies at these centers.

- *Prepare for primary service interruptions.* Since a megadisaster is likely to interrupt essential services such as water and power, it is critical to install alternatives such as portable toilets and power generators. Sendai City is planning to equip its designated facilities with renewable energies, such as solar panels, as a backup power source.

- *Evacuees should take part in managing activities and services at evacuation centers.* They are not guests who are simply receiving food and materials, but capable enough to manage the evacuation centers themselves.

- *Anticipate different needs in evacuation centers.* Evacuees consist of diverse groups of people who have different needs and wants: women and children, the elderly, the disabled, and some foreigners. Those in charge of managing evacuation centers should be sensitive to this diversity. It is also critical that women are included in management and leadership positions at these facilities.

- *Creative management pays.* Some local governments have come up with innovative arrangements for managing evacuation centers and supporting evacuees. These governments should share their experiences and learn from one another so that good practices may be replicated in the future.

- *Providing the information that disaster victims need* is not only critical to their well-being but also comforting. It is important to listen to evacuees to understand what kinds of information they need and want, and to continue listening as their needs change over time.

RECOMMENDATIONS FOR DEVELOPING COUNTRIES

Most of the lessons described above are applicable to developing countries. Evacuation centers are needed after most natural and industrial disasters, including not only earthquakes and tsunamis but also floods, landslides, volcano eruptions, and so on.

Plan ahead. In disaster-prone developing countries, evacuation centers should be safely located. Schools and community centers should be designed and built to also serve as evacuation centers. They should also be stocked with essential supplies such as food and drinking water, and equipped with emergency power generators. In developing countries, rainwater harvesting systems in schools and other public facilities, and renewable energies such as solar panels may also serve well in emergency situations. Political and financial support for predisaster investment in evacuation centers and supplies should be mobilized.

Support community organizations. One of the biggest challenges to managing evacuation centers in developing countries is weak local government capacity. Evacuees should, therefore, get organized to help themselves as illustrated by the Japanese experiences. In many developing countries this effort could perhaps be supported by nongovernmental organizations.

Integrate gender considerations into planning. Gender sensitivity and an ability to serve diverse groups of evacuees are required in any country. Communication among these groups and governments should be established at evacuation centers. Developing countries would be well advised to learn from the Japanese experience, especially with respect to gender.

NOTE

Prepared by Toshiaki Keicho, World Bank.

BIBLIOGRAPHY

Committee of "Alive, Living and Life." 2012. *Alive, Living, and Life: The 169 Days at the Fukushima Big Palette Evacuation Shelter.* Amu Promotion.

Committee on Disaster Emergency Response in the Great East Japan Earthquake. 2011a. *Documents from the 5th Meeting: Evacuation Center Operation* [in Japanese]. http://www.bousai.go.jp/3oukyutaisaku/higashinihon_kentoukai/5/naikakufu.pdf.

———. 2011b. *Interim Report* [in Japanese]. http://www.bousai.go.jp/3oukyutaisaku/higashinihon_kentoukai/cyukan_torimatome.pdf.

Committee for Technical Investigation on Earthquake Disaster Management in Regional Cities. 2011. *Documents from the 7th Meeting: Examples from the Great East Japan Earthquake* [in Japanese]. http://www.bousai.go.jp/jishin/chubou/toshibu_jishin/7/sub2.pdf.

Fire and Disaster Management Agency. 2011. *Enhancement and Strengthening of Earthquake and Tsunami Countermeasures in the Regional Disaster Management Plan* [in Japanese].

Fukushima University. 2012. *2011 Report of Basic Data Collection for Disaster Reconstruction Survey in Futaba 8 Towns.* 2nd ed. [in Japanese].

Imai. 2011. "The Great East Japan Earthquake and Public Policies of the Municipalities—Response to Nuclear Disaster." *Kokyoseisaku kenkyu* [Public Policy Research] 11.

Yomiuri Newspaper. 2011a. "Forty Shelters Receive Elderly and Disabled" [in Japanese]. April 2. http://www.yomiuri.co.jp/national/news/20110402-OYT1T00745.htm.

———. 2011b. "Issuing Newsletter Everyday" [in Japanese]. May 9. http://www.yomiuri.co.jp/e-japan/iwate/feature/morioka1304174360304_02/news/20110509-OYT8T00076.htm.

CHAPTER 19

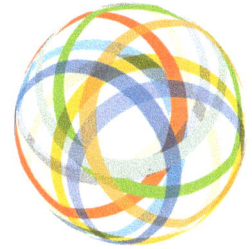

Ensuring Sensitivity in Response and Equity in Recovery

As in every disaster, certain groups were more vulnerable than others to the effects of the Great East Japan Earthquake. Two-thirds of those who lost their lives were over 60 years old. Response efforts to the catastrophe reflected other existing inequities, including that of gender. The special needs of children and the disabled were not always met. Women, the elderly, and the disabled—and experts sensitive to the needs of all vulnerable groups—should be engaged in the planning, design, and implementation of relief-and-recovery activities to ensure a more effective and efficient recovery. Such efforts promise to contribute to the sustainability and resilience of communities in the long term.

FINDINGS

Vulnerability to the impacts of natural hazards normally varies by social and demographic group. The Great East Japan Earthquake (GEJE) was no exception, with the elderly proving to be the most vulnerable. Two-thirds of the deaths occurred among people over 60 years old, who accounted for some 30 percent of the total population in the affected areas (figure 19.1). They were physically weaker than other groups and could not run fast enough to reach higher ground.

Nine hundred and eighty-five children and young people (0–19 years old) lost their lives in

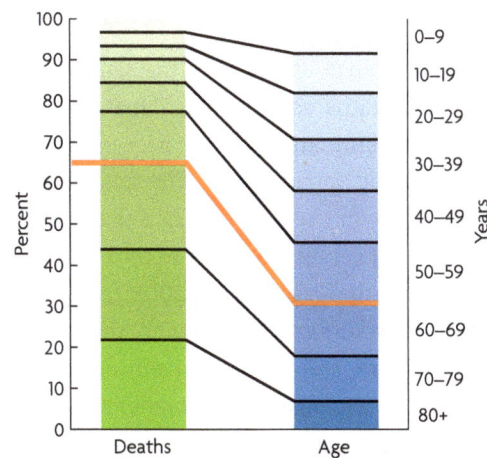

Figure 19.1 Age distribution of people killed in the GEJE

Source: Cabinet Office.

the GEJE (as of March 31, 2012). As of October 31, 2011, 1,327 children had lost one parent and 240 children had lost both their mother and father. Of these, 160 were adopted by relatives. A survey conducted by Ashinaga ("daddy longlegs," a scholarship organization for orphaned students) revealed that households with disaster-affected children, in particular those headed by females, face difficulties paying their bills. The details are as follows:

- Half of the affected children are in female-headed households.

- Forty-five percent of the heads of households have permanent full-time jobs, while 30 percent are unemployed or looking for work.

- Among female-headed households, 24 percent are employed full time, while 47 percent are unemployed or looking for work.

- The homes of 70 percent were damaged; 30 percent are living in their own homes, with the remainder living with relatives (29 percent) or in evacuation or transition centers.

CHALLENGES FACED DURING RESPONSE AND RECOVERY

Gender

Women in Japan do not have the same socioeconomic status as men, participate less in decision making, and have less access to social and economic opportunities. The relative poverty rate of women is higher than that of men (28.1 percent versus 22.9 percent in 2007). The average hourly wage rate in 2008 for female full-time workers was 69 percent of the rate for male workers, and the proportion of women in positions equivalent to or higher than section manager in private corporations was 6.5 percent. The prefectures affected by the event belong to a medium range of gender equality in Japan: rankings on the gender equality index

for Iwate, Miyagi, and Fukushima prefectures were 11th, 27th, and 17th, respectively, out of 47 prefectures. The GEJE relief-and-response efforts reflected and reinforced these preexisting inequalities. Most evacuation centers were managed by men. In fact, throughout Japan, 96 percent of the leaders of neighbors associations (*Jichikai*), many of whom served as the leaders of evacuation centers, are men.

Privacy and security

Privacy for women was rarely available at evacuation centers, which added greatly to their stress (figure 19.2). A survey conducted by the Cabinet Office in April 2011 revealed that only 26 percent of the centers had private spaces for women; at many centers women had to change their clothes under blankets or in bathrooms.

Women were hesitant to voice their needs to the male leadership of the centers, even when basic needs relating to hygiene were overlooked or handled in an insensitive manner. For example, in one center, male staff distributed a sanitary napkin to each woman and said: "If you need another one, please let me know." In centers where women were engaged in management, those items were made readily available in bathrooms. Male leaders at evacuation centers considered skin lotions and other cosmetic items to be luxury goods, while for women they contributed to a sense of normalcy. When a cosmetic company sent makeup kits to several centers, women were able to put on makeup for the first time since the disaster, which raised their spirits and encouraged them to be more active.

It is difficult to obtain verifiable estimates of sexual harassment incidents since they can take many forms—from sexual taunting to physical harassment—and often go unreported. In May 2011, there were two reported cases of rape confirmed in the three affected prefectures after the disaster, compared to nine reported incidents at the same time in 2010. There were 13 reported cases of forcible indecency compared to 32 cases in the previous year. The minister of state said these incidents did not occur in the affected areas. At evacuation centers, personal alarms were distributed to protect women and children, and they were cautioned to avoid going to the outdoor toilets alone, especially at night.

In one center, a grievance desk was set up; however, since there were no partitions in the facility, everyone could see and hear who was registering a complaint. This made women reluctant to report any concerns or incidents. In another center, a private, soundproof space was set up where women felt more confident about reporting grievances.

Domestic violence is also difficult to track, as it is typically considered a family matter and seldom discussed or reported. Of the cases that police responded to in the three affected prefectures from March 11 to December 31, 2011, 98 were recognized as having a clear linkage to the disaster. Many of these involved violent acts by husbands who had increased their alcohol consumption after the disaster.

The Gender Equality Bureau of the Cabinet Office recognized that gender perspectives were not sufficiently considered in managing evacuation centers, and on March 16, 2011, issued an ordinance on "Disaster Response Based on the Needs of Women and Women with Children" to provide guidance to relevant agencies. They also initiated consultation services for women dealing with distress or violence. But conditions on the ground made it difficult to reach the evacuees and people managing the centers.

At the Fukushima Big Pallet, a major evacuation center accommodating more than 2,000 evacuees, spaces for women were set up in collaboration with local women's organizations. The organizations provided advice to women and referred them to experts when necessary. They provided a safe space for women to gather and share their thoughts and concerns with others, and also held events such as cooking and handicraft classes. Women said that they felt relaxed and comfortable in these spaces.

Figure 19.2 An evacuation center, one month after the earthquake

Source: © Mikio Ishiwatari. Used with permission. Further permission required for reuse.

Maternal care

Many nursing mothers did not have privacy to breast-feed. Some went outside in the cold in search of privacy and others gave up nursing and changed to powdered milk. A number of maternal care clinics and hospitals offered temporary evacuation facilities free of charge for families with pregnant women and infants. But the Japan Primary Care Association reported that many pregnant women refused to move because they were concerned that their neighbors would no longer consider them to be community members if they moved to a separate place.

The Japan Primary Care Association set up several programs to support pregnant women and families with infants, and sent an obstetrician and gynecologist to the affected area.

Workload and livelihoods

Women in many evacuation centers were requested to prepare meals for the evacuees three times a day, in addition to caring for the elderly and children while the men were out looking for work. This placed a heavy burden on them. In some centers, a rotation system was established to alleviate the pressure on any specific person or group. Moreover, while men were engaged in cash-for-work programs cleaning up debris from the disaster, women were not compensated for their work in the centers.

Men's needs

Integrating a gender-sensitive approach to relief-and-recovery efforts means understanding and addressing the needs of men and boys in addition to those of women and girls. While data still need to be collected in the affected area, there are indications of a need for counseling for men to deal with alcoholism and domestic violence. Moreover, men may need special counseling for child rearing if they have become single parents (box 19.1) or if they have lost their means of livelihood.

Children

The GEJE left children feeling frightened, confused, and insecure. The number of incoming calls to "Childline," a free counseling service for children, increased fourfold in the Fukushima, Miyagi, and Iwate prefectures following the event. The government made plans to deploy some 1,300 mental health counselors to all public schools in affected areas.

The government expanded its support to foster parents caring for relatives' orphans, and recommended that the children's previous connections with friends and with their home region should be maintained. Governments and various organizations, such as Ashinaga and

the Fund for the Future of Children affected by the GEJE, started providing financial support or scholarships to orphans.

Because of the accident at the Fukushima Daiichi Nuclear Power Station, children in Fukushima Prefecture have stopped playing outside or swimming in pools, and have suffered from the stress of remaining indoors. In 74 percent of Fukushima households, children have decreased the time they play outdoors to 13 minutes per day to avoid the effects of radiation. These children demonstrate signs of increased stress, in many cases acting out twice as much as children in other areas. The government organized a few days of "refresh camp" where children can play sports and engage safely in outdoor activities. Some 6,000 children participated in the program.

The elderly and the disabled

A lesson learned from the Great Hanshin-Awaji (Kobe) Earthquake in 1995 was that special centers should be established for elderly people and the disabled. In 2008 the Ministry of Health and Welfare issued guidelines stating that welfare evacuation centers for special care needs should be established within seven days of a disaster emergency, but only 20 percent of municipal governments in the three affected prefectures did so. Many disabled people faced challenges accessing evacuation centers; there were some reports of mentally ill and autistic people leaving centers because they were not properly cared for.

People over 60 made up 30 percent of the population in the affected area, but local authorities were unprepared to respond to their needs. The evacuation of elderly people with dementia and their family members was challenging. While long-term care facilities organize regular evacuation drills, local governments had limited knowledge about the elderly with dementia who lived in their communities and were not well prepared to support them. Older people also faced accessibility issues at evacuation sites and temporary

---

BOX 19.1

Single Father Japan

Single Father Japan was established before the GEJE to support single fathers. After the disaster, the organization requested the Japanese government to extend bereavement pensions for men who had lost their wives in the event. Their main activities are providing counseling and open lectures, awareness raising, and research on single-parent families. See http://zenfushiren.jp (in Japanese).

housing sites. A number of older people in need of soft food and diapers went with their needs unmet. Older people are prone to withdrawal when disconnected from friends and family, which is an issue for many people in temporary housing who have lost their social networks.

The elderly residents in care facilities that were damaged in the GEJE were relocated to evacuation centers such as school gymnasiums, where they faced difficulties living without nursing care. Finding nursing care staff was a challenge because many of them had suffered from the GEJE: 52 out of 1,165 eldercare facilities in the Iwate, Miyagi, and Fukushima prefectures were damaged by the event, and 173 staff members were reported dead or missing. In April 2012, the Ministry of Health, Labour and Welfare (MHLW) issued an ordinance to local governments to prepare for large disasters by arranging for the evacuation of the elderly living in care facilities, supporting staff sent to devastated areas, and providing support to the elderly who needed care at home.

One elder-care facility became an evacuation site by default. Designed as a group home for 20 people, the building was equipped with an accessible kitchen, bathrooms, bedrooms, and a living room for individuals with physical and cognitive impairment. While large-scale multilevel elder-care facilities could not function without electricity and running water because of the GEJE, this small-scale group home was able to provide basic services and an accessible environment for over 100 people of all ages from the community.

Coordination challenges among agencies may have hindered the collection of data and the provision of support to disabled people affected by the GEJE. For example, disaster risk management (DRM) staff at local governments could not have access to information on the disabled in the affected area because of privacy policies, and a housing facility that provided income-generation activities for disabled people did not fall under the purview of the MHLW and so did not receive assistance.

Such bureaucratic mismatches resulted in certain groups falling through the cracks.

In an effort to ease the burden on vulnerable groups, the MHLW temporarily suspended the collection of national insurance system premiums for long-term nursing care. It also simplified procedures for claims, allowed affected people to receive services without showing their insurance identification cards, and reduced or waived service fees.

EMPOWERING MARGINALIZED GROUPS FOR LONG-TERM RECOVERY

Recognizing its importance, a number of groups have acted to enable marginalized groups to participate meaningfully in medium- and long-term recovery efforts.

The first meeting of the Government's Reconstruction Design Council was held on April 11, 2011. No mention was made of gender or of issues related to the disabled in the council's reconstruction principles, and only one woman was appointed to the 15-member council. This is a nationwide problem, reflected in the following figures:

- On the National Disaster Prevention Council, only one out of the 25 committee members is a woman.

- At disaster prevention councils at the prefectural and municipal levels, the participation rate of women is only 4 percent.

In response to the GEJE, there was an appeal led by several women leaders, including Akiko Domoto, former governor of Chiba Prefecture, and Hiroko Sue Hara of Josai International University, to establish the Japan Women's Network for Disaster Reconstruction and Gender. In June 2011, on the three-month anniversary of the disaster, the network held a symposium on gender equality in the GEJE reconstruction process. The network's advocacy efforts have been successful, and have contributed to the inclusion of the following text in the Basic Act for Reconstruction in response to the GEJE, which was passed on June 20, 2011: "Opinions of the residents in the disaster-afflicted regions shall be respected and opinions of a wide range of people including women, children and disabled persons shall be taken into account." There were also accompanying guidelines issued on promoting the participation of women, children, and the disabled in all aspects of the reconstruction process. The real challenge in the coming months will be the implementation of the law and guidelines, as so far the capacity and will to engage and address the needs of vulnerable groups and women has been quite limited.

A number of United Nations organizations and civil society organizations (CSOs) are also supporting children. Four organizations—the United Nations Children's Fund (UNICEF), Save the Children (box 19.2), the General Research Institute of the Convention on the Rights of the Child, and Childline—established a network for supporting children affected by the GEJE, with the objective of coordinating among governments, CSOs, experts, and the private sector. Through the network, information is shared on support activities, damages incurred, and the progress of recovery; also, children's messages are issued to the public and recovery policies are recommended. As of November 2011, 29 organizations were participating in the network.

UNICEF is providing assistance to the children of Japan for the first time in nearly half a century with a budget of ¥4 billion. The assistance covers emergency support supplies; health and nutritional support; educational support; psychosocial support (psychological care); protection of children in harsh environments, such as if they are orphaned, in need, or in impoverished families; and child-friendly reconstruction plans.

Older people are more often thought of as a vulnerable group in need of care rather than as a resource to support younger generations. When marginalized, elders lose opportunities for interaction and the ability to contribute to society, and young people lose the wisdom and talents that elders can offer. After the GEJE, a nongovernmental organization (NGO) called Ibasho, which focuses on the issues faced in aging societies, visited the affected area and heard many stories about elders who saved younger people's lives by telling them where to escape or by teaching them how to survive with extremely limited resources. Older people also expressed a great deal of gratitude for all the foreign aid they had received, and wanted to give back. "I want to be useful to others but I do not know how," was heard numerous times.

To empower elderly survivors of the GEJE, Ibasho is building a café adjacent to a large temporary housing site in Ofunato City, Iwate Prefecture. The Ibasho café is being designed in partnership with the community as a place where people of all ages can gather and share conversation and refreshments in an informal setting. It is envisioned that elders will plan, manage, and operate the

Save the Children

One key lesson Save the Children has learned over many years of responding to emergencies is that while children are more adversely affected by disasters, they also have a great capacity to recover quickly, provided they are given the proper support and are directly engaged in supportive dialogues. Children can inform families, school officials, and local officials of their needs, and of how they can help their communities recover. When asked about what would be of most support to them, children generally expressed their desire to return to normal routines and living situations— and to help their communities recover. Save the Children surveyed more than 11,000 children in the affected area on what type of role they would like to play in the recovery process, and how they would like to see their towns rebuilt. Close to 90 percent said they wanted to contribute in some way to rebuilding their communities. Save the Children is strengthening children's participation in the recovery process by ensuring their views are part of the planning for rebuilding their towns and communities, and assisting children to convey their thoughts and ideas to their communities and to local and national government officials.

café. Everyone—including people with physical disabilities or cognitive illnesses such as dementia—will be encouraged to participate to their fullest ability. It is hoped that this intergenerational exchange and interaction will create stronger social capital in the community, resulting in strengthened resilience to natural hazards and the risks associated with the rapid growth of an aging population.

LESSONS

Lessons learned from the GEJE include

- *Data collection disaggregated by gender and age, and including the disabled, is needed* to understand the relief and recovery needs of all affected people, and particularly those groups that have special needs. Arrangements and agreements to share data across agencies in case of an emergency are key.

- *Once an emergency occurs, it is already too late to start advocating for gender-sensitive perspectives.* It is crucial to involve women in center management, and to make plans that ensure women's privacy and safety.

- *The livelihoods of women also need to be supported;* opportunities for income generation during relief and recovery should be provided to both women and men.

- *Children are in particular need of support* in regaining a certain sense of security and normality; they can also be meaningfully engaged in rebuilding their communities.

- *Think of vulnerable people first.* When planning evacuation sites, it may be beneficial to reexamine how care facilities for the elderly and disabled are designed and integrated into neighborhood and city planning.

- *Engaging marginalized groups actively in the design and implementation of recovery* efforts contributes to their recovery and to the future resilience of the community.

RECOMMENDATIONS FOR DEVELOPING COUNTRIES

Plan for diverse needs. The needs and impacts of different groups can be quite varied. Assess and understand the different needs of women, girls, boys, men, the elderly, the disabled, ethnic groups, the very poor, and other marginalized groups to respond effectively. Those working in the informal economy may face particular difficulties, for example, where the loss of housing also means the loss of workplaces, tools, and supplies. It is important to formally recognize and compensate such cases.

Adopt rights-based approaches. Women should be encouraged to participate in disaster management committees, camp management, and risk assessment. National and local disaster management policies and strategies should be reviewed for their gender sensitivity.

Involve women and children in decision making. Establish specific monitoring mechanisms (for example, Continuous Social Impact Assessments) to ensure that women and children can access recovery resources, participate publicly in planning and decision making, and organize to sustain their involvement throughout the recovery process.

Protect the vulnerable. Sexual harassment and domestic violence comes in various forms. It is necessary to create safe and secure spaces for women, children, and other marginalized groups. Protection shelters and consultation services for victims should be established in collaboration with NGOs, governments, and the police.

Support marginalized groups. For longer-term recovery, support can be designed to help upgrade the living standards of the poor, to enable the most marginalized to participate, and to establish mechanisms that promote an inclusive, more resilient society. Supporting marginalized groups requires a solid understanding of the broader societal and policy contexts (for example, labor market practices).

NOTE

Prepared by Yoko Saito, Disaster Reduction and Human Renovation Institution, International Recovery Platform; Hironobu Shibuya, Save the Children Japan; and Margaret Arnold and Mikio Ishiwatari, World Bank. Valuable contributions were also provided by Emi Kiyota, Ibasho; and Akiko Domoto and Hiroko Sue Hara, Japan Women's Network for Disaster Reconstruction and Gender.

BIBLIOGRAPHY

Asahi Newspaper. 2012. "The Disaster for Women" [in Japanese]. February 28.

Ashinaga. 2012. "Report on the Status of 2,005 Children in 707 Households Who Lost Parents or Guardians in the Great East Japan Earthquake and Tsunami." Press Release, February 28. http://www.ashinaga.org/en/news/press/entry-378.html.

Benesse. 2012. "Survey on Childcare: Effects of GEJE." Press Release, February 24 [in Japanese]. http://www.benesse.co.jp/jisedaiken/pdf/shinsai_311_release2.pdf.

Cabinet Office. 2011. *Reality Check for Evacuation Centers in Three Prefectures* [in Japanese]. http://www.cao.go.jp/shien/2-shien/6-zentyosa/1-result-1th.pdf.

Gender Equality Bureau. 2011a. *In Response to Disaster Management and Rehabilitation from the Viewpoint of Gender Equal Participation* [in Japanese]. Cabinet Office, Tokyo. http://www.gender.go.jp/pdf/saigai_22.pdf.

———. 2011b. *Women and Men in Japan.* Cabinet Office, Tokyo.

ISS-GCOE. 2011. "Gender Equality and Multicultural Conviviality in the Age of Globalization." *ISS-GCOE Research Series* 4, *ISS Research Series* 46 [in Japanese].

Japan Primary Care Association. 2011. *Primary Care for All Team.* Tokyo: Ground Publishing. http://www.pcat.or.jp.

Kyodo Press. 2011. *Supporting Affected Women from the Viewpoint of Women; Sending Make-Up Sets to Evacuation Shelters* [in Japanese]. http://www.47news.jp/feature/woman/womaneye/2011/05/post_20110523170949.html.

Osawa, M., A. Domoto, and K. Yamaji, eds. 2011. "Proceedings on Gender Equality and Disaster." Rehabilitation 6.11 Symposium—Perspectives of Gender Equality in Disaster and Rehabilitation [in Japanese]. June 11, 2011.

Yoshida, Hiroshi. 2010. "Development of Gender Equality Index in Japan: Referring Gender Equality Index of Norwegian Statistic Bureau" [in Japanese]. *GEMC Journal* 3.

Yoshida, Honami. 2011. *The Prenatal and Postpartum Care Support Project in Tsunami-Affected Areas after 311,* Final Report Submitted to AmeriCares, Ground Publishing, Tokyo.

PART IV

Reconstruction Planning

CHAPTER 20

Infrastructure Rehabilitation

Social infrastructure and public utilities are critical for quick and effective disaster response and recovery. Japan's rigorous seismic reinforcement of infrastructure has greatly reduced the effort required to restore essential facilities. Identification of priority infrastructure, legislation of financial arrangements for rehabilitation, and establishment of predisaster plans alongside the private sector have enabled prompt emergency response operations and facilitated a quick rehabilitation.

FINDINGS

The Great East Japan Earthquake (GEJE) caused tremendous damage to infrastructure and public utilities in the eastern region of Japan. According to the Cabinet Office, damages to public utilities and social infrastructure were estimated to be about ¥1.3 trillion (approximately $16 billion) and ¥2.2 trillion ($27 billion), respectively (chapter 30).

Damage to infrastructure

Since damage to the road network was limited, and rehabilitation work was efficient, (figure 20.1) the main highways and roads to the affected areas were repaired within one week. Bullet train service was resumed within 49 days of the event. These developments, in turn, facilitated full-scale relief activities in the devastated areas. All of this was a huge improvement compared to the aftermath of the Great

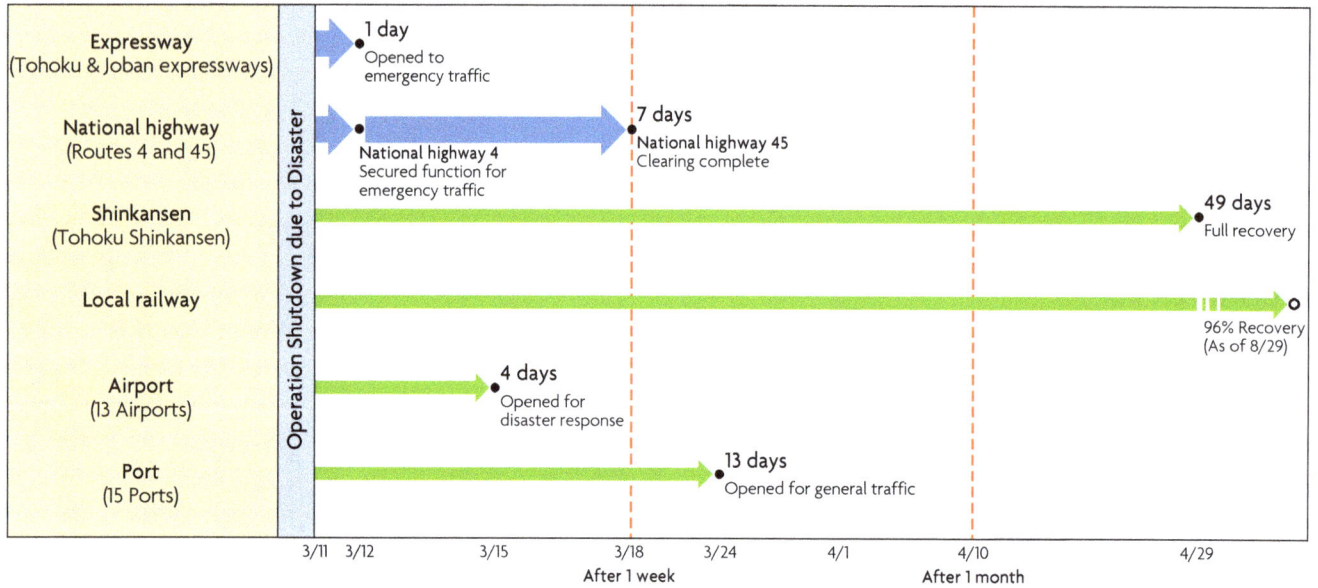

Figure 20.1 Securing emergency transportation

Source: Ministry of Land, Infrastructure, Transport and Tourism (MLIT).

Hanshin-Awaji Earthquake (Kobe earthquake) in 1995, when it took over 1.5 years for highway reconstruction and 82 days for the bullet train line to be repaired.

Roads

Some 15 expressway routes and 69 sections of the national highway system, mainly in the Tohoku region, were closed immediately after the earthquake. Many prefectural and municipal roads were also closed. Because they had been retrofitted, bridges on national highways or expressways were not damaged, but 20 bridges on prefectural and municipal roads collapsed or were severely damaged (chapter 2).

The subsequent tsunami flooded approximately 100 kilometers (km) of national highway, and submerged three expressway interchanges and junctions. The tsunami also washed away five national highway bridges. Massive amounts of debris brought in by the waves left many of the coastal roads unusable (map 20.1).

Railways

Railway facilities were also severely damaged, but various earthquake countermeasures, including the seismic reinforcement of railway

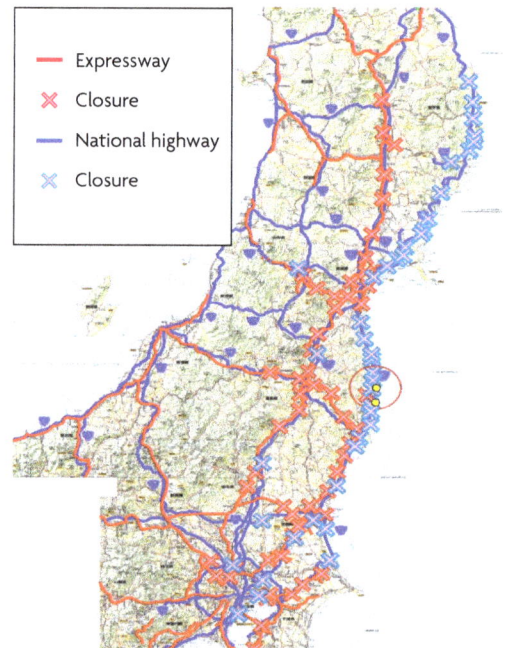

Map 20.1 Status of expressways and national highways immediately after the earthquake

Source: MLIT.

facilities, prevented most of them from breaking down and causing fatalities. Some 325 km of railway were damaged, mostly by the tsunami. Damage included the displacement or washing away of railroad tracks, power poles, bridges, and stations; the collapsing of earthen embankments; and damage to platforms.

Airports

The Sendai Airport, the major airport in the Tohoku region, is located about 1 km from the Pacific coast at an elevation of 4 meters above sea level. The tsunami hit the airport and flooded the runway, the first floor of the terminal building, and the airport access railways (figure 20.2).

Ports

Fourteen international and other major ports and 18 local ports were severely damaged by the tsunami and unable to function. Numerous ports that support the region's fishing industry were also destroyed. The tsunami and the earthquake together destroyed much of the port infrastructure. Debris from the tsunami washed into the port area, preventing ships from entering.

Damages to public utilities

Public utilities were severely damaged by the earthquake and tsunami. About 2.3 million houses were left without water supply after the earthquake, and the sewerage systems were destroyed in the coastal cities and towns in an area spanning some 550 km.

Wastewater treatment plants were damaged at 63 locations, 48 of which had to stop operating because of tsunami inundation. The condition of six wastewater treatment plants near the Fukushima Daiichi Nuclear Power Station is still unknown because of access restrictions. In Urayasu city, Chiba Prefecture, sewerage systems were severely damaged by liquefaction (figure 20.3).

The number of houses left without electricity reached 8.5 million. Several nuclear and conventional power plants, including the Fukushima Daiichi Nuclear Power Station, went offline after the earthquake, reducing the region's total power generation and supply capacity. The capacity of the Tokyo Electric

Figure 20.2 Sendai Airport after the tsunami
Source: MLIT.

Figure 20.3 Manhole raised by liquefaction in Urayasu City
Source: © Urayasu City. Used with permission. Further permission required for reuse.

Power Company (TEPCO) was reduced by about 40 percent from 50 gigawatts (GW) to about 30 GW, not enough to meet the typical 40 GW peak demand for that season.

Infrastructure rehabilitation planning and implementation

Concerned organizations were able to start rehabilitation work immediately after the earthquake and tsunami, to a large extent subsidized by the national government under the National Government Defrayment Act for

Reconstruction of Disaster Stricken Public Facilities (enacted in 1951). This act applies to a variety of transport systems and other infrastructure such as rivers, coastal facilities, *sabo* facilities, roads, ports and harbors, parks, and sewerage systems. The typical course of rehabilitation project implementation is illustrated in figure 20.4. In the aftermath of a disaster, local governments report their infrastructure damage to the national government, usually within 10 days of occurrence, with a request for a national subsidy. Upon receipt of the application, the national government conducts a disaster assessment within two months of the disaster and approves the subsidy. To ensure quick rehabilitation, local governments can begin implementing their projects immediately after the disaster occurs, even before applying for the subsidy.

The national government subsidizes two-thirds of the project costs, and much of the local government's share is covered by issuing local bonds. Thus, local governments actually cover only 1.7 percent of the costs at most. This local government share decreases as the severity of the disaster increases. In the case of the GEJE, the costs were so large that the local government share was minimal.

To ensure the quick rehabilitation of infrastructure, the national government enters into predisaster agreements with the private sector, ensuring that in the event of a disaster, the needed workforce will be mobilized quickly, without burdensome contracts and paperwork. Such arrangements are made between government field offices and private companies or private sector associations, and they cover such postdisaster activities as construction, engineering consulting, surveying, telecommunications, and broadcasting.

Roads: Operation Toothcomb

Transportation infrastructure is critical for delivering relief supplies. After the GEJE, roads were recovered early on to secure an emergency transportation network. Immediately after the earthquake on March 11, the Ministry of Land, Infrastructure, Transport and Tourism (MLIT) deployed a strategic initiative to make sure that the entire length of the Tohoku Expressway and National Highway 4 was passable to traffic. This major artery runs south to north from Tokyo to Aomori along the inland part of the region, which suffered relatively little damage. Next, 16 routes were opened up, stretching out from various points on this major north–south artery and reaching east to the coastal areas that were worst hit by the tsunami. The plan was called *Kushinoha Sakusen,* or Operation Toothcomb, because of the shape of the road network (map 20.2 and figure 20.5). From the next day, the operation began clearing debris from the emergency roads that run eastward from the inland arterial highway—National Route 4 (running north–south)—connecting them to the Pacific coast. By March 15, four days after

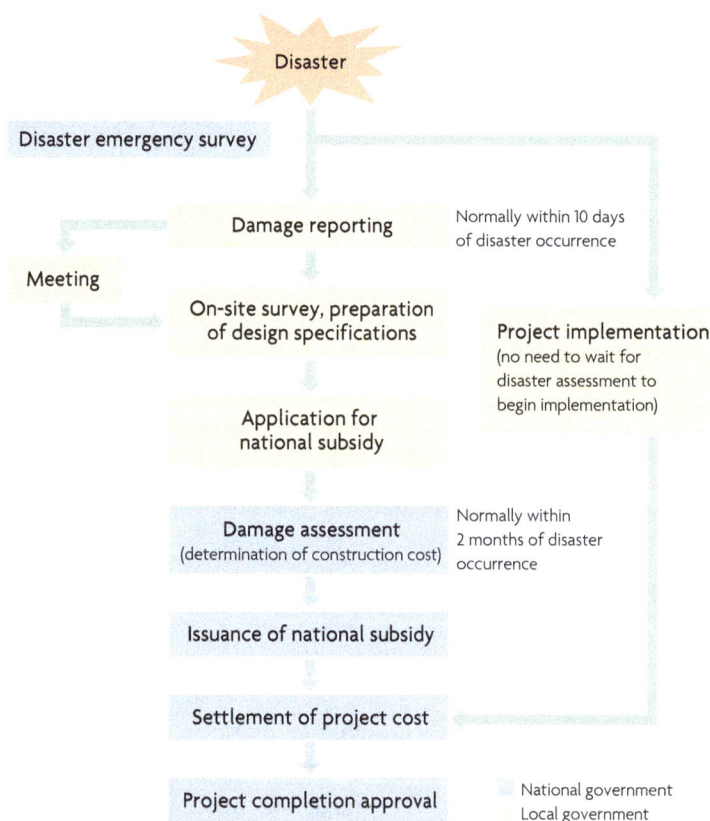

Figure 20.4 Steps in infrastructure rehabilitation

Source: MLIT.

the earthquake, 15 eastward access roads were usable, and by March 18, 97 percent of the national coastal highways were accessible.

Furthermore, 13 days after the earthquake the entire Tohoku Expressway, the main expressway connecting the Tohoku region to central Japan, was open to general traffic.

The quick rehabilitation of roads was possible for a number of reasons:

- The seismic reinforcement of road structures had helped minimize damage.

- There was a clear focus on opening up the 16 eastward routes by concentrating the workforce on them first.

- The authorities used their predisaster agreements to mobilize contractors immediately after the disaster.

Ports and navigation passages

The MLIT requested contractors to begin clearing navigation passages so that disaster relief vessels could enter ports. The operations began in 14 principal ports on March 14, the day after the lifting of the tsunami warnings. This included removing debris as well as ensuring the safe passage of emergency relief

Figure 20.5 Clearing of roads

Source: MLIT.

First step	Second step	Third step
March 11 Earthquake occurred	Establish the vertical artery	**March 15** Establish the horizontal lines	**March 18:** National highway 45 & 6 were 97% rehabilitated (operation completed)

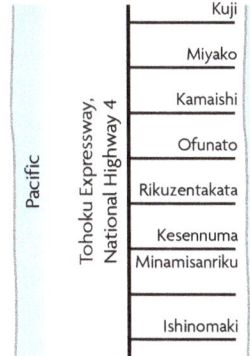

Kuji
Miyako
Kamaishi
Ofunato
Rikuzentakata
Kesennuma
Minamisanriku
Ishinomaki

Tohoku Expressway, National Highway 4

Pacific

Available for traffic

Closed

Map 20.2 Operation Toothcomb

Source: MLIT.

Figure 20.6 Clearing of navigation passages

Source: MLIT.

vessels (figure 20.6). By March 15, four days after the earthquake, all 14 ports were either entirely or partially usable and began accepting vessels delivering emergency supplies and fuel. At Sendai's Shiogama Port in Miyagi Prefecture, the first oil tanker entered 10 days after the earthquake, reducing the fuel shortage in the disaster-affected areas.

Railways

The Tohoku Shinkansen (bullet train) resumed operations between Tokyo and Nasushiobara (the southern section) on March 15, and between Shinaomori and Morioka (the northern section) on March 22. By April 29, the entire Tohoku Shinkansen line was in operation, as were most of the other railways except for those along the coast. The rehabilitation of the coastal railways, especially the Joban Line that runs through an area 20 km from the Fukushima Daiichi Nuclear Power Station, has still not happened. Many are currently being evaluated for possible rehabilitation along with the reconstruction of the towns and cities. The Sanriku Railway, which runs along the coast, resumed its operation in April 2014.

Sendai Airport

The Sendai Airport rehabilitation operation began two days after the earthquake, and by March 15, four days after the earthquake, the airport was being used by rescue and emergency supply rotorcraft. Fixed-wing aircraft were able to use it by the following day,

allowing the U.S. Army to bring in emergency supplies. The airport was available for commercial services on April 13.

Water supply systems

Although water supply services were resumed for about 90 percent of residents within one month of the disaster, the aftershocks on April 7 and 11 temporarily increased the number of households without water (figure 20.7). The Japan Water Works Association (JWWA) set up emergency headquarters to arrange for relief teams. The Ministry of Health, Labour and Welfare; JWWA; and 400 water utilities nationwide provided assistance to the affected areas by dispatching emergency teams with water supply trucks and machinery. They also helped conduct investigations for the restoration and reconstruction of water works.

Sewerage systems

Of the 120 disaster-affected wastewater treatment plants, those with minor damage (95 facilities) were rehabilitated and have recovered their predisaster capacities. Sixteen treatment plants are still inoperable because the tsunami destroyed their infrastructure and equipment. The 13 facilities that are accepting influent sewage have been providing primary treatment only, consisting of settlement and disinfection (box 20.1).

The reconstruction planning for the sewerage systems is the responsibility of the local municipalities. However, some 6,575 personnel have been dispatched from national or local municipalities in other regions to support their rehabilitation efforts. Sanitation is a major challenge in a disaster. Higashimatsushima City in Miyagi Prefecture did not have enough toilets for the people staying at evacuation centers. The city installed "manhole" toilets, paid for by a national subsidy system for promoting earthquake proofing of sewerage systems across the country. These toilets, which can be easily and quickly installed, were well received, especially by the elderly.

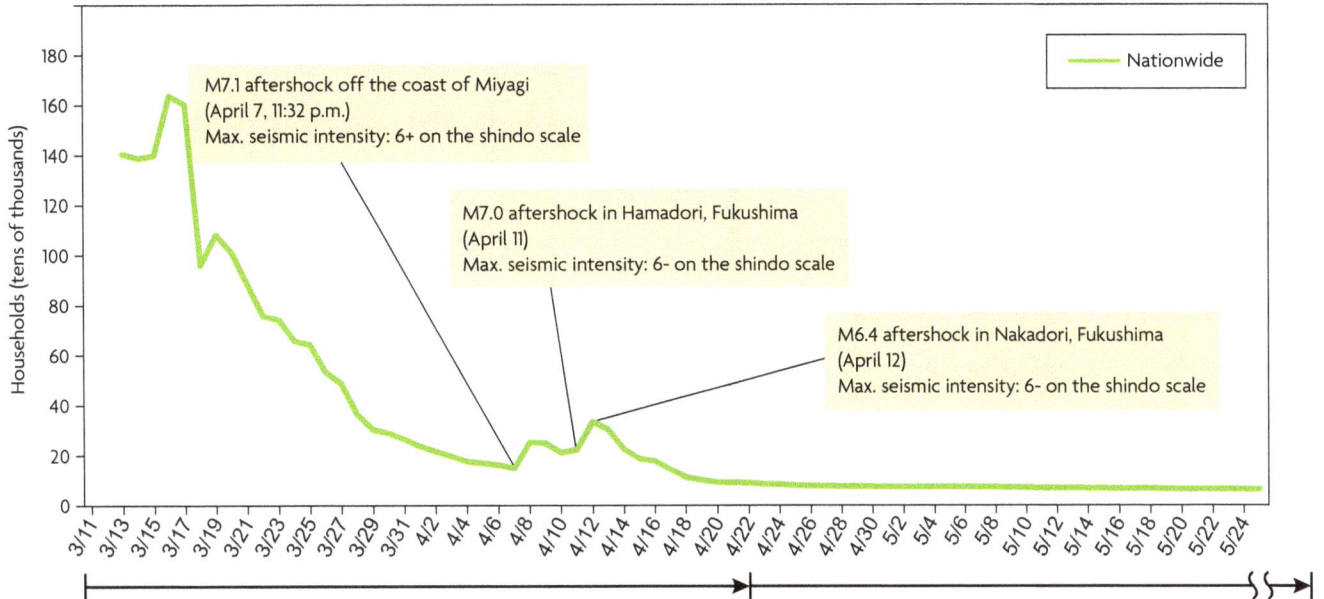

Figure 20.7 Water works rehabilitation

Source: Cabinet Office.

Note: As a point of reference, in the Great Hanshin-Awaji Earthquake, water supply was cut off to 1,270,000 households. Temporary recovery was completed 42 days after the earthquake, and water supply to all households resumed 91 days after the earthquake. Figure excludes areas located within the Fukushima restricted area where surveys could not be conducted.

Electricity services

About 90 percent of the power services were recovered within one week of the disaster; however, the aftershocks on April 7 and April 11 temporarily increased the outages (figure 20.8). Because of its reduced power supply capacity, TEPCO implemented rolling blackouts in its service areas, including Tokyo, between March 14 and 28.

LESSONS

- *Act fast.* Quick emergency response initiatives, such as Operation Toothcomb, contributed greatly to the prompt rehabilitation of transportation networks and the starting of relief activities.

- *Identifying the routes to be recovered first, and prioritizing resources and manpower accordingly,* was an effective approach to rehabilitating transportation networks.

- *Pre-agreements with the private sector.* Agreements, made with the private sector

Rapid rehabilitation of sewerage system in Rikuzentakata City

In Rikuzentakata City in Iwate Prefecture, the wastewater treatment plant was severely damaged by the tsunami. But within its service area, 400 houses located on higher ground had survived the tsunami. When water supply services resumed, the sewage generated by these 400 houses had nowhere to go. Following a proposal by a private company, the city decided to introduce a movable membrane bioreactor unit, which was quickly installed and began operating within a month.

Source: MLIT.

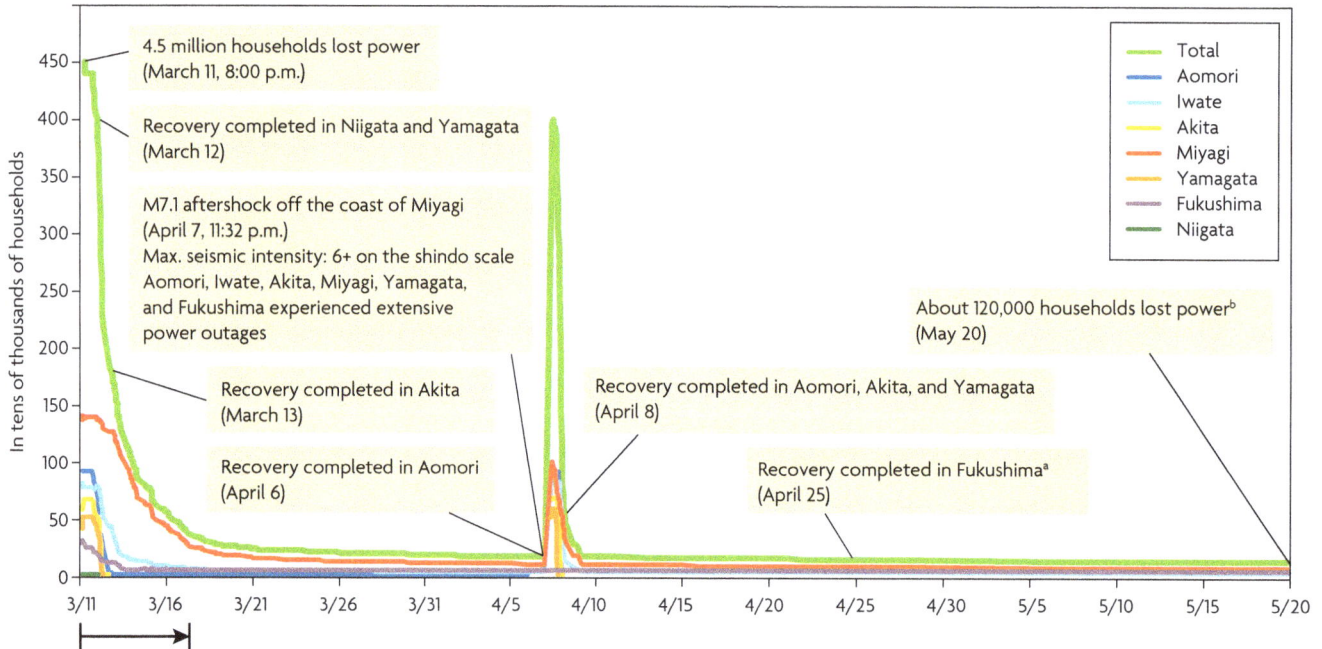

Figure 20.8 Electricity rehabilitation

Source: Cabinet Office.

Note: As a point of reference, in the Great Hanshin-Awaji Earthquake, about 2.6 million households lost power and fully recovered 6 days from the earthquake, excluding houses that were destroyed in the earthquake.

a. Excludes cases where service was suspended despite recovery when safty of indoor wiring could not be verified due to absence of residents; where public infrastructures, houses, etc., had been lost due to tsunami or other damage; and where households were located within the Fukushima restricted area.

b. The number of households without power, excluding a. above, totaled 1,452.

before the disaster, to provide emergency response operations were effective in quickly mobilizing the needed workforce and resources.

- *Pre-agreements with national and other local governments.* Experts and equipment dispatched from national and local governments contributed to prompt rehabilitation.

- *Building codes enforcement reduced damages.* Rigorous implementation of the seismic reinforcement of infrastructure prevented excessive damage to structures, minimizing the effort required to restore their functions.

- *Restoring utilities/service functions is a priority.* At the time of a disaster, sanitation can

be a major challenge. Resumption of water supply services without adequate sanitation led to sanitation and hygiene problems.

RECOMMENDATIONS FOR DEVELOPING COUNTRIES

Effective emergency and rehabilitation operations depend on social infrastructure and public utilities. The following arrangements are required if rehabilitation works are to be started and completed promptly.

Establish financial arrangement mechanisms. Budget-sharing mechanisms between local governments and the central government should be established in advance (chapter 31). Negotiating between governments only after a disaster has occurred will delay rehabilitation work. Such negotiations should cover

- Procedures for applying for a subsidy to the central government

- The cost-sharing ratio of rehabilitation works, shared between national and local governments

- Criteria for which types of disasters—and at what scale—require which mechanisms

- Establishment of a body of experts and responsible organizations at the central government level

- Team formulation and procedures for damage assessment

Arrange predisaster agreements with the private sector. Prearranged agreements with the private sector allow for quick mobilization of the needed rehabilitation workforce. Government agencies can skip the procurement process and start work immediately. These agreements should include (1) the designated responsibilities of governments and private companies for rehabilitation work, (2) a government guarantee of payment for the work involved, and (3) procedures for project requests from the government.

Arrange support teams. Emergency support teams should be established during normal times (chapter 14). Rehabilitation requires enormous additional resources from local governments, which are already burdened by the aftermath of disaster. Emergency teams from other government agencies can assist those local governments affected by disaster.

Develop disaster-resilient infrastructure. If infrastructure and utilities are planned and developed to mitigate potential disaster damage, the effort and time required for rehabilitation can be minimized. Retrofitting bridges can reduce both damage and rehabilitation efforts (chapter 2).

Identify key infrastructure. Transportation or communication networks that are critical to emergency operations should be identified before the disaster and given priority during the rehabilitation efforts (chapter 5).

NOTE

Prepared by Mikio Ishiwatari, World Bank, and Junko Sagara, CTI Engineering

BIBLIOGRAPHY

Central Disaster Management Council. 2011. *Report of the Committee for Technical Investigation on Countermeasures for Earthquakes and Tsunamis Based on the Lessons Learned from the 2011 Earthquake off the Pacific Coast of Tohoku.* Cabinet Office, Tokyo.

East Japan Railway Company. 2011. *Safety Report 2011* [in Japanese]. Tokyo.

Ishiwatari, M. 2014. "Institution and Governance Related Learning from the East Japan Earthquake and Tsunami." In *Disaster Recovery: Used or Misused Development Opportunity,* edited by R. Shaw, 77–88. Tokyo: Springer Japan.

MLIT (Ministry of Land, Infrastructure, Transport and Tourism). 2011. *Yuso chosei nitsuite, Higashinihon daishinsai niokeru kotsuro no kakuho* [Logistics coordination and securing of transportation routes in the Great East Japan Earthquake]. http://www.bousai.go.jp/3oukyutaisaku/higashinihon_kentoukai/3/kokudokoutu2.pdf.

CHAPTER 21

Reconstruction Policy and Planning

The unprecedented damage caused by the Great East Japan Earthquake affected multiple locations, posing severe challenges for local governments. Based on advice from an independent council, the government acted quickly and issued a basic policy and regulation framework within four months, laying the foundation for an inclusive process of recovery and reconstruction. This note documents the interactive process of reconstruction planning, as conducted by various levels of government with the active engagement of affected people, experts, volunteers, and the private sector.

FINDINGS

The Great East Japan Earthquake (GEJE) was Japan's first major multilocation disaster in recent history. With over 200 municipalities affected, it required both a national-level response as well as inclusive and participatory local planning. By adopting early policy and regulatory guidance and releasing several budgetary supplements, the government supported the evolution of effective recovery and reconstruction plans, including coordination at the prefecture and municipal levels. Overall,

the policy and planning process involved three stages:

- *Stage I (0 to 4 months):* The government established a disaster headquarters, chaired by the prime minister and an independent Reconstruction Design Council (RDC). Basic guidelines and an act were issued within 4 months, based on the council's recommendations. The first supplementary budget was passed within 1.5 months of the disaster.

- *Stage II (4 to 11 months):* The provisional reconstruction headquarters was established. Prefectures and municipalities prepared basic recovery plans in close consultation with disaster-affected people. Two other supplementary budgets were adopted to fund the recovery.

- *Stage III (11 months to 10 years):* A reconstruction agency and special zone for reconstruction were formed, and a fourth supplementary budget was passed. The reconstruction was envisaged to last 10 years and to be implemented through flexible grants and policies in support of the municipalities.

Although challenges remain—particularly with respect to the role of the new reconstruction agency—the GEJE reconstruction planning process can be seen as a model for other megadisasters. Prior to the GEJE, Japan already had a sound institutional and policy framework for disaster response and mitigation, based on lessons learned from past disasters. Building on this foundation, Japan acted rapidly to establish a reconstruction planning framework based on mutual trust, respect, and collaboration among stakeholders. At the same time, the fact that the GEJE required a new agency and reconstruction act shows that megadisasters, by their very nature, tend to overwhelm existing institutional arrangements. The chronology of policy and planning followed during the GEJE is summarized in figure 21.1 and explained in further detail below.

Basic principles, guidelines, and legal framework for reconstruction (March to June 2011)

The government set up headquarters for emergency disaster control less than an hour after the disaster. At the same time, building on lessons learned from the Great Hanshin-Awaji Earthquake (Kobe earthquake) in 1995, the government sought to broaden the recovery strategy by setting up an RDC. This advisory panel was composed of a team of highly respected intellectuals, academics, religious figures, and elected officials. Within two months of the disaster, the council issued "Seven Principles for the Reconstruction Framework," a consultative vision for the reconstruction. By the end of June 2011, a final report was given to the prime minister, which in turn became the basis for the government's *Basic Guidelines for Reconstruction* and *Basic Act for Reconstruction* (GOJ 2011a, 2011b), issued 3.5 months after the disaster. Thus, the initial process of national consultation set the stage for the entire recovery and reconstruction effort.

The Basic Guidelines set in place several innovative policies (box 21.1). It placed municipalities and residents at the center of the reconstruction; it promoted the concept of multiple defenses and people-oriented measures in

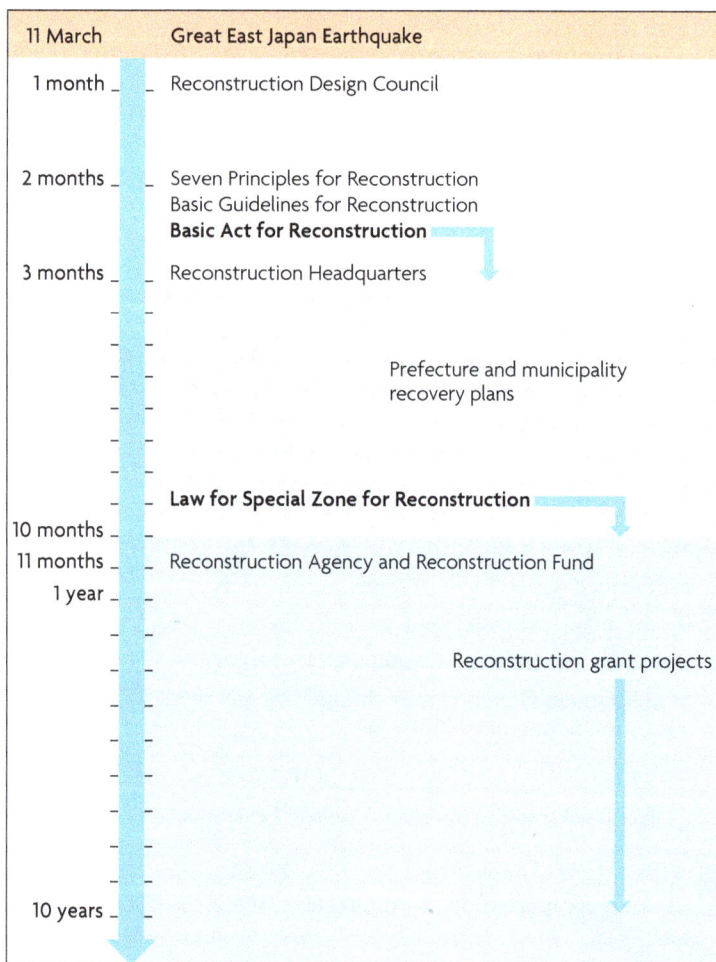

Figure 21.1 **Chronology of key policy and planning measures after the GEJE**

disaster reduction (departing from past reliance on defensive structures); and it encouraged land-use planning as a way to balance safety considerations with the need to preserve links between communities and infrastructure.

The recovery and reconstruction period was estimated to last 10 years and cost ¥23 trillion (approximately $290 billion), with the bulk of the effort focused on the first 5 years. The financial resources were to be secured through reconstruction bonds, reduction of public expenditures, increase in nontax revenues, and temporary taxation. As of early February 2012, the government had passed four supplementary budgets, worth a total of ¥21.9 trillion ($274 billion). The budgets were issued over a period of several months and served to support different stages of recovery and reconstruction.

The *Basic Guidelines* also provided for the establishment of a special zone for reconstruction containing financial and regulatory incentives, and a central one-stop reconstruction agency to respond to, and help coordinate, the needs of local governments (see section below titled "Reconstruction").

Recovery planning process (July 2011 to March 2012)

Prefecture-level planning

Based on the national guidelines, the most affected prefectures and municipalities—Iwate, Miyagi, and Fukushima, with more than 120 affected municipalities among them—developed their own recovery plans. These plans were not intended to be comprehensive, but rather to reach consensus among residents on the vision and key principles to be followed, the proposed land-use planning (including potential relocation of communities), and the implementation program (figure 21.2). It was understood that the plans would evolve over time through further consultations with ministries and elected officials, and eventually result in more detailed reconstruction plans (and cost estimates).

BOX 21.1

Basic guidelines for reconstruction after the GEJE

Key policies

- Recognize the challenges of an aging and declining population by promoting adequate public transportation and support services.

- Promote a strategy of multiple defenses through both soft and hard (structural) measures, putting people at the center of disaster reduction.

- Promote a "new public commons" through social inclusion of a wide range of stakeholders in the reconstruction.

- Make municipalities in disaster areas the main actors accountable for reconstruction, aided by financial and technical support from the central government and prefectures.

- Promote rapid reorganization of land use, to stimulate investment and prevent speculation.

- Prioritize providing stable residences for the affected, through favorable housing loans and low-rent public housing.

- Assist municipalities with reconstruction planning through external experts.

- Promote employment of affected people through recovery and reconstruction investments under the "Japan as One" project.

- Prioritize rehabilitation of key transport and logistics infrastructure and revival of local economic activities.

- Open reconstruction to the world through active international cooperation and lesson sharing.

- Create a special zone for reconstruction to support local projects through flexible procedures and financing.

Source: GOJ 2011a.

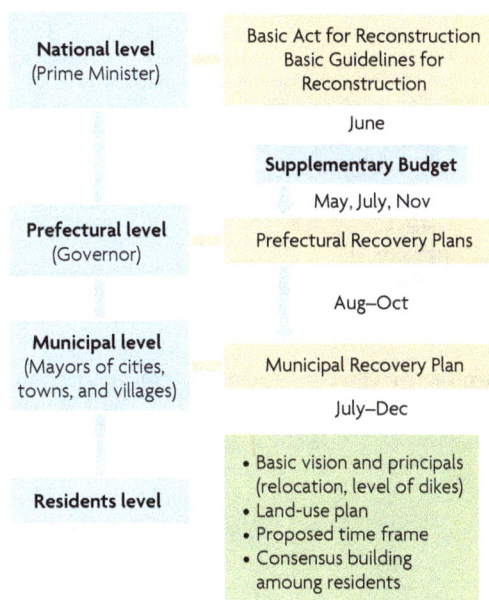

Figure 21.2 Recovery plans after the GEJE

National level (Prime Minister)	Basic Act for Reconstruction Basic Guidelines for Reconstruction
	June
	Supplementary Budget
	May, July, Nov
Prefectural level (Governor)	Prefectural Recovery Plans
	Aug–Oct
Municipal level (Mayors of cities, towns, and villages)	Municipal Recovery Plan
	July–Dec
Residents level	• Basic vision and principals (relocation, level of dikes) • Land-use plan • Proposed time frame • Consensus building amoung residents

The three most affected prefectures benefited substantially from a partnership arrangement supported by the Union of Kansai Governments (a grouping of prefectural governments in Western Japan), which provided expert personnel to assist with the emergency and relief efforts. This twinning experience, which also proved beneficial after the 2008 Sichuan earthquake, is outlined further in chapter 17.

To formulate the prefecture recovery plans, task force meetings were held with experts and citizens to collect public comments. In general, prefectural-level plans allowed local stakeholders to make decisions on infrastructure and other issues (such as debris disposal) that required intermunicipal coordination.

Fukushima, for example, faced a special problem due to the nuclear accident, which restricted access to contaminated areas and led to the evacuation of large numbers of residents. The Miyagi Prefecture recovery plan, in turn, developed a detailed tsunami protection plan, including structures resistant to a 100-year tsunami, elevated structures, population relocation to higher altitudes, an accessible evacuation plan, and the promotion of a culture of disaster prevention.

Municipal-level planning

Planning processes at the municipal level tackled such issues as risk assessment, financing, land tenure and land use, transportation infrastructure, and the role of the government in building consensus and providing relevant information to communities. Recovery plans had a positive tone, reflecting the municipalities' confidence in the nation's ability to assist affected people in improving their lives.

Similar to the prefectural recovery planning process, municipalities established recovery planning committees involving experts, residents, and community representatives. Generally, they used surveys and workshops to incorporate residents' opinions into the plans. For instance, in Minamisanriku (in Miyagi), a residents' committee played a key role in proposing "symbolic projects" that were then integrated into the town recovery plan (figure 21.3). Similarly, Ofunato City (in Iwate), held residents' workshops and students' reconstruction meetings involving more than 3,000 residents. In Sendai (in Miyagi), the largest city in the Tohoku region, the mayor herself visited residents' workshops and talked directly with victims. About 80 workshops were held to share information between residents and the city government, and residents submitted more than 2,000 comments on the draft recovery plan.

The central government supported municipal efforts by deploying two professional private sector consultants per municipality to provide technical services linked to damage assessment and engineering analysis. Experts

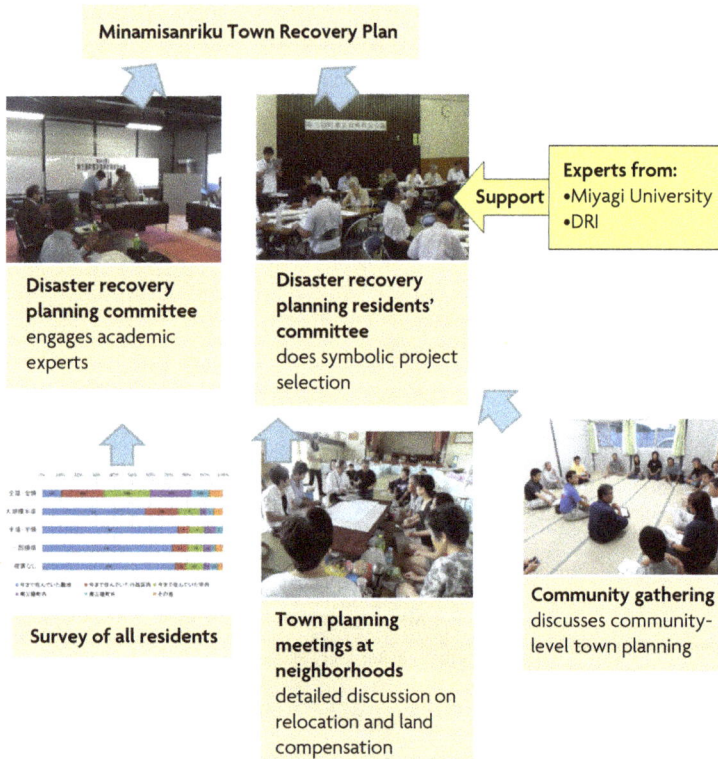

Figure 21.3 Community involvement in recovery planning in Minamisanriku Town (Miyagi Prefecture)

Source: International Recovery Platform (IRP).

Note: DRI = Disaster Recovery Institute.

such as university faculty members, architects, engineers, lawyers, and members of nongovernmental organizations (NGOs) also participated actively and voluntarily in the municipal planning process, according to their field of expertise. Thus, the process of participatory planning was widely supported by governmental and nongovernmental actors across all administrative levels in Japan. Chapter 33 covers updated information.

Two issues were particularly challenging in recovery planning: land-use planning and demographic trends.

Land-use planning

Municipalities used land-use planning as a tool to reach consensus on the strategy for reconstruction. This was based on a tsunami simulation conducted by the prefectural governments.

The simulation assumed two different levels of a tsunami (map 21.1): a maximum-level tsunami such as the GEJE (a 1,000-year event) and a frequently occurring tsunami (a 100-year event). The height of the coastal seawall is usually planned to protect from a frequently occurring tsunami. If a maximum-level tsunami hit the area, water may overtop the seawall and inundate the town. However, because of land-use planning—such as relocation of residential areas, land elevation, and multifaceted protection using forests and/or roads—the water level is projected to be less than 2 meters high in residential areas (making it unlikely for houses to be washed away). Low-lying areas would be reserved for parks, commerce, and industry (figure 21.4). In case of a maximum-level tsunami, people would have to evacuate, and early warning systems and evacuation routes would become crucial.

In the coastal areas of Iwate and the northern part of Miyagi, there was not enough land space available for relocation since steep mountains line the coast. In Minamisanriku Town, for example, many fishing villages that were

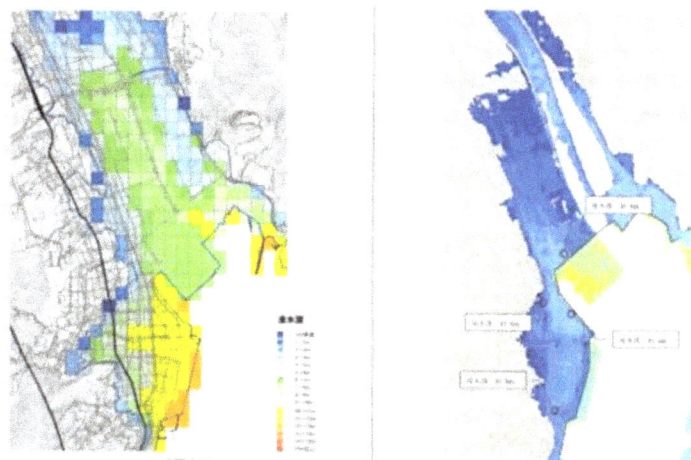

Map 21.1 Tsunami simulations
Source: Ofunato City.

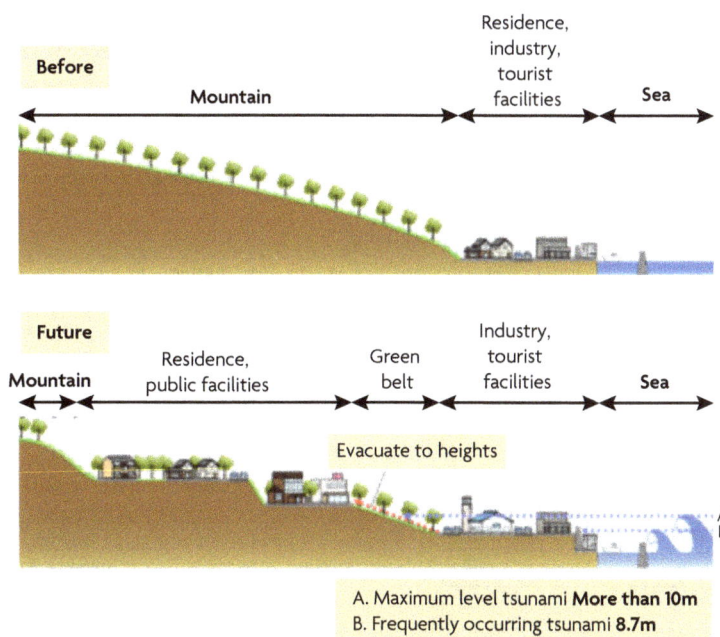

A. Maximum level tsunami **More than 10m**
B. Frequently occurring tsunami **8.7m**

Figure 21.4 Recovery concept of Minamisanriku Town
Source: Minamisanriku Town.

located adjacent to the coast were severely affected by the tsunami and had to be relocated. However, residents wanted to live close to their original location and to the fishing port to maintain their livelihoods. A policy of *separate relocation* was therefore proposed, whereby each village would move to a small hillside space close to its original location (box 21.2).

BOX 21.2

Land-use and population relocation strategies

There are generally three land-use strategies to address tsunami events (see upper figure): (1) avoiding risk, (2) separating risk, and (3) controlling risk. In the **risk avoidance** strategy, residential uses are prohibited or restricted in high-risk areas, although nonresidential purposes (for example, recreational) may be allowed. This strategy is being considered in several municipalities in Tohoku and has been adopted within 20 kilometers of the nuclear power station in Fukushima. It requires a relocation plan and identification and planning for the relocated infrastructure and population at the new site.

In a **risk separation strategy**, some areas are restricted, some are elevated, and others are used to divert the tsunami to controlled directions. The **risk control strategy** uses multiple defenses (such as elevated areas/infrastructure, seawalls, and levees). This

type of strategy was adopted in Otsuchi Town in Iwate and is proposed for parts of Sendai. It requires knowing the optimal height and location of multiple defenses.

Population relocation can also follow different strategies (lower figure). In a **separate relocation** plan, each community is relocated separately to a higher location. In a **collective relocation**, separate (original) communities are relocated to a common (safer) area. A third **combination** strategy uses variants of the above.

In the wide coastal plains, such as near Sendai, the city government adopted a *controlled risk* strategy, whereby house rebuilding would be restricted in areas where water levels could rise above 2 meters. The government also intends to raise the height of the roads to act as breakwaters, as well as use green belts.

Type	Avoid risk	Separate risk	Control risk
Aim (in the event of a huge tsunami)	To protect lives and resources	To protect lives and protect many resources	To protect lives and prevent catastrophic damage to resources
Image	Relocation to higher altitudes	Regrading/relocation to higher altitudes	Regrading/relocation to higher altitudes

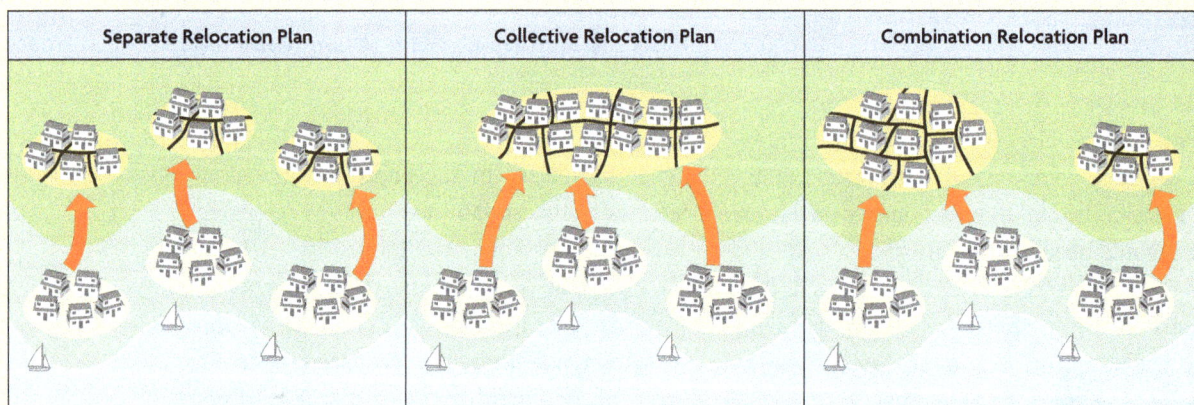

Sources: Siembieda, Chen, and Maki 2011; and Minamisanriku Town.

Residents plan to establish community development associations to facilitate relocation planning.

Population movements

According to government statistics, a large number of people moved out of the affected municipalities following the disaster. The gap between out-migrants and in-migrants relative to the total population in 2011 was particularly high for coastal municipalities—9.4 percent in Minamisanriku, 8.9 percent in Yamamoto, and 8.5 percent in Otsuchi. That gap was also large among young people (less than 15 years old)—up to 14.6 percent in Minamisanriku and 13.2 percent in Onagawa, further raising concerns about the aging population. In Minamisanriku, some residents gave up rebuilding altogether due to lack of funds and planned to either leave town or move to public housing (figure 21.5 and map 21.2).

By contrast, Sendai City experienced a net population inflow (6,633 in 2011). Urbanization in Sendai has therefore accelerated and the population gaps between urban and rural areas are widening. Thus, preexisting trends of aging and declining populations in rural areas and small towns have been exacerbated since the disaster and must be taken into account in the reconstruction planning.

Reconstruction (2012–20)

On February 10, 2012, 11 months after the tsunami, the Japanese cabinet established a national Reconstruction Agency for a period of 10 years (figure 21.6). The agency—headed by the prime minister—aims to promote and coordinate reconstruction policies and measures, and support affected local governments in the Tohoku region. It will serve as a "one-stop

Households

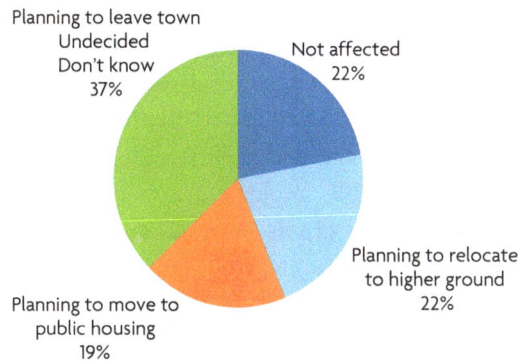

Companies

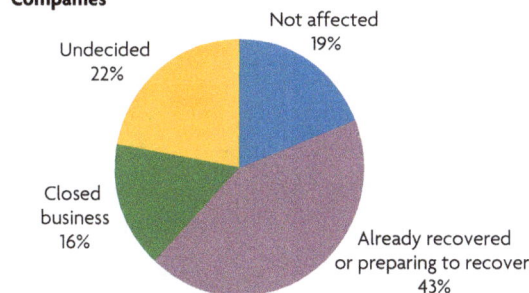

Figure 21.5 Population decrease in disaster areas and survey of population and businesses in Minamisanriku (December 2011)

Source: Ofunato City.

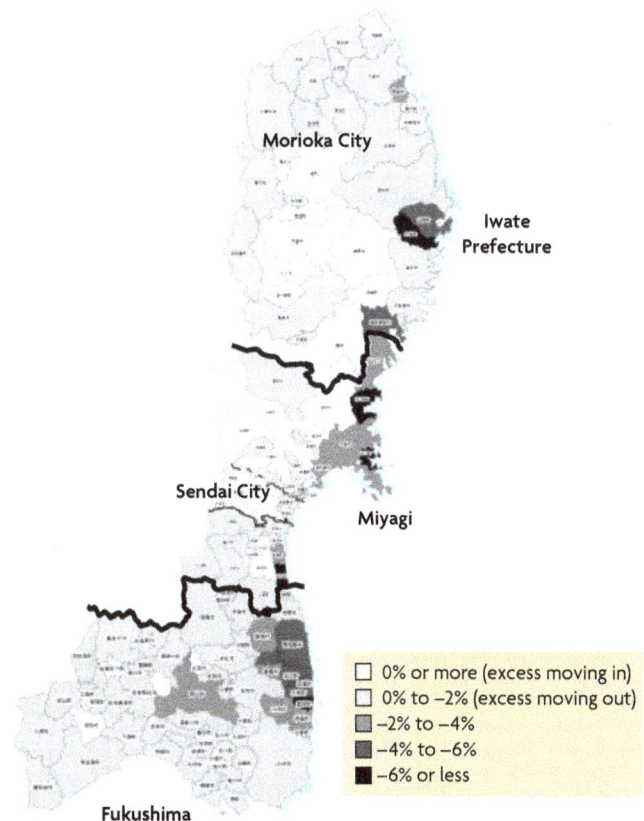

Map 21.2 Gap between people moving in and people moving out as a share of the population

Sources: Statistics Bureau, Ministry of Internal Affairs and Communications, and Minamisanriku Town.

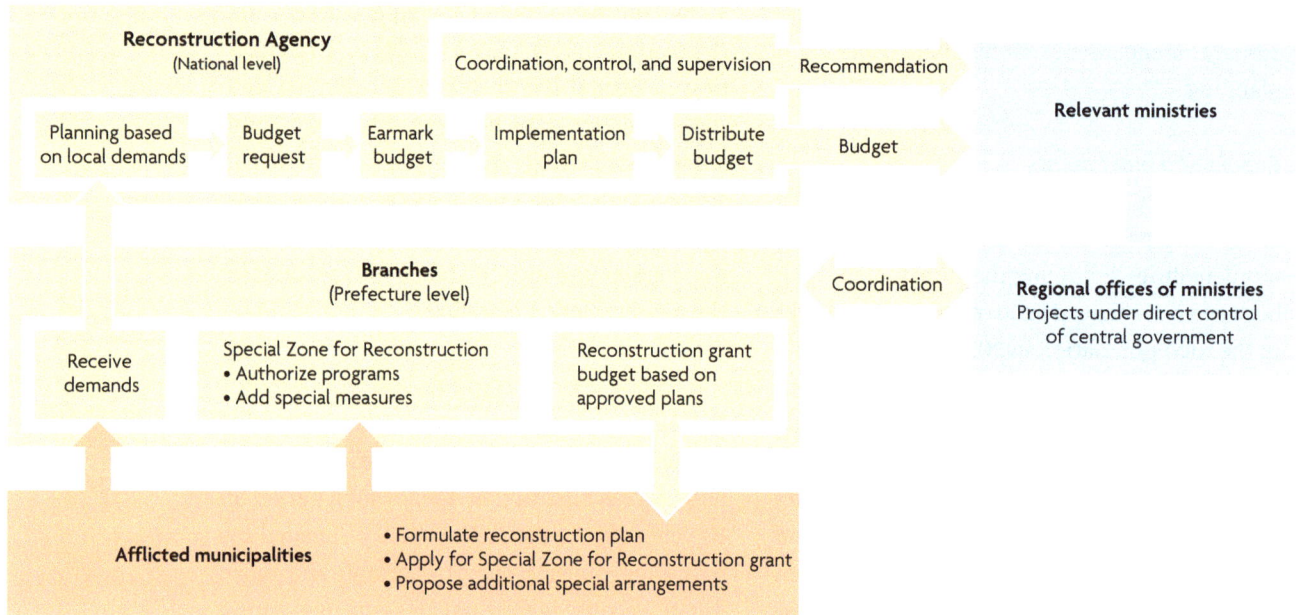

Figure 21.6 Coordination framework for the Reconstruction Agency in Japan

Source: Reconstruction Agency.

shop" for local authorities. Although it is based in Tokyo, it includes three regional branches in the most-affected prefectures (Iwate, Miyagi, and Fukushima).

As envisaged under the *Basic Guidelines,* the government also created a Special Zone for Reconstruction, benefiting 222 municipalities in the disaster-afflicted zones. These municipalities were allowed to submit specific reconstruction plans and apply to the government for funding, as well as a package of special arrangements—such as concessions for land-use planning, creation of new systems related to land use, tax incentives, and special deregulation and facilitated procedures for housing, industry, and services. This strategy supports flexible implementation over time. Reconstruction grants and plans for special measures are submitted to the prime minister, whereas special arrangements for land use are subject to public hearings and inspections.

The process of reaching an agreement on detailed project plans has just begun in most municipalities. In Minamisanriku, for example, total reconstruction costs are estimated at a few hundred billion yen, a vast sum compared to the annual budget of the town (¥8 billion a year). Two projects are being proposed: a land readjustment project for recovery and a group relocation project (map 21.3). An application for a Special Zone for Reconstruction will also be submitted to the central government to relax regulations and attract businesses. Implementation capacity remains a worry, however, as

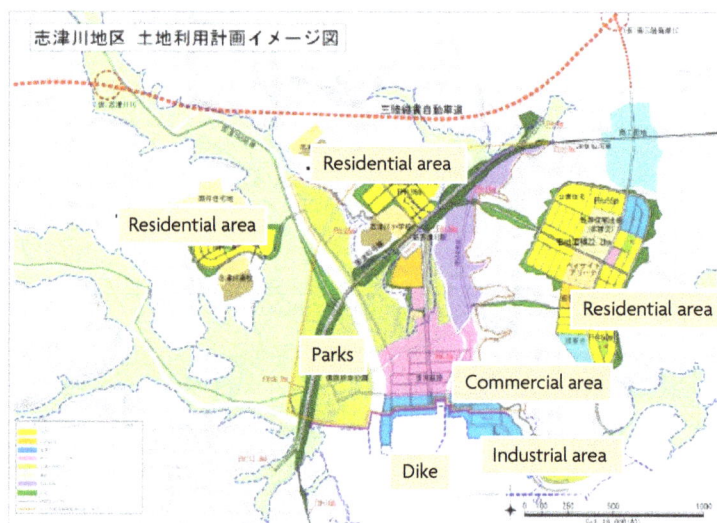

Map 21.3 Land-use planning and projects in Minamisanriku

Source: Minamisanriku Town.

40 out of the 170 town officials (administrative posts) died or went missing during the disaster.

The creation of the Reconstruction Agency and the Special Zone for Reconstruction are designed to respond to reconstruction timelines and facilitate a high number of reconstruction projects at increased speed. They represent a major step forward; after the Great Hanshin-Awaji Earthquake (Kobe earthquake) in 1995, a Reconstruction Agency and Special Zone were not put in place. But it remains to be seen how these new systems will be able to coordinate the various recovery plans, turn them into effective projects, and—significantly—overcome a highly sectoral government structure. Already there are indications that prefectures and municipalities are finding ways to bypass the structures and access funds directly. To succeed, the system must be able to adapt and adjust.

Similarly, it remains to be seen whether the innovative policy of the Special Zones for Reconstruction will be able to help slow or reverse preexisting economic and demographic trends, such as struggling industries and declining and aging rural populations in the affected areas.

LESSONS

- *To be effective, recovery planning and policies must be based upon local conditions and culture.* As such, the highly participatory recovery-planning process followed in Tohoku has proven to be a solid model for megadisaster recovery.

- *A role for independent institutions.* In disasters of this magnitude, a well-respected and independent advisory council can play a key role in setting the blueprint for the recovery.

- *New reconstruction agencies are needed when a disaster compromises institutional functions.* Even though municipalities were responsible for disaster response, they became effectively dysfunctional in the aftermath of the disaster due to the destruction of their offices and the large numbers of dead or missing (a situation also faced in Haiti). Such destruction is one of the main factors slowing recovery. Furthermore, the implementation of a large number of projects and the outpouring of volunteer support posed a significant burden for smaller municipalities, where financial and human resources are constrained, even at the best of times. This has been one of the principal justifications for the establishment of the Reconstruction Agency.

- *The large scale and diversity of the recovery make information and communication management more challenging* and more critical to a successful recovery. Systematic information on victims, for example, was a challenge for many smaller municipalities who lost both records and staff. As a result, prefectures have begun to centralize such information for use by local governments.

- *Support from experts contracted by the central government for damage assessment and logistics.* The affected municipalities also benefited from the support of expert consultants contracted by the central government, who had the expertise to quickly carry out damage and needs assessments and provide logistical support. Damage assessments were completed quickly, as the central government relied on private engineering companies who had readily available information on infrastructure replacement costs.

- *Twinning arrangements with local governments.* Similar to the provincial pairing system employed in China after the Sichuan Earthquake of 2008, and to staff secondments following the Nargis cyclone in Myanmar, twinning arrangements with local governments outside the disaster-affected areas proved very effective for prefectures and municipalities facing a shortage of expertise and manpower (chapter 17).

- *While recovery projects may secure the safety of residents' lives, they will be costly.* The population of most disaster-affected areas is sharply decreasing, and it will be a challenge to balance the needs of aging survivors with long-term financial efficiency.

- *Pre-disaster recovery plans are useful.* The design of new residential areas could have been facilitated had a predisaster recovery plan been in place to preselect suitable areas. Taking into consideration the likelihood of large-scale disasters in Japan, enactment of new legislation should be considered to not only facilitate postdisaster response, but also predisaster recovery planning.

RECOMMENDATIONS FOR DEVELOPING COUNTRIES

Involve community members in planning. Megadisasters in developing countries often involve a multiplicity of humanitarian agencies, donors, and NGOs. As such, it is even more critical to develop, early on, a shared vision for recovery and reconstruction that recognizes local cultural and life values and is perceived as legitimate by key stakeholders. Failure to do so can result in a proliferation of external-driven plans and strategies, as seen recently in Haiti.

Make recovery plans before disasters strike. Predisaster planning can help promote a more resilient recovery. This was the case following the 1995 Bangladesh floods, where the response benefited considerably from the level of disaster preparedness introduced after the 1985 floods. In Gujarat, by contrast, a lack of proactive planning despite past disasters hampered recovery efforts following the 2001 earthquake.

Balance central and local control of resources. Every megadisaster is different, and the necessity for a dedicated reconstruction

agency depends on postdisaster governance and coordination capacity. The Agency for the Rehabilitation and Reconstruction of Aceh and Nias (BRR), established 3.5 months after the tsunami, was generally effective largely due to a strong mandate, national commitment, and external financial support. Concerns about slow recovery, however, led the BRR to take over implementation responsibilities, posing a potential conflict of interest with its oversight function. In later years, the BRR progressively devolved implementation to local governments. Another example of an agency with both coordination and operational functions (albeit not in a developing country) was the Victorian Brushfire Recovery and Reconstruction Authority established after the 2009 brushfires in Australia. Using a successful model based on people, economy, environment, and reconstruction, the authority completed its mandate in 30 months. In other disaster contexts, however, a hybrid model may be more appropriate, where a centralized agency coordinates reconstruction, but implementation capacity continues to be delegated to government agencies.

Integrate many viewpoints into recovery plans. In general, recovery planning is most effective when it uses participatory methods and directly integrates the views of experts with those of affected people. Response to numerous megadisasters (for example, the GEJE, 2006 Yogyakarta earthquake, and 2010 Pakistan floods) attest to the merits of this approach. Community members' participation in planning workshops should be arranged. Also, community leaders should be assigned as members of planning committees. The 2008 Wenchuan earthquake provides an alternative model, where centralized, top-down planning led to rapid reconstruction. At the same time, there was a weak focus on local capacity building and community preparedness, issues that could hamper future disaster response.

Use recovery to improve spatial planning in general. Governments in developing countries have a very narrow window of opportunity to decide whether to rebuild in situ or relocate populations to safer areas. The government of Thailand, for example, considered seriously whether to relocate parts of the capital to higher grounds following the 2011 floods, but this opportunity was quickly lost due to social and political pressures. While moving entire cities has proven historically difficult to achieve, megadisasters can still provide opportunities to improve spatial planning—as demonstrated after the 2011 tsunami in Samoa, when affected coastal communities agreed to relocate further inland.

Relocation may be needed to preserve public safety, but it often removes people from their sources of livelihood. In a disaster response, both *safety* and *livelihood* have to be well balanced, and nowhere is this delicate balance more difficult than in developing countries. In such countries, affected people are often poor and marginalized, having settled in unsafe areas often because they offer the only land available. When disaster strikes, land speculation and security problems are often rampant; residents quickly rebuild in their original neighborhoods out of fear that someone else may move in. As house insurance markets tend to be nonexistent, governments are left with very few instruments to promote relocation: they can resettle people involuntarily (which is seldom successful), or they can promote voluntary relocation by investing in alternative "growth centers" (for example, by building social infrastructure in safer areas). Often, relocating people as close as possible to their original homes and livelihood sources proves to be the most sustainable solution.

Open and transparent information sharing is a key prerequisite to successful planning. This can be a major constraint in developing countries, where information on key issues such as land tenure and historical exposure tends to be scarce or inaccessible. Since Haiti, development partners working in megadisasters have promoted the use of crowdsourcing and other open data platforms, often with great success. The challenge now is to mainstream such processes effectively into local planning, so that they can provide vulnerable people with a greater voice in mitigating future disasters. The processes should be formulated considering local conditions, since relationships between governments and civil societies vary from country to country.

NOTE

Prepared by International Recovery Platform; Yasuo Tanaka, Yoshimitsu Shiozaki, and Akihiko Hokugo, Kobe University; and Sofia Bettencourt, World Bank.

BIBLIOGRAPHY

Beck, T. 2005. "Lessons Learned from Disaster Recovery: The Case of Bangladesh." Disaster Risk Management Working Paper Series 11, World Bank, Washington, DC.

Ge, Y., Y. Gu, and W. Deng. 2010. "Evaluating China's National Post-Disaster Plans: The 2008 Wenchuan Earthquake's Recovery and Reconstruction Planning." *International Journal of Disaster Risk Science* 1 (2): 17–27.

GFDRR (Global Facility for Disaster Reduction and Recovery). 2010. "Haiti Earthquake Reconstruction—Knowledge Notes from the DRM Global Expert Team for the Government of Haiti," World Bank, Washington, DC.

GOJ (Government of Japan). 2011a. "Basic Guidelines for Reconstruction, June 2011." Reconstruction Headquarters in Response to the GEJE. http://www.reconstruction.go.jp/english/topics/2012/12/basic-act.html.

——. 2011b. *Basic Act on Reconstruction in Response to the Great East Japan Earthquake.* June 24. http://www.reconstruction.go.jp/english/topics/Basic%20Act%20on%20Reconstruction.pdf.

IRP (International Recovery Platform). http://www.recoveryplatform.org/.

Ramalingam, B., and S. Pavanelio. 2008. "Cyclone Nargis: Lessons for Operational Agencies." Active Learning Network for Accountability and Performance in Humanitarian Action. http://www.alnap.org/.

Reconstruction Agency Web site. http://www.reconstruction.go.jp/english/.

RDC (Reconstruction Design Council). 2011a. "Seven Principles for the Reconstruction Framework." Resolution of the Reconstruction Design Council in Response to the Great East Japan Earthquake, May 10, 2011.

———. 2011b. *Towards Reconstruction: "Hope Beyond the Disaster."* Report to the Prime Minister of the Reconstruction Design Council in response to the Great East Japan Earthquake, June 25, 2011.

Siembieda, W., H. Chen, and N. Maki. 2011. *Multi-Location Disaster: Shaping Recovery in the Great East Japan Earthquake and Tsunami of March.* International Association for China Planning (IACP) Conference, Wuhan, June 17–19.

Silva, J. 2010. *Lessons from Aceh—Key Considerations in Post Disaster Reconstruction.* Practical Action Publishing.

Shiozaki , Y., Y. Tanaka, and A. Hokugo. 2012. "Reconstruction Policy and Planning." Presentation.

Tokyo Metropolitan Government Disaster Prevention. http://www.bousai.metro.tokyo.jp/english/index .html.

World Bank. 2012. "Current State of Reconstruction and the Way Forward." Presentation at the Workshop on Reconstruction, January 18, Kobe, Japan.

CHAPTER 22

Transitional Shelter

Transitional shelter can play a crucial role in housing reconstruction following a megadisaster. Reconstruction of permanent housing cannot move forward until a number of complex issues are settled, such as relocation planning and removal of debris. Even after plans are agreed on and reconstruction begins, it may take several years for permanent housing to be completed. In this context, affected people may need to rely on transitional shelter for extended periods of time, and this will have a significant effect not only on their housing, but also on their overall recovery, including livelihood rehabilitation.

FINDINGS

The Great East Japan Earthquake (GEJE) led to the total collapse of some 108,000 residential houses. An additional 117,000 houses suffered damage to more than half of their structure (chapter 2). As a result, more than 450,000 people had to be evacuated to evacuation centers. Within four months of the disaster, 75 percent of the centers had closed, as people were moved gradually to transitional shelters (chapter 18).

Lessons learned from the Great Hanshin-Awaji Earthquake (Kobe earthquake) of 1995 and other disasters led the Japanese government to promote the concept of networked relocation following the GEJE, when an attempt was made to preserve, to the extent possible, existing social networks. The government also offered multiple options for transitional shelter, depending on geography, reconstruction planning, and local preferences. These included temporary housing, mostly prefabricated; government-owned accommodations and public housing; and private rental apartments, which proved popular due to lower prices, higher comfort, and greater versatility. Local governments, volunteers, and nongovernmental organizations (NGOs) provided complementary support, including counseling. As

Figure 22.1 The housing recovery process in Japan

relocation into transitional shelters proceeded, several innovations were introduced, including physical upgrades to improve comfort, wooden housing (easier to convert into permanent use), and multiple-story accommodations. Key challenges have been the lack of sufficient land due to the volume of remaining debris, as well as logistical difficulties in keeping track of disaster survivors to ensure ongoing support. This note discusses the GEJE experience and offers lessons learned with application to developing countries.

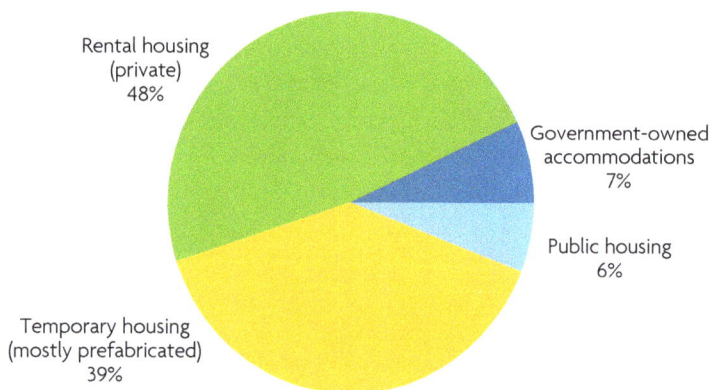

Shelter type	Number of houses allocated or chosen	Number of houses supplied
Temporary housing (mostly prefabricated)	52,182	52,620
Government-owned accommodations	9,832	38,464
Public housing	8,238	24,505
Private rental housing	65,692	—
Total	135,944	115,589

Figure 22.2 Characteristics of transitional shelters used after the GEJE (as of December 27, 2011)

Source: Reconstruction Agency.

Note: — = not available.

The Japanese framework for transitional shelter

Prefectural governments are responsible for transitional shelter according to the provisions of the Japanese Disaster Relief Act (1947), with funds allocated from the central government. The prefecture, outside of exceptional cases, can choose the type and form of housing as well as hire private construction companies. Municipal governments coordinate with prefectures for the selection of sites, distribution of affected people, and maintenance of shelters. Affected people are expected to move into permanent accommodations within a period of two years (the time normally allowed by Japanese law), and at their own cost, although they receive up to ¥3 million (approximately $37,500) in compensation from the government, depending on the housing damage (chapter 34). Alternatively, they can rent public housing at subsidized rates. The usual flow of the housing reconstruction process is shown in figure 22.1.

Basic types of transitional shelters used after the GEJE

The government adopted three main programs of transitional shelters in the aftermath of the GEJE (figure 22.2):

- Newly constructed temporary housing (mostly prefabricated by private contractors)

- Private rental apartments

- Existing public housing and government-owned accommodations (previously built to house government officials)

The type of transitional shelter was influenced by geographic and demographic considerations (map 22.1).

- Temporary housing was commonly used in the ria coastal areas north of Sendai (including part of the Miyagi Prefecture and most of the Iwate Prefecture), where most of the resident houses suffered major destruction. This area is characterized by steep and fjord-like topography, and both small fishing villages and larger towns located near the ocean; there is little available land near the ocean fit for building.

- Private rental apartments predominated in Sendai City and urban areas in the coastal plains, much of it undamaged.

- The towns in Fukushima Prefecture presented a unique case: due to the radiation hazard, residents had to be evacuated for an uncertain length of time. Facing the prospect of having to provide long-term transitional shelter (possibly for many years), the Fukushima Prefecture decided to construct more than 4,000 units of wooden temporary housing, including larger-size units for larger families. As of March 2012, about 60,000 residents had evacuated the Fukushima Prefecture to other prefectures (chapter 36).

Temporary housing

Temporary housing, typically one-story prefabricated row houses built by private companies (29 square meters), is the most common type of transitional shelter used in Japan (figure 22.3). Typical construction costs have ranged from $5.7 million to $6.6 million (approximately $71,000–$80,500 per unit), slightly more than double the price of similar units during the 1995 Kobe earthquake. As of early 2012, some 52,000 housing units have been built.

Many prefectures have preexisting agreements with construction companies to build

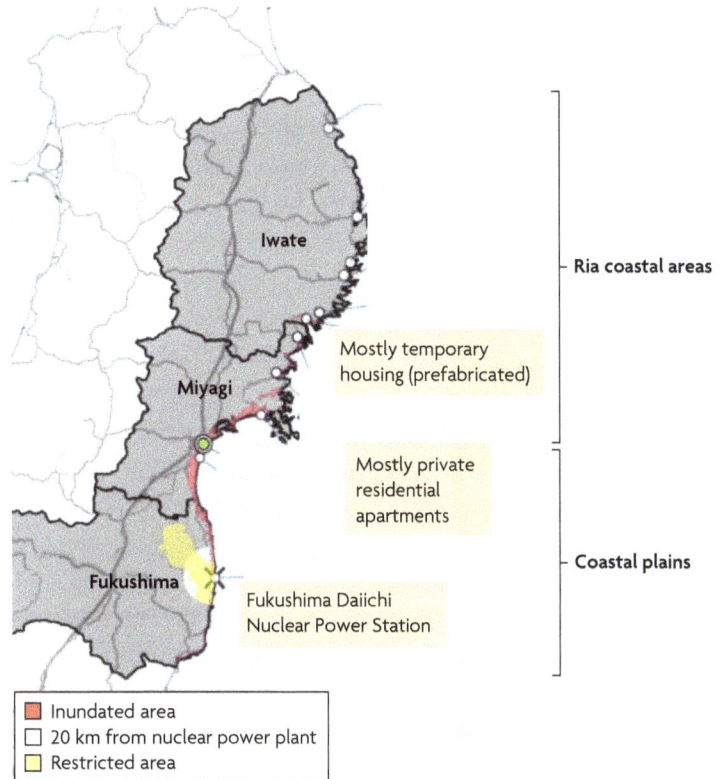

Map 22.1 Predominant transitional shelter in affected areas
Source: Kobe University.

Small group of temporary houses forms a new neighborhood

Temporary houses in Ofunato, Iwate; and Onagawa, Miyagi

Figure 22.3 Typical prefabricated temporary houses
Source: © International Recovery Platform (IRP). Used with permission. Further permission required for reuse.

prefabricated temporary housing during emergencies. But even with these agreements in place, it was not possible for construction companies to build all the units needed immediately, due to shortages of construction materials and workers. Because of such shortages and a lack of coordination across companies, the quality and level of construction of temporary houses varies across the disaster area.

Government policy requires that temporary housing be built on publicly owned land, outside high-risk areas. This posed a significant challenge for much of the disaster area, particularly along the ria coastline north of Sendai, where there was almost no available land—a major reason for the initial delays in the construction of temporary housing. The first residents moved in April/May, one to two months after the disaster (figure 22.4).

In many municipalities, however, a high percentage of temporary housing remained empty, as prospective residents found them inconvenient (too distant from their original villages), uncomfortable, and much smaller than their original houses. The houses were constructed using low-quality, bare-minimum standards, and were not suited to the cold climate of the Tohoku region. Problems included gaps between walls and roofs, drafts,

and the absence of noise or temperature insulation, shelves or storage areas, places to sit outside, an awning or enclosure around the front door, and a veranda outside the sliding door (which made it dangerous for the elderly hanging laundry, or small children). Moreover, as allocations were determined by lottery, residents complained that they did not know their neighbors and lost their community connections. Some people preferred to stay in evacuation shelters as long as possible because food and utilities were provided (a trend also observed following other megadisasters).

Private rental apartments

Although not widely used during the Kobe earthquake, privately owned rental housing became the preferred form of transitional shelter after the GEJE, with about 66,000 units used by disaster victims. Rents were paid directly by the government. Such apartments were widely used in the urban areas of Tohoku, including Sendai City.

As also observed in Haiti, private rental units offer many advantages over conventional temporary houses: they are considerably cheaper—about ¥0.7 million–¥1.5 million ($9,000–$18,000) per year per unit or for a two-year average stay, which makes them two to three times less costly than temporary housing. They also allow affected people to move into transitional shelters quickly (people started moving in less than a month after the disaster, compared to one to two months for the prefabricated units). In addition, regular apartments are considered more comfortable and livable for residents.

Nonetheless, private rental apartments are not a viable option for areas that suffer extensive destruction of existing housing stock. In addition, the fact that affected residents are scattered across existing housing units makes it difficult for government and relief workers to track them to provide the necessary information and support. It also makes disaster

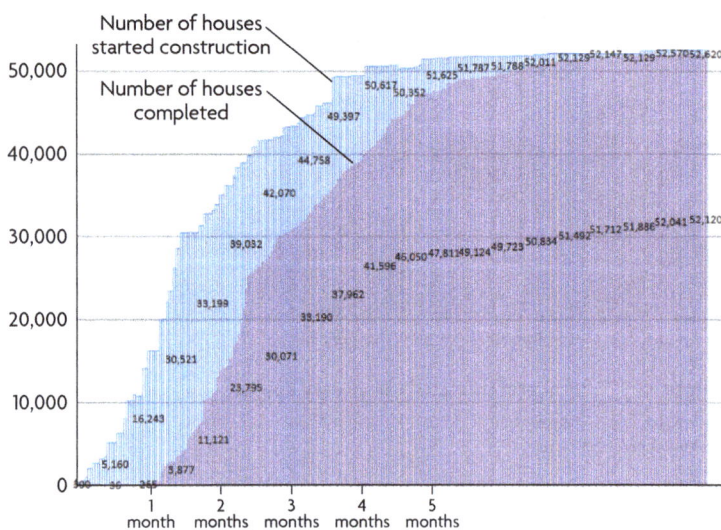

Figure 22.4 Number of temporary houses completed

Source: Ministry of Land, Infrastructure, Transport and Tourism (MLIT).

survivors more prone to losing social connections than when they are grouped together in conventional temporary housing.

Public housing and government-owned accommodations

Some disaster survivors moved into public housing managed by government entities, as well as into other government-owned residential facilities. Public housing shares many of the positive features of private rental housing, although it can also lead to residents' isolation, with limited access to the information and social networks found in the more aggregated temporary housing.

Support systems

Community building and emotional care

Throughout the disaster region, local governments, volunteers, and NGOs started numerous support initiatives to help disaster victims at transitional shelters. These included both physical (provision of furniture, building of additions or improvements, provision of community spaces, buses) and nonphysical support (social events, counseling, health checks, visits, shopping and support for elderly and children).

One example is the Disaster Victims Support Center, started by the town government of Minamisanriku (Miyagi Prefecture) through the National Government Emergency Employment Fund. The center hired about 100 disaster victims to visit other affected people in temporary shelters, counsel them, and provide support to the most vulnerable. It also established one satellite location in each of the four regions of the town to be closer to the temporary housing residents. This initiative built upon the earlier example of the community centers established in the aftermath of the Kobe earthquake (box 22.1).

The Japanese Red Cross Society provided six electric household appliances (televisions, refrigerators, washing machines, cooking pots, microwave ovens, and hot water pots) to those

---

BOX 22.1

The case of community centers at transitional shelter sites after the Kobe earthquake

A total of 232 community centers were opened as bases to support residents, established by an association of local organizations:

- Volunteers and nonprofit organizations manage the centers.
- Life support advisors visit each house to confirm safety and provide advice.
- Events and gatherings are held by volunteers to promote communication among residents.
- Establishment of community-based organizations is supported.

Source: © IRP. Used with permission. Further permission required for reuse.

---

families who moved to new but empty prefabricated houses and apartments. By June 2012 the number of beneficiary families reached over 130,000 throughout Japan, from Okinawa to Hokkaido, including those families displaced by the Fukushima nuclear accident.

Transportation

One of the key difficulties faced by residents of transitional shelters is the distance from work, schools, hospitals, and shopping. Providing adequate transportation to support these residents is therefore an important challenge.

Livelihood support

Many support groups have started projects to assist residents of transitional shelters in generating side incomes. Examples include the friendship bracelet "Tamaki" produced by wives of fishermen, and hammocks produced by fishermen (both from fishing nets). Other women's groups have started making and selling products such as key chains, fabric bags, and slippers. The link between transitional shelter and livelihoods has proven important not only to help improve the socioeconomic status of affected people, but also their psychological recovery (see chapter 24).

The evolution of transitional shelters following the GEJE

Networked (group) relocation

Given the shortage of publicly available land in disaster-stricken areas, the government allowed some temporary housing units to be built on privately owned land.

Lessons were also learned from Kobe. Many elderly residents had died a "solitary death" (*kodokushi*) after being separated from their social networks by lottery systems that dispersed them into transitional shelters. In the GEJE, a lottery system was also used during the initial stages of the recovery as the number of temporary houses were much fewer than the number of affected people wanting to move out of the emergency shelters. In Minamisanriku Town, for example, some 62 percent of the temporary shelters followed the lottery system.

As more temporary houses became available, municipalities made an effort to support

Figure 22.5 Improvements to temporary housing—adding insulation to the walls and double-pane windows

Source: © IRP. Used with permission. Further permission required for reuse.

community building and design group housing units that encouraged interaction between neighbors. In Minamisanriku, therefore, two models of temporary group housing were adopted: large group sites built on public land (schools or athletic facilities) and smaller group sites built on private land. On the larger group sites (built earlier), prospective residents were chosen by lottery, which prioritized senior citizens, families with small children, and other vulnerable residents. Affected people were also given the choice to go to a large group site sooner, or wait a little longer and be relocated collectively into one of the smaller group sites, closer to their former neighborhoods. Smaller group sites were built specifically to support collective group relocation from nearby neighborhoods, to keep affected communities relatively intact.

Physical improvements

The close network of support to affected people enabled local governments and NGOs to do some improvements to the poor physical condition of the temporary housing units by adding awnings, balconies or verandas, and insulation or soundproof materials and by providing benches, shelves, and other indoor furniture (figure 22.5). But problems of basic construction persisted over the entire disaster area, and it was very difficult to improve the situation for all residents.

Multiple-story temporary housing made from stacked containers was introduced in Onagawa Town to compensate for the scarcity of available land. Stacking the containers to form two- and three-story group temporary housing also helped reduce overall construction time (figure 22.6).

Wooden temporary housing has been used extensively in Fukushima Prefecture, where long-term, temporary residency is required, as well as in Sumita Town, Rikuzentakata City, and Tono City. The main advantage is that it can be used for longer periods than the prefabricated houses, and can potentially be converted and reused for the construction of permanent housing. It is also more comfortable and warmer and has the advantage of being disposable. But it is not as standardized as the prefabricated type and cannot easily be produced in large quantities offsite. In addition, in megadisasters such as that in Aceh, the extensive use of wood resources has contributed to deforestation of already fragile environments.

Temporary to permanent housing

In common with other megadisasters (for example, those in Haiti, Aceh and Yogyakarta in Indonesia, and Chuetsu and Kobe in Japan), it is expected that owner-built transitional shelter will start to emerge. Like wooden temporary housing, it can be reusable and converted to permanent use.

In the 2006 Central Java earthquake in Yogyakarta, the government promoted a "roof first" concept to transitional shelter, allowing residents to incrementally finish the structure. The 2001 Gujarat earthquake in India and the "Katrina Cottages" built following the 2005 Hurricane Katrina (United States) provide further examples where materials and/or semipermanent structures were provided to residents to gradually rebuild their homes (box 22.2). This process, however, needs to be carefully monitored to ensure that residents rebuild according to safer standards and do not settle on disputed land.

A relatively unanticipated challenge to the general recovery and reconstruction has been the vast quantity of debris left by the tsunami. Collecting and disposing of such a large amount of debris requires time, large spaces, and resources—impeding other aspects of recovery (chapter 23).

LESSONS

- As discussed in this note, the GEJE experience demonstrates the *importance of providing multiple options for transitional*

Figure 22.6 Multiple-story temporary housing made with stacked containers

Source: © IRP. Used with permission. Further permission required for reuse.

BOX 22.2

International examples of creative, temporary-to-permanent housing

The "roof first" concept of temporary shelter was adopted in Yogyakarta following the Central Java earthquake (2006). It prioritized putting a roof over the heads of residents, who could then incrementally finish the structure. For permanent housing recovery, a core house was used to provide a structurally safe permanent shelter as soon as possible for a large number of beneficiaries, who could then expand their housing incrementally over time.

Source: © International Federation of Red Cross and Red Crescent Societies (IFRC). Used with permission. Further permission required for reuse.

shelter. It also shows the importance of allowing local governments and affected communities to have a voice in the location, type, and services provided. This leads to flexible housing solutions that better match the needs of residents. Table 22.1 summarizes some of the advantages and disadvantages of the various types of transitional shelter, based on the GEJE as well as international experience.

- *The design of transitional shelters was built upon experiences with past disaster recovery in Japan.* In Kobe a great deal of temporary housing was constructed far from the city center and former neighborhoods, with residency determined by a lottery system. These conditions exacerbated the feeling of loss for affected people, and there were many cases of "solitary deaths" (*kodokushi*), where no one even knew that the individual had passed away. The GEJE model tried to prevent this to a certain extent by promoting group relocation and preservation of improved social networks.

- *Community-based organizations support evacuees in transitional shelters.* Community-based organizations (such as *jichikai*) and support groups can play important roles in assisting affected people to understand and resolve issues by themselves during their stay at transitional shelters.

- *Transitional shelters must be designed with efficiency and sustainability in mind.* The design of transitional shelters should be better from the start to promote efficient recovery—for example, by taking into consideration climate conditions and transportation and livelihood needs. It is also important to consider the special needs of vulnerable groups—including the elderly, children, and disabled. Transitional shelters need to be accessible to them, and complementary care services planned and provided. To facilitate this, local governments in highly vulnerable areas should select a suitable construction site for temporary housing and coordinate the works and services needed *before* a disaster

Table 22.1 Advantages and disadvantages of various types of transitional shelter

TYPE	ADVANTAGES	DISADVANTAGES
Temporary housing (prefabricated)	• Standard specifications • Can be built in large quantities offsite • Easy to keep track of relocated people • Can be used for collective relocation (preserving social networks)	• Requires available, safe, and undisputed land • Slower relocation than rental units (needs to be constructed) • Low quality and lack of comfort • Often built in inconvenient locations, far from original homes • If use is prolonged, risks degrading to a slum
Temporary housing (owner built)	• Can evolve to permanent housing • Flexibility in location, materials, style	• Requires available, safe, and undisputed land • Principles of "building back better" (or in nonrisk areas) may not be followed
Private rental housing	• Cheaper • Fast relocation (already constructed) • Flexibility and comfort	• May not exist in affected areas • Difficult to keep track of and provide services for relocated people, who are more scattered • Can reinforce social isolation
Public and government-owned housing	• Cheaper • Fast relocation • Comfort	• Can reinforce social isolation • More difficult to preserve social networks and provide services than temporary housing

occurs. Neighborhood groups should also be trained in network relocation.

- *A better information database of disaster survivors is necessary in order to provide suitable support to the affected population.* For example, such data can help in the planning of how many houses to build as affected people move out of the area into surrounding cities, as well as help forecast demographic changes over the long term. This information is also critical for more efficient and economic reconstruction planning.

RECOMMENDATIONS FOR DEVELOPING COUNTRIES

The timeline and costs of transitional shelters must be considered carefully. In developing countries, affected people often start rebuilding their homes immediately after a disaster, and often according to poor safety standards. As such, transitional shelters may not be needed for long periods (as was the case during the 2010 Pakistan floods), and resources should be shifted toward permanent reconstruction.

Long periods in transitional shelters may also make it more difficult for beneficiaries to move to permanent housing (such as in the Marmara earthquake, Turkey) and encourage the growth of slums or ghettos.

In general, megadisasters in developing countries require transitional shelters that are upgradable, reusable, and recyclable, allowing shelter materials to be gradually used for permanent housing. Salvageable materials from debris can often be used to build or complement shelters, and their salvage can be a temporary boost to local livelihoods.

Owner-built shelters or units built with strong beneficiary participation are often best (for example, 2001 Gujarat, 2006 and 2008 Yogyakarta, and 2010 Haiti), but care must be taken to oversee the quality of the construction or provide incentives for better standards (such as conditional cash transfers). Cash or voucher programs, such as those used in Haiti

(2010) and Wenchuan (2008), can promote flexible solutions and allow families to pool resources and rebuild together.

Transitional shelters must be planned together with strategies supporting daily life (shopping, health care, social life, schools, infrastructure, psychosocial support) as well as livelihoods. To the extent possible, affected people themselves should participate actively in these services, helping rebuild a sense of community and a quick return to normalcy.

The location of temporary housing is particularly important, especially where land is scarce. Sites with uncertain tenure should be consistently avoided. The preparation of a "land bank"—preselected areas that can be quickly converted to be used as transitional shelters or permanent relocation—should therefore be a critical component of any predisaster contingency plan in highly vulnerable areas. In places where public land is scarce, this may require that the government prenegotiate the use of the land with private landowners to prevent subsequent land speculation.

To the extent possible, the distance between transitional shelters and former homes should be minimized to allow displaced people to maintain social networks and livelihoods, and protect their land and property.

Community cohesiveness should be ensured by providing timing and site options for temporary shelter. This, however, requires high levels of government capacity and costs, and could slow down shelter transitions. Community members should provide one another mutual help.

A systematic communication and monitoring strategy is critical to avoid harmful rumors, keep affected people informed, and allow for beneficiary feedback.

Governments have an important role to play. Civil society and the private sector may not be robust and resilient enough to face the disaster, and may not have the necessary relations with their governments in some countries. In these countries, government initiatives are crucial.

NOTE

Prepared by International Recovery Platform; Yoshi-mitsu Shiozaki, Yasuo Tanaka, and Akihiko Hokugo, Kobe University; and Sofia Bettencourt, World Bank.

BIBLIOGRAPHY

CRS (Catholic Relief Services). 2012. "Learning from the Urban Transitional Shelter Response in Haiti." Catholic Relief Services, Baltimore. http://www.crsprogramquality.org/storage/pubs/emergencies/haiti_shelter_response.pdf.

Dercon, B., and M. Kusumawijaya. 2007. "Two Years of Settlement Recovery in Aceh and Nias. What Should the Planners Have Learned?" 43rd ISO-CARP (International Society of City and Regional Planners) Congress, September 19–23, 2007.

Frederica, L., J. Reed, and H. Gloor. Undated. *Transitional Shelter Evaluation in Pakistan.* International Organization for Migration.

GFDRR (Global Facility for Disaster Reduction and Recovery). 2010. *Haiti Earthquake Reconstruction—Knowledge Notes from the DRM Global Expert Team for the Government of Haiti.* http://www.gfdrr.org/gfdrr/node/149.

IFRC (International Federation of Red Cross and Red Crescent Societies). 2010. *Owner-Driven Housing Reconstruction Guidelines.* http://www.ifrc.org/PageFiles/95526/publications/E.02.06.%20ODHR%20Guidelines.pdf.

IRP (International Recovery Platform). *Guidance Notes on Recovery: Shelter.* http://www.recoveryplatform.org/resources/guidance_notes_on_recovery.

Shelter Center. *Shelter after Disaster.* http://www.sheltercentre.org/library/shelter-after-disaster.

———. *Transitional Shelter.* http://www.sheltercentre.org/transitional-shelter.

Trohanis, Z., and G. Read. 2008. "Housing Reconstruction in Urban and Rural Areas. Knowledge Notes." Disaster Risk Management in East Asia and the Pacific Working Paper Series No 9, World Bank, ISDR (United Nations International Strategy for Disaster Reduction), and GFDRR, Washington, DC.

UNDP (United Nations Development Programme). 2006. *Early Recovery Assistance/ERA. Programme for D.I. Yogjayakarta and Central Java—Call for Proposals for Small Grants for NGO/CSO Shelter Activities.* United Nations Development Programme.

CHAPTER 23

Debris Management

Some 20 million tons of waste resulted from the Great East Japan Earthquake. The amount of debris in Iwate Prefecture was 11 times greater than in a normal year, and in Miyagi Prefecture 19 times greater. Appropriate treatment and disposal depends on the type of debris or waste, while recycling should also be considered. Authorities should prepare for disasters by designating temporary storage sites and routes for transporting waste. Japan's existing debris management plans are being revised to include methods for estimating the amount of disaster waste generated by tsunamis and appropriate measures for dealing with it.

FINDINGS

The many causes of disaster

Disasters have a variety of causes including earthquakes, tsunamis, typhoons, floods, and fires. Over the past decade, several major disasters have destroyed social infrastructure all over the world: Sumatra's Andaman earthquake in 2004, Hurricane Katrina in 2005, the Sichuan Earthquake in 2008, and the earthquakes in New Zealand and Turkey in 2011, to name a few. Differences in the nature and geographical extent of the environmental effects, and other waste-related problems that may arise, are dictated by many variables including the cause of the disaster, types of local industry, building densities, and so forth. In other words, big differences exist and it is extremely difficult to generalize.

The amount of disaster waste and its classification

The Great East Japan Earthquake (GEJE) generated large amounts of disaster waste. Japan's Ministry of the Environment estimated 20 million tons of waste as of May 21, 2012. This number is very large even when compared with the 15 million tons from the Great Hanshin-Awaji Earthquake (Kobe earthquake), the 20 million tons from the 2008 Sichuan earthquake, or the

10 million cubic meters (m³) found in Indonesia alone following the 2004 Indian Ocean tsunami (Brown, Milke, and Seville 2011).

Estimates for the Kobe earthquake in 1995, based on the unit waste generation intensity for totally destroyed structures, were 61.9 tons/household and 113 tons/building. Although there are few reports on the per-unit-floor-space amount, one value reported for the Kobe earthquake was 0.62–0.85 tons/square meter (m²), and a more recent review put it in the range of 0.20–1.44 tons/m² (Takatsuki, Sakai, and Mizutani 1995).

Tsunami sediment deposits and their properties

Tsunami sediment deposits consist mainly of sand, mud, and other bottom material, but their properties and compositions vary widely. Some examples of deposits causing concern are those mixed with the ruins of homes crushed by tsunamis, those containing oils, and those that release offensive odors or dust due to putrefaction or drying. Deposits may also be mixed with substances such as pesticides, acids, alkalis, and other hazardous chemicals from industries in the disaster-stricken areas. Doing nothing about such substances raises public health concerns. The tsunami from this earthquake left heavy deposits. To estimate the amount, we multiplied the tsunami-inundated area by the average thickness of the deposits and a volume-to-weight conversion factor, and obtained a total estimated 11,990,000–19,200,000 m³ and 13,190,000–28,020,000 tons for the six disaster-stricken prefectures of Aomori, Iwate, Miyagi, Fukushima, Ibaraki, and Chiba (JSMCWM 2011). The deposit height is between 2.5 and 4 centimeters.

The gist of the chemical analysis results is as follows. Ignition loss (600°C, 3 hours) had a spread of 1.2 percent to 16.3 percent, and there were some samples influenced by the organic matter and oils in the seabed mud. Hexane extracts exceeded 0.1 percent in a number of samples, and on the high end oily mud was at 9.8 percent. While tests for heavy metals did not detect much, lead was detected in many samples in the milligram per kilogram (mg/kg) range. Leaching amounts of heavy metals (using a method based on Ministry of the Environment Notification No. 46) were found in some instances to exceed environmental quality standards for soil contamination from lead, arsenic, fluorine, and boron. In the cases of lead and arsenic, it is conceivable that natural sources were responsible for exceeding leaching standards. Because concentrations of fluorine and boron are high in the seawater of this area, the influence of seawater is a possibility. There were no samples in which the content of persistent organic pollutants (POPs) such as dioxins, polychlorinated biphenyls (PCBs), or pesticides exceeded the standards (for example, for PCBs the standard is the destruction target of 0.5 parts per million [ppm] for PCB treatment, for dioxins it is the environmental quality standard for soil and for sediment in bodies of water, and for other substances it is the established reference guidelines). The levels found were generally the same as the results of environmental monitoring surveys of sediment and soil that were performed in recent years by the Ministry of the Environment in nearby water and land areas. Because our investigation is based on 62 samples and a limited study, a more detailed study may be carried out in the future, but it is safe to say that at this point no serious contamination in particular has been found.

Essentially, the guidelines for disposing of tsunami deposits call for removing pieces of wood and other materials, detoxifying them, and then using them as fill in landfills or for embankments. In urban areas, where hydraulic excavators are hard to use, removal is performed by people with shovels or other tools. After being gathered, deposits are carried away by heavy machinery, while septic tank pumper trucks can be used for sludge, which has a high water content. After removal, the deposits are put in temporary storage sites; pieces of wood and concrete, which can be used as civil engineering materials, are separated out. If the deposits contain hazardous substances, they are detoxified by washing and/or physical/ chemical treatment, and then either likewise used as material, or taken to a municipal solid waste disposal site if they cannot be effectively used. It was decided that if tsunami deposits contain no pieces of wood or other matter and are not contaminated with hazardous substances, they could be left in place after making arrangements with landowners.

Hazardous waste separation and disposal

The types of waste that present dangers, and the methods of handling them, require various cautions, particularly if operations are on-site. There are hazardous wastes such as gas cylinders, building materials containing asbestos, and transformers and capacitors containing PCBs. The Japan Society of Material Cycles and Waste Management (JSMCWM) has prepared a disaster-waste quick reference chart, and it is desirable that personnel performing waste removal should use this (or others like it) to learn about hazardous wastes.

Here is an example from Sendai City of how to treat hazardous waste: such waste, ranging from household cleaners, paints, lead-acid automobile batteries, and emergency power supply systems used by industries, is all being stored separately in a space about the size of a baseball field. Of these types of waste, a decision has been made only about gas cylinders and fire extinguishers—which should be treated by the related industries—while the treatment and disposal of other materials is still undecided. A high level of caution is needed in daily dealings with household hazardous waste, and further detailed measures are required to tackle this issue when establishing plans to deal with disasters.

LESSONS

Basic framework for dealing with disaster waste

On April 5, 2011, the Science Council of Japan issued the "Urgent Proposal Related to Measures for Earthquake Disaster Waste and Prevention of Environmental Impact." The proposal's overall framework was drafted by the JSMCWM, and then issued in collaboration with the Japan Society of Civil Engineers and the Japan Society on Water Environment. The medium- and long-term response was also taken into consideration in formulating a basic policy for the disposal of earthquake waste and the minimizing of environmental impacts. The essential points are given below:

- *Waste is to be treated and disposed of quickly,* while keeping in mind the securing of public health and the handling of hazardous waste. Priority is to be given to dealing with putrefied organic matter and quickly removing it from cities and streets, or—while taking measures such as spreading lime to delay putrefaction—to determining locations of hazardous wastes such as medical waste, asbestos, and PCBs, and trying to process each waste type in the proper manner.

- *Temporary storage sites are to be created* (which take the water environment into consideration) and waste is to be uniformly separated. Waste collection locations are to be decided on immediately, and putrefied materials including sludge-type items, flammable materials, and hazardous wastes should not be mixed. Care is to be taken not to create huge piles, to prevent fires and other such events, and not to cause contamination of water, soil, or groundwater.

- *Recycling should be considered* to help put resources to use in recovery and reconstruction. Concrete debris might be recycled in the recovery and rebuilding phases, wood scraps could substitute for fossil fuels in power generation and other applications, and various other types of recycling could be conceived.

- *Local employment and wide-area cooperation should be facilitated in disaster-waste recycling.* It was determined that in this case what is promoted internationally as "cash for work" could be effective. On dealing with disaster waste in the Tohoku region, even if wastes were to be recycled, the region would not have sufficient treatment and disposal capacity, which raises the possibility of widespread cooperation. A case can be made for taking a nationwide response: integrating industry, government, academia, and the citizenry.

Figure 23.1 shows the basic flow involved in operating temporary storage sites and preliminary waste storage sites to facilitate the local management of municipal solid waste. These storage sites play a major part in the smooth removal of debris from disaster areas. For instance, it was known that since much of

the disaster-stricken area in the Tohoku region comprises narrow coastal zones and also because of the urgent need for land for temporary housing and other purposes, it was not easy to secure land for temporary storage sites. In all geographical areas, authorities should prepare for disasters beforehand by designating places for temporary storage sites, traffic routes for waste transport, and other related needs.

In situations such as when a tsunami has scattered individuals' private possessions and mixed them with disaster waste, removal and processing must proceed while also determining who owns what. At the end of March 2011, the government issued "Guidelines on the Removal and Other Treatment of Collapsed Homes and Other Property after the Tohoku Region Pacific Coast Earthquake" (Ministry of the Environment 2011), which contained the following three points:

- Make sure everyone knows in advance the plans for where operations will be conducted, schedules, and other particulars.

- Before removal, take photographs and make other records of buildings, automobiles, motor scooters, and boats.

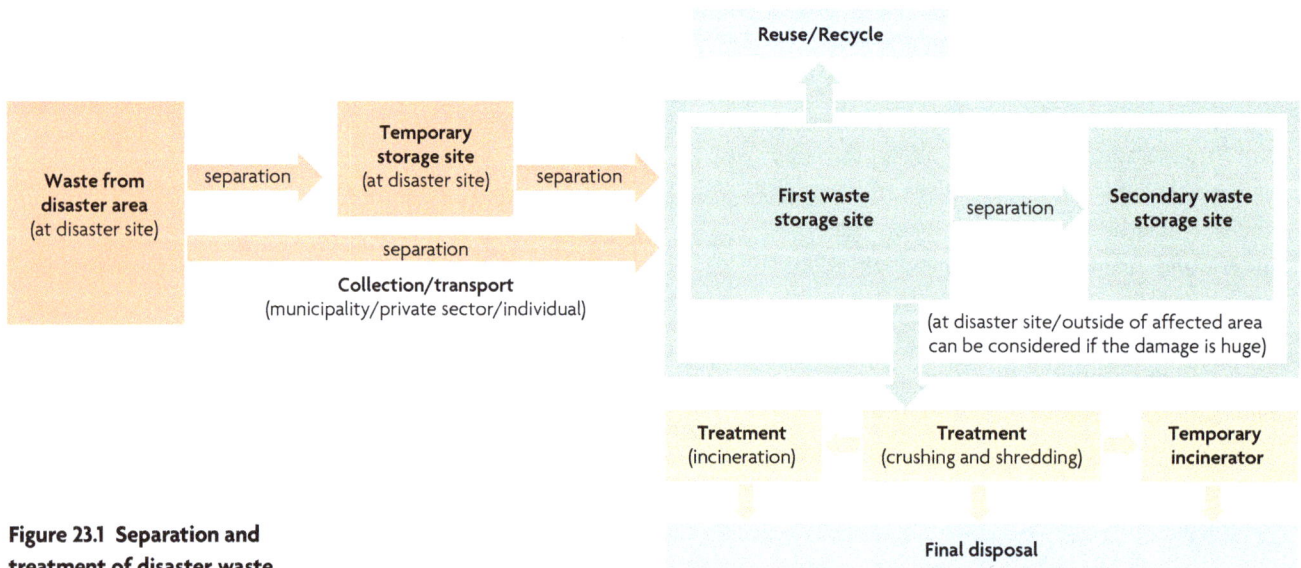

Figure 23.1 Separation and treatment of disaster waste

- For ancestral tablets, photo albums, and other items that are valuable to owners and other persons, as well as chattels, provide opportunities to return them to the respective owners and other persons.

Valuables such as precious metals and safe boxes should be put into temporary safekeeping. Efforts should be made to contact the owners or relevant parties in the event they are identified, and the valuables should be returned when the owners or relevant parties so request. When the owners or other relevant parties are unknown, the guidelines call for the valuables to be processed as directed by the Lost Property Act.

Separation and recycling: The Sendai City model

Following is one conceivable classification scheme for the composition of disaster wastes from earthquakes and tsunamis:

- Waste consumer electric appliances and electronics, and various household effects

- Waste wood, concrete rubble, tiles, and so on

- Plants, trees, and other natural items

- Large structures and so on

- Deposits (silt, bottom sediment, and so on)

- Wrecked vehicles and boats

- Hazardous wastes (asbestos, pesticides, PCBs, and so on)

- Evacuation center waste

- Infectious waste, human corpses, and animal carcasses

Depending on the composition of each type, it is necessary to identify and carry out the appropriate treatment and disposal methods, while keeping in mind the possibilities for recycling. Table 23.1 lists the specific types of waste that fall under the above categories, and their recycling and disposal methods.

Although people tend to concern themselves with removing disaster waste quickly, they should from the outset consider how wastes could be recycled to reuse valuable resources and preserve landfill space.

Disaster waste and tsunami deposits generated in Sendai City were estimated to be around 1.35 million tons and 1.3 million tons, respectively. As of April 2012, concrete, which accounts for about half of the 1.35 million tons of disaster waste, could possibly be reused as material for reconstruction. Strategies for waste other than tsunami deposits were near completion.

The city had already estimated the amount of disaster waste only three weeks after the March 2011 earthquake, and set up a target of disposing of it within three years. Realizing that it was impossible to treat the waste using only existing facilities, the city decided to set up additional temporary incinerators, which were constructed in autumn 2011. Three temporary incinerators (one stoker furnace and two rotary kilns; 480 tons/day of total disposal capacity) were installed in three designated temporary storage sites along the coastal area. The following items were separated and recycled: wood lumber (for fuel use), metals, tires, four items designated in the Home Appliance Recycling Law, automobiles, and motorcycles.

Including wastes that are supposed to be landfilled, the amount of waste collected and moved to temporary storage sites is measured by a huge weighing scale, and in some cases the results are recorded.

Financial support

To facilitate disposal of disaster waste, half the cost is covered by government subsidies, and 80 percent of the remaining cost is covered by issuing municipality bonds (that is, a local government has to pay only 10 percent of the total cost). Additional measures are being taken to reduce the burden on local governments, considering the size of the enormous damage caused by the GEJE.

Table 23.1 Segregation of disaster waste and recycling and treatment methods

CATEGORY	OUTLINE	TYPE OF WASTE	RECYCLING AND DISPOSAL METHOD
Waste from household goods	Household goods destroyed by earthquake and tsunami	Valuables and mementoes	Each item stored for return to owner
		Home appliances (TVs, refrigerators, air conditioners, washing machines)	Home appliance recycling system
		Other home appliances	Metal recycled after dismantling and crushing; organic material incinerated, inorganic material disposed of in landfill
		Tatami mats, mattresses	Shredded and used as fuel or incinerated
Waste from collapsed houses	Collapsed houses and buildings (including furniture) destroyed by earthquake and tsunami	Timber from houses, furniture	Desalted if necessary. Potential usages include: 1) particle board, charcoal, and reuse of material; 2) use as fuel in cement kilns; 3) energy recovery from incineration
		Concrete, asphalt, waste tiles	Crushed and used as aggregate for roadbed material and in construction
		Asbestos-containing building materials	Controlled management: disposed of in landfill, melted
		Plasterboard	Controlled management: disposed of in landfill
Wood	Scattered and accumulated garden trees, pine wood, and other trees	Garden trees, live trees, etc.	Desalted if necessary. Potential usages after chipping include: 1) particle board, charcoal, reuse of material, papermaking material; 2) use as fuel in cement kilns; 3) energy recovery from incineration
Bulky waste	Large-size and unusual waste from factories and infrastructure	Tanks, power poles, feedstuffs, fertilizer, and fishing nets that each require a specific disposal method	Crushed and separated and then recycled, incinerated, or disposed of in landfill Caution is required for hazardous substances such as asbestos
Deposits generated by the tsunami	Gravel and mud left in disaster area after the tsunami. Most is bottom sediment from water bodies, but sometimes organic materials and contaminants are included	Sediments mixed by the tsunami with the debris of collapsed houses and other debris. Some include oil. Odor and dust could arise on putrefaction or drying. Hazardous chemicals such as acids, alkalis, and pesticides from the disaster area could be included	Used as fill for landfills or embankments after removing woody debris and detoxifying. Detoxified by washing or incineration when material contains hazardous substances. Nonrecyclable items are taken to final disposal site and disposed of as general waste. Where there is no wood debris and no contamination with a hazardous substance, they could be left in place after making arrangements with landowners
Vehicles/ships	Automobiles/ships	Automobiles, motorbikes, tires, ships, etc.	Automobile recycling system. Tires chipped and used as a supplemental fuel. Ships are dismantled, recycled, and disposed of. Caution required for asbestos materials
Hazardous waste	Asbestos, PCBs, etc.	Batteries, fluorescent lamps, fire extinguishers, gas cylinders, waste oil, waste liquids, transformer oil, etc.	Controlled management undertaken as necessary for each type of waste

RECOMMENDATIONS FOR DEVELOPING COUNTRIES

Prepare a disaster waste management plan in advance

It is essential to make disaster waste disposal plans beforehand to help reduce the need for decision making with insufficient information in the wake of a disaster (box 23.1). Guidelines on measures to manage disaster waste and on measures to treat waste from flooding were established in Japan in 1998 and 2005. Both sets of guidelines require that any plan should specify how to

- Establish basic policies for waste management.

- Construct and manage the system that deals with waste management.

- Classify disaster waste and secure necessary equipment and temporary storage sites for disaster waste.

In 2010, 72 percent of municipalities across the country (a rather high rate), had disaster waste management plans in place. But they are now being revised to include the following:

- Estimation method for the amount of disaster waste generated by tsunamis, and countermeasures for dealing with the waste

- Multiple predictions for disasters of different scales

Accordingly, periodic review of disaster waste management plans is indispensable.

Build cooperative structures with various organizations and institutions

When disasters occur, cooperative ties with various organizations and institutions are key to the smooth management of disaster waste. This is because many problems and administrative needs arise, while the number of appropriate policy experts is limited, and the waste disposal sites in the affected areas are often

---

BOX 23.1

Preliminary findings of the United Nations Environment Programme's (UNEP) expert mission on Japan's earthquake waste

- The contingency plans put in place by some prefectures before the earthquake allowed them to respond more quickly to the waste management challenge (for example, in Sendai City, which had contingency plans, three incinerators were already in place processing 460 tons of waste a day).

- While Japan has done much to advance global best practices on handling disaster debris, there is still scope for substantial optimization so as to lower the costs of postdisaster debris management and reduce its environmental impacts.

- Commendable emphasis has been placed on waste segregation and recycling. Waste is divided into several categories such as wood, metals, electrical items, tatami mats, fishing nets, vehicles, plastics, and so on. Some segregated materials are already being reused: for instance, tree trunks are being sent to a paper mill, shredded wood is being sent to a cement company for use as fuel in the manufacturing process, and building rubble is being recycled as building material, landfill, or in road construction.

- Maximizing the possibilities for waste recovery and recycling while minimizing the need for transportation are priorities for effective debris management.

- Under Japanese law, the manufacturers of cars and white goods (refrigerators, washing machines, and so on) are responsible for the final disposal of their products. But the volume of disaster debris generated is likely to overwhelm their intake capacity, which may need to be expanded.

- Despite the magnitude of the challenges, and their own personal tragedies, the officials in the various Japanese cities are doing systematic and dedicated work to manage the debris in a time-bound fashion.

- Opportunities exist for learning from best practices in various cities, and a systematic approach to capturing them and disseminating them would be beneficial.

- The national guidelines produced for disaster debris management could be locally adapted, with input from academic experts to reflect local circumstances. This will lead to more environmentally optimal outcomes.

- There is scope for improved monitoring and communication of the waste management activities in the disaster-impacted areas, which will enable everybody to appreciate the challenges faced and the efforts made.

Source: http://www.unep.org/newscentre/Default.aspx?DocumentID=2676&ArticleID=9067&l=en.

damaged. Above all, much more waste is generated in these circumstances. Developing cooperative relations between local governments in the surrounding affected areas and with communities far from the stricken areas should be considered. Sendai City, for example, which was affected by the GEJE, over the course of a year received 58 staff members from eight organizations to help promote its waste management plans. For waste collection, the city received help from 7,510 staff members from 10 organizations, as well as 88 vehicles.

In addition to cooperating with industries and local municipalities, building and making effective use of cooperative relationships with academic organizations, other expert groups, and civil society organizations are also recommended.

Customizing the removal process to local contexts

Each country has its own environmental safeguards, technology, and recycling practices. Utilizing these local practices are crucial in effective debris management.

NOTE

Prepared by Shinichi Sakai, Kyoto University, and the International Recovery Platform.

BIBLIOGRAPHY

Brown, C., M. Milke, and E. Seville. 2011. "Disaster Waste Management: A Review Article." *Waste Management* 31: 1085–98.

JSMCWM (Japan Society of Material Cycles and Waste Management). 2011. "Guidelines for Treatment of Tsunami Deposits" (proposed). July 5. http://eprc .kyoto-u.ac.jp/saigai/archives/files/Sediment ManagementGL%20by%20JSMCWM.pdf.

———. 2012. *Manual on Separation and Disposal of Disaster Waste: Based on Experience from the Great East Japan Earthquake* [in Japanese]. Gyosei.

Ministry of the Environment. 2011. "Guidelines on the Removal and Other Treatment of Collapsed Homes and Other Property after the Tohoku Region Pacific Coast Earthquake" [in Japanese]. March 2011. http://www.env.go.jp/jishin/sisin 110326.pdf.

Sendai City. 2012. "Disposal Processes of Disaster Waste in Sendai City" [in Japanese]. February.

Takatsuki, H., S. Sakai, and S. Mizutani. 1995. "Disasters and Waste Properties: Disaster Waste Intensity and Change in the Composition of Municipal Solid Waste" [in Japanese]. *Waste Management Research* 6 (5): 351–59.

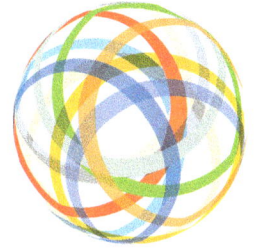

CHAPTER 24

Livelihood and Job Creation

Livelihood and job creation have long been critical challenges to disaster recovery. Following the Great East Japan Earthquake, the Japanese government launched an innovative cash-for-work project, hiring more than 31,700 jobless people to work not only on reconstruction, but also on clerical and support work for affected people. This allowed it to reach out to women and the elderly, vulnerable groups that were traditionally excluded from schemes focusing primarily on manual work.

FINDINGS

The Great East Japan Earthquake (GEJE) caused some 140,000–160,000 people to lose their livelihoods and jobs. By February 2012, in part as a consequence of an innovative emergency job-creation project initiated by the government, 143,820 people had found employment in the three most affected prefectures. Of these jobs, 22 percent (31,700) were jobs directly created by the emergency job-creation project. Despite gaps between sectors, regions, and types of employment available, the government-initiated job-creation policy has generally been effective in sustaining employment in disaster-affected areas.

Record of livelihood and job creation in Japan following catastrophic disasters

Livelihood and job creation has long been a critical issue in disaster response and recovery, both in Japan as well as worldwide. Fundamentally, it plays three critical roles:

- *Economic.* It serves as a key—and in some cases the only—source of income for the population affected by disaster.

- *Social.* It encourages affected people to participate in the recovery process, thus strengthening their social ties.

- *Psychological.* It helps those who lose their jobs regain their self-esteem and look forward to the future.

Historically, job-creation policies benefiting those affected by disasters have not been particularly successful in Japan, despite their recognized importance and long record—even dating back to the 1854 Ansei Nankai earthquake disaster. After the 1923 Kanto earthquake, the Ministry of the Interior encouraged local governments and private firms to hire affected people for disaster response and recovery work; however, this attempt was unsuccessful, as the work provided was mainly manual while affected people aspired to non-manual, "white collar" labor. The national government instead encouraged jobless people to move to suburban areas of the cities from which they originated.

Livelihood- and job-creation attempts were also unsuccessful following the Great Hanshin-Awaji (Kobe) Earthquake of 1995. As a result of the disaster, some 40,000–100,000 people were left jobless. The national government issued a law in March 1995 forcing public projects in affected areas to reserve up to 40 percent of their workforce for affected people rendered jobless by the earthquake. A year later, however, only 30 people had been hired under the policy. Contractors continued to make employment decisions based on profitability and efficiency, and there were no penalties for noncompliance. As a result, the employment of affected people was limited to simple and unskilled public work tasks. During the recovery process, 254 people died in transitional shelters without the care of family members or neighbors. Some study reports point out that most of the people who died alone (*Kodokushi*) were jobless, suggesting that they were isolated from society and had no contact with others.

Damage caused by the GEJE and its impact on livelihoods and jobs

The GEJE could be the most severe of recent disasters in Japan. The Japan Research Institute estimates that 140,000–160,000 people lost their livelihoods and jobs in May 2011 due to earthquake and tsunami damage. Moreover, most of the tsunami-hit cities lost the bulk of their infrastructure.

Ishinomaki City, for instance, was one of the largest cities hit by the tsunami (population 160,000). The Ishinomaki fishery port is the third largest in Japan in terms of total landings. Fishery and seafood processing were the main industries of the city, engaging hundreds of companies and employing several thousand people.

The tsunami washed out nearly the entire central business district of the city. Aside from damage to buildings and facilities, the earthquake lowered soil levels by approximately 1.4 meters, allowing seawater to penetrate the area at full tide. To restart the industry it will be necessary to first elevate the soil, something very few companies can afford to do given the burden of existing loans. Over three years have passed since the earthquake and tsunami, and the national government has included the elevation costs under its third supplementary budget (fiscal year 2011). But it will take several years to complete such a large reconstruction project and, therefore, job recovery in Ishinomaki City is expected to be slower than what followed the 1995 Kobe earthquake.

In Fukushima, the national government designated the area within a 20-kilometer (km) radius of the Fukushima Daiichi Nuclear Power Station accident as a restricted area, affecting some 78,000 people. Areas with relatively high radiation levels, even outside the 20-km radius, were designated as Deliberate Evacuation Areas, affecting an additional 10,000 people (most of whom lost their jobs).

Although the national government is planning to remove restrictions in areas with relatively low radiation levels, the recovery of livelihoods and jobs in these areas will be difficult to address. A questionnaire of evacuees from these areas conducted by Fukushima University indicates that only 4 percent intend to return to their homes immediately after the lifting of the restrictions. Of the respondents, 25 percent have already decided not to return at all, citing lack of jobs as one of the major reasons. Close to 46 percent of respondents under the age of 35 say they will not return. Since the power station was the main source of economic activity in the area, there are now very few job opportunities left. Thus, livelihood and job creation will also be critical to recovery in these areas. The survey results further indicate that 16 percent of the respondents say that recovery of the infrastructure will be necessary, while 21 percent argue for a concrete plan for radium decontamination (chapter 36 covers updated information).

Livelihood and job creation under the GEJE
Government initiatives
Following the GEJE, the Japanese government's response involved both cash transfers to the most vulnerable, as well as an emergency job-creation project.

To help secure the well-being of the most vulnerable (such as the elderly and any handicapped not regarded as employable), the government provided cash transfers through the regular social security system based on the Public Assistance Act, amounting to around ¥50,000 to ¥250,000 (approximately $550 to $26,000) per month. In addition, the Japanese disaster management system provided up to ¥3 million (approximately $37,500) to households that lost their houses to assist them with reconstruction efforts (chapter 20). Cash was also individually distributed to the most vulnerable people in the form of donations received from all over Japan.

To promote job creation, the Ministry of Health, Labour and Welfare (MHLW) launched the "Japan as One" Work Project immediately after the earthquake. The project had three major policy objectives:

- Steadily create jobs through reconstruction projects

- Develop a system to match disaster victims with jobs

- Secure and maintain securing employment among disaster victims

The first policy objective built upon an earlier emergency job-creation fund created in 2008 after the global financial crisis. Following the GEJE, the government spent ¥50 billion ($625 million) to enlarge the fund, expanding its eligibility to disaster-related job losses.

Examples of activities supported by the project included:

- *Evacuation center management and administration,* such as food distribution, cleaning, procurement, and the delivery of food and other materials

- *Safety management and life-support services* such as patrolling, caring for the elderly and disadvantaged, babysitting, supplementary lessons for students, and bus driving

- *Office-work support for local governments* such as issuing resident cards, operating the call center, guiding visitors, distributing donations, and monitoring and performing needs assessments at evacuation centers

- *Reconstruction and recovery work* such as debris removal, the cleanup of houses of the elderly, parks and public building maintenance, planting of flowers in parks, and public relations activities for sightseeing promotions

The basic thrust of this policy was very similar to that of a cash-for-work (CFW) program (see box 24.1), but it differed substantially from typical CFW programs in developing countries. The range of work created by this project was so diverse that women and elderly could also work, whereas other CFW programs have tended to provide mostly manual labor (for example, infrastructure reconstruction).

One of the constraints faced by the job-creation project was that employers had to

Livelihood options in humanitarian assistance

International humanitarian assistance has typically used two instruments to promote livelihood recovery after disasters: cash transfer and public works programs cash-for-work (CFW) programs.

Cash transfers are typically used to provide short-term assistance to the most vulnerable affected people. To be effective, cash grant programs must be well targeted (for example, aimed at the elderly, widows, refugees), be transparent, have sound mechanisms for monitoring and evaluation, and have a clear exit strategy. Typical programs implemented during the 2005 Pakistan earthquake and 2004 Sri Lanka tsunami involved a transfer of $50 per month per target household for a period of four to six months. Often, cash transfer programs coexist with or graduate to become CFW programs.

Cash-for-work (CFW) programs have been common tools for humanitarian assistance. These programs provide cash to affected people in return for their work on various recovery projects, such as debris removal and the repair or reconstruction of damaged infrastructure. They have been used in many disaster situations, including the 2004 Indian Ocean tsunami, the 2008 Myanmar cyclone, and the 2010 Haiti earthquake.

CFW programs were developed as an alternative to food-for-work (FFW) programs, in which affected peoples could receive food in return for their disaster-recovery and mitigation work (during droughts and famine). Cash has several advantages over food as a worker incentive: (1) related logistics are less complex and management costs are lower; (2) workers can choose what they buy, thus empowering them; and (3) cash has a large market impact when it is spent locally. At the same time, CFW programs must avoid crowding out the normal job market and, like cash transfers, require close monitoring.

comply fully with domestic labor laws. For example, employers had to compel workers to take compensation, employment, and social insurance. Paperwork accompanying employment procedures proved a bottleneck during job creation. Although many of the government agencies, nongovernmental organizations (NGOs), and private contractors were major sources of job opportunities, they were reluctant to hire the jobless since they were otherwise occupied with the emergency response.

Public-private partnerships were an effective solution to this problem. The Fukushima Prefecture government, for example, requested private staffing agencies to hire affected people for the work of disaster-response organizations (including municipal governments). This scheme was very effective since the organizations involved did not have the burden of paperwork or personnel management.

Public-public partnerships were also used. The CFW activity in Ofunato City was partially undertaken by the Kitakami City government. Kitakami City received emergency job-creation funding from the Iwate Prefecture government, and entrusted a private staffing agency to hire affected people to care for those staying in transitional shelters in Ofunato City.

To meet the second policy objective of the "Japan as One" project—matching disaster victims with jobs—the government intended to fully activate and empower public employment exchanges in the affected areas. This was effective to some degree but not adequate to the significant burden of the aim, which was why (as mentioned above) private staffing agencies played a significant role in job creation.

The third objective—to secure and maintain employment among disaster victims—was supported by two activities. Some ¥727 billion ($9 billion) was distributed as an employment adjustment subsidy to affected industries, as an incentive for them to secure employment. In addition, the government provided ¥294 billion ($3.7 billion) to extend benefit terms of unemployment insurance. This helped protect workers in the formal sectors. Without this assistance, the burden of the job-creation project would have been much higher.

Nongovernmental organizations and the private sector

NGOs and the private sector also played important roles in the aftermath of the GEJE. The International Volunteer Center Yamagata, for example, launched a CFW project in which jobless affected people were hired for debris removal and cleaning activities. Their salaries were financed by donations from all over Japan as well as overseas. The work was eventually expanded to community-support activities. The project ended on March 31, 2012, having hired 112 jobless people. Although it was a

Figure 24.1 Minamisanriku shopping village

Source: © International Recovery Platform (IRP). Used with permission. Further permission required for reuse.

Figure 24.2 A poster promoting the friendship bracelet (*tamaki*)

Source: Source: © Shingo Nagamatsu. Used with permission. Further permission required for reuse.

typical CFW scheme, it was not as large as programs seen in developing countries.

Another example was the Sanriku-ni Shigoto-wo Project in the Sanriku area, driven by a nonprofit alliance of Iwate Hakuhodo Co. Ltd., Iwate Menkoi TV, and Sendai Television Inc. This project provided livelihoods to fishermen's wives previously engaged in seafood processing. While affected fishermen had benefited from an emergency job-creation project promoted by the Fishery Agency for debris removal and fishing port clean-up efforts, their wives had been left jobless.

Thirty new shops were opened in the Minamisanriku shopping village, inaugurated on February 25, 2011, for the purpose of temporary job creation following the disaster (figure 24.1). The Ministry of Economy, Trade and Industry, through its "Small Medium Enterprise Support, JAPAN Program," facilitated the establishment of this temporary shopping village. Souvenir items produced by local residents, particularly women, were sold in some shops.

The project promoted a new handicraft made by women: a friendship bracelet called *tamaki* ("ring") made of fishing-net materials (figure 24.2). Approximately 50 percent of the sales went to the women producers. This project was covered extensively by television and the social media, and for several months production could not keep abreast of sales. As of February 29, 2012, 298 producers had received

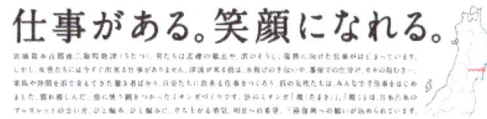

as much as ¥83 million ($1 million), according to the project website. The success of this project triggered many other kinds of handicraft production.

The Security Support Fund, operated by Music Securities Inc., was an e-commerce citizen aid initiative that matched prospective investors with small businesses affected by the GEJE to help restart them. Those who needed financial support submitted proposals via the fund's website. In turn, prospective donors could visit the website and find projects for their potential investment. Thus, it worked as a microfinance project where prospective donors were matched directly to the recipients.

This fund has two important features: (1) one unit of investment can be as small as ¥10,500 ($131) and (2) investors do not expect an economic return from their investment. About half (¥5,000) of the single unit of

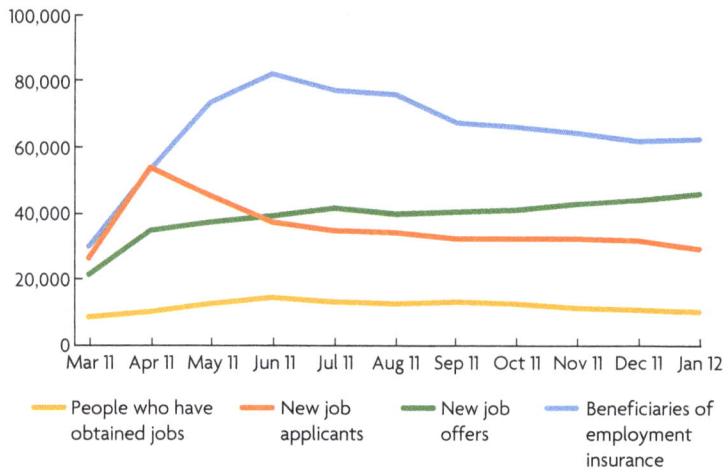

Figure 24.3 Recovery process of labor markets in Iwate, Miyagi, and Fukushima prefectures

Source: Ministry of Health, Labour and Welfare (MHLW).

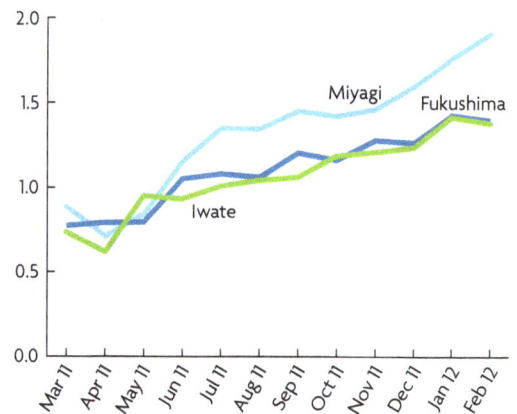

Figure 24.4 New-job-opening ratios of the Iwate, Miyagi, and Fukushima prefectures

Source: MHLW.

investment is considered a donation. Most of the investors enjoy communicating through the website with the businesses they are supporting. The fund had grown to ¥700 million (approximately $8.8 million), attracting more than 20,000 investors as of 2012.

Policy results and outstanding challenges
Partially as a result of the government policy, the labor market recovered rapidly in the affected areas. The number of beneficiaries of employment insurance leapt to 81,179 in June 2011 from 29,931 the previous March. Since June 2011, moreover, job offers exceeded the number of new applicants, and this gap had grown in 2011 (figure 24.3).

Although the job situation is surely improving in general terms, recovery is not yet complete, and there are gaps in four major areas: (1) differences between job offers and applicants (mentioned above), (2) gaps among regions, (3) gaps among sectors, and (4) gaps in employment patterns.

In common with other disasters, job opportunities have disproportionally been concentrated in urban areas. Figure 24.4 illustrates trends in new-job-opening ratios by prefecture. Miyagi Prefecture—where Sendai City, the capital of the Tohoku region, is located—has been attracting more jobs than the other

two prefectures. Even within the Miyagi Prefecture, job opportunities were concentrated in the Sendai metropolitan area (a new-job-opening ratio of 1.17 in February 2012), while Ishinomaki and Kesennuma, both of which are located on the coastal areas severely affected by the tsunami, offered relatively scarcer job opportunities (ratios of 0.77 and 0.55, respectively).

Additional gaps are seen among job sectors. With rising reconstruction demand, many new job offers come from construction and related industries, with relatively fewer offers in the manufacturing and distribution industries. Job applicants, on the other hand, appear to be seeking occupations more focused on food processing and clerical work.

A final gap is seen in employment patterns. In spite of an increase in job offers, most involve part-time or short-term employment. The job-opening ratio for full-time, regular workers in Miyagi Prefecture in February 2012 was only 0.49. The situation for those who are looking for regular, full-time work is therefore not as favorable as the general statistics suggest.

Part of the reason why a large proportion of job openings involve so much short-term employment relates to the government-supported emergency job-creation project. Between March 2011 and February 2012, 31,700

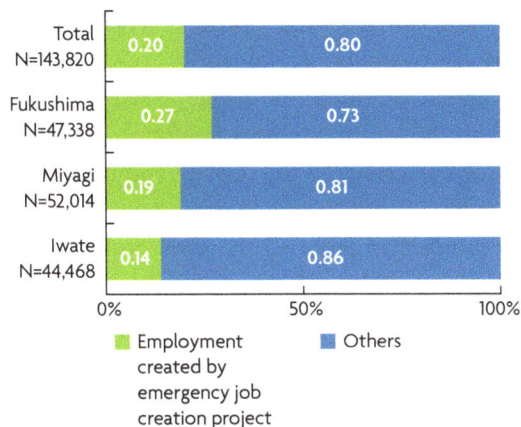

Figure 24.5 Ratio of employment sustained by government policy: March 2011–February 2012

Source: MHLW.

workers—or 22 percent of all job offers in the Iwate, Miyagi, and Fukushima prefectures—stemmed from the emergency job-creation project (figure 24.5).

This finding has key two implications. First, the government-initiated job-creation policy is effective in sustaining the job market in disaster-affected areas. In its absence, unemployment issues would have been far more severe. At the same time, the transition from CFW jobs to regular jobs is a difficult challenge for the economic recovery process.

CFW programs in developing countries typically assist in the process of economic recovery and even economic growth: this is plausible since disasters in developing countries tend to affect growth rates positively. As such, CFW programs fill an important employment gap immediately following a disaster, after which economic growth creates sufficient permanent jobs to take over.

But economic recovery in developed countries does not necessarily follow this trend: the populations of the three prefectures affected by the GEJE had been declining since before the earthquake. As an economy shrinks, it does not necessarily generate sufficient permanent jobs to take over the role of emergency job-creation programs. Japan could well be facing this problem.

LESSONS

- *Dedicated emergency job-creation programs, complemented by cash transfers to the most vulnerable, can be effective* ways to assist disaster-affected people during a recovery. At the same time, they need to be adjusted progressively to emerging job markets, and avoid cluttering them in the process. More prolonged assistance may be needed when local economies are contracting.

- *The livelihood needs of disaster-affected people are diverse, and thus require diverse solutions.* The most vulnerable may need cash transfers, whereas those already benefiting from pensions (for example, the elderly) may need primarily an occupation to make them feel needed. Others—such as widows with young children—require regular employment with insurance benefits.

- The experience of the GEJE shows how *learning from past disasters has been used effectively to design the emergency job-creation project.* Regulatory measures and market forces alone did not succeed in creating jobs following the Kobe disaster. The GEJE helped launch a more proactive government project, which promoted diverse employment and partnerships with NGOs and the private sector, while retaining the means to monitor its overall progress.

- *The GEJE job-creation program has been innovative in facilitating public-private and public-public partnerships.* In particular, hiring staffing agencies helped reduce the administrative burden, which would otherwise have prevented many employers from engaging the victims of the disaster.

- *Matching jobs with the needs of the jobless* is a very important but difficult task. Most of the affected areas have seen excess labor demand and labor supply simultaneously, but in different sectors, and urban areas have clearly benefited over rural areas.

Interventions such as continuous monitoring of job supply and demand, job retraining, and further integration with municipal plans are necessary to effectively complete the recovery.

- *Unemployment insurance can be effective in securing the incomes of those affected.* But there are several limitations: (1) unemployment insurance does not cover self-employed workers and those who run private enterprises, and (2) the national government has twice had to extend the beneficiary period of insurance, allowing even those covered for the shortest period to benefit from the program until January 2012. Unemployment insurance therefore needs to be seen as part of a broader livelihood recovery program following a disaster.

RECOMMENDATIONS FOR DEVELOPING COUNTRIES

Match jobs with worker skills. To the extent feasible, CFW and employment programs following a disaster should expand the range of work opportunities, from simple manual labor for infrastructure reconstruction to nonmanual work. While in developing countries most of those affected are poor and unskilled, mega-disasters such as the Haiti earthquake of 2010 also affected skilled workers. It is important that all be given opportunities to contribute meaningfully to the recovery and reconstruction of their neighborhoods, although priority for external assistance must naturally be given to the poorest and most vulnerable. In particular, the jobs created should be

- Appropriate to the workers' skills and abilities

- Help boost the morale and self-esteem of those affected

- Build upon the workers' skills, to help them secure their next occupation

Consider the bigger picture. The balance between quality and quantity needs to be planned carefully in developing countries, where the primarily goal is often to provide rapid cash relief to the poorest and most vulnerable of the disaster victims. As a rule, the proportion of labor to the total costs of the activity should therefore remain high (for example, 50–80 percent). CFW schemes also need to be designed with a view to providing a smooth transition to long-term jobs, and avoid attracting people back to vulnerable urban areas. As such, prevailing wages should be set just below the market rate for unskilled manual labor, thus ensuring that programs attract only those without other alternative means to earn income, and do not crowd out more permanent job creation.

In the above context, CFW schemes in developing countries differ from those promoted under the GEJE. Under the GEJE, the beneficiaries of the job-creation project were paid market wages, as there was no possibility of circumventing minimum wage regulations. In addition, as they had the option of claiming unemployment insurance, it was important to set the wages at levels sufficiently attractive to motivate them to work. Statistics in the GEJE prefectures do not show that this approach—at least in Japan—caused wage inflation. Thus, it was not supposed to prevent a transition to normal employment.

Integrate job-creation initiatives with other social protection systems. Similar to the experience of Japan, CFW programs in developing countries need to be part of a broader social protection program which can include cash transfers to the most vulnerable, such as was done in the aftermath of the Pakistan earthquake or Sri Lanka tsunami. If so, the eligibility, amount, and duration of payments and the cash-delivery mechanisms must follow transparent procedures.

Continue evaluating progress. Periodic evaluations are essential to determine whether livelihood programs are reaching their goals, and allow for corrections among program partners. In the case of Haiti, for example, preliminary evaluations pointed to the need to better target the most vulnerable, while avoiding prolonged aid dependency. A particularly neglected aspect tends to be seasonal competition between CFW and agriculture or fishing occupations, as well as assistance to people who, while not direct victims of the disaster, may be under traditional obligations of sheltering family members, with consequent strains on food supplies.

Involve the private sector. Job-creation programs in Japan tend to be smaller than those in developing countries—most hire fewer than 100 people each. Although this model is not necessarily an efficient way to maximizing employment, it helps integrate CFW programs with long-term job opportunities, as employers are directly responsible for supervising and caring for employees.

Use social media. The case of the Security Supporting Fund in Japan proves the effectiveness of e-commerce in directly linking affected people with potential benefactors. This has also been observed in other recent megadisasters (for example, the Pakistan and Bangkok floods), where social media increasingly played an important role in disaster recovery (see also chapter 21).

Continue supporting regular employment. While CFW programs are effective schemes for the short term, the transition from CFW jobs to regular jobs is a difficult challenge. Job opportunities for construction works will complete within a few years. Government support for creating regular jobs—such as arranging jobs, building factories, rehabilitating facilities of irrigation and fishery harbors, and resolving double debt—is essential in devastated areas (chapter 31).

NOTE

Prepared by Shingo Nagamatsu, Kansai University, with contributions from Sofia Bettencourt, World Bank.

BIBLIOGRAPHY

Albara-Bertrand, J. M. 1992. *Political Economy of Large Natural Disasters: With Special Reference to Developing Countries.* Oxford, U.K.: Oxford University Press.

Doocy, S., M. Gabriel, S. Collins, C. Robinson, and P. Stevenson. 2006. "Implementing Cash for Work Programmes in Post-Tsunami Aceh: Experiences and Lessons Learned." *Disasters* 30 (3): 277–96.

Echevin, D., F. Lamanna, and A-M. Oviedo. 2011. "Who Benefits from Cash and Food for Works Programs in Post Earthquake Haiti." MPRA Paper No. 35661, Munich Personal RePEC Archive, Munich.

GFDRR (Global Facility for Disaster Reduction and Recovery). 2010. "Haiti Earthquake Reconstruction—Knowledge Notes from the DRM Global Expert Team for the Government of Haiti," GFDRR, World Bank, Washington, DC.

Harvey, P. 2007. *Cash-based Responses in Emergencies.* HPG Report 24, Overseas Development Institute.

Mercy Corps. 2007. *Guide to Cash-for-Work Programming.* Portland, OR: Mercy Corps.

MHLW (Ministry of Health, Labour and Welfare). 2011. *"Japan as One" Work Project.* Conclusion on the countermeasures Phase 1, Ministry of Health, Labour and Welfare.

Music Securities Inc. Web page for Japan Earthquake Security Support Fund. http://oen.securite.jp/.

Myanmar Red Cross Society. 2009. *Myanmar: Cyclone Nargis Operations, Cash for Work (CFW) Program Project Progress Report.* Yangon, Republic of the Union of Myanmar: Myanmar Red Cross Society.

Recovery Research Institute. 2011. "Questionnaire Survey for Evacuees from Futaba 8 Municipalities" [in Japanese]. Fukushima University, Fukushima City.

Vishwanath, T., and X. Yu. 2008. "Providing Social Protection and Livelihood Support during Post-Earthquake Recovery." Knowledge Notes on Disaster Risk Management in East Asia and the Pacific, Working Paper Series No. 15, World Bank, ISDR (United Nations International Strategy for Disaster Reduction), and GFDRR, Washington, DC.

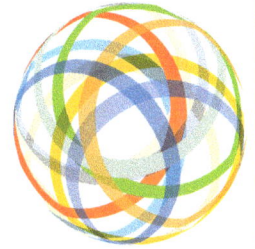

PART V

Hazard and Risk Information and Decision Making

CHAPTER 25

Risk Assessment and Hazard Mapping

Hazard and risk assessments are the crucial first step in disaster risk management and the basis for formulating relevant policies. They must take into account worst-case scenarios in the event of the largest possible hazard, while recognizing that hazard assessments of earthquakes and tsunamis will always have their limitations and associated uncertainties. In Japan, so-called hazard maps, which combine hazard information with evacuation routes and locations of evacuation centers, are effective tools for promoting evacuation procedures and risk awareness among the public. However, in the case of the Great East Japan Earthquake, these hazard maps, created before the event, may have given people a false sense of security by underestimating the disaster's potential impact. Hazard maps should be designed to guide and facilitate prompt evacuation. They should be easy to understand and readily available.

Risk assessment involves estimating the hazard levels of possible earthquakes and tsunamis to be considered when formulating disaster management policies. It is the first step in developing disaster risk management (DRM) plans and countermeasures. In Japan, the responsibility for risk assessment rests with government agencies at multiple levels. Implementing agencies at the national, prefectural, and municipal levels normally conduct risk assessment to inform their planning and the design of preventive measures. The national government is responsible for providing information and technical assistance to help prefectural and municipal entities assess risks properly and to reflect these risks in DRM measures.

FINDINGS

Megadisaster hazards considered in risk assessment

In Japan, countermeasures against earthquakes and tsunamis have been based on the risks associated with five large earthquakes that have occurred over the past several hundred years (map 25.1, box 25.1). The Central Disaster Management Council has set up a

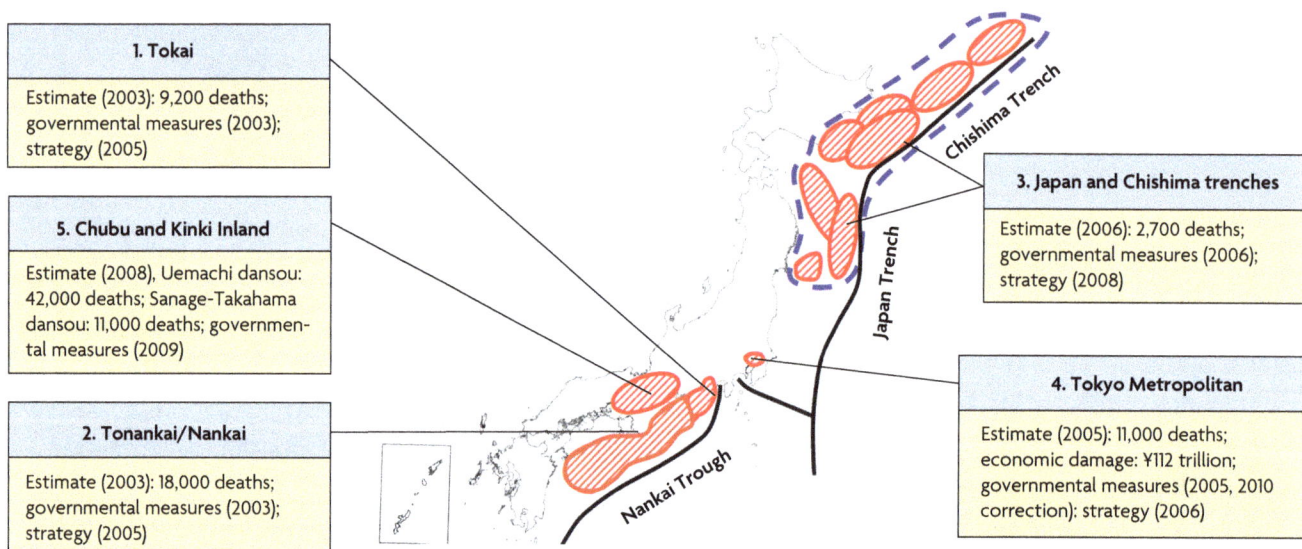

1. Tokai

Estimate (2003): 9,200 deaths; governmental measures (2003); strategy (2005)

5. Chubu and Kinki Inland

Estimate (2008), Uemachi dansou: 42,000 deaths; Sanage-Takahama dansou: 11,000 deaths; governmental measures (2009)

2. Tonankai/Nankai

Estimate (2003): 18,000 deaths; governmental measures (2003); strategy (2005)

3. Japan and Chishima trenches

Estimate (2006): 2,700 deaths; governmental measures (2006); strategy (2008)

4. Tokyo Metropolitan

Estimate (2005): 11,000 deaths; economic damage: ¥112 trillion; governmental measures (2005, 2010 correction): strategy (2006)

Chishima Trench

Japan Trench

Nankai Trough

Map 25.1 Five mega-earthquakes used as basis for risk assessment

Source: Cabinet Office (CAO).

committee to investigate and assess the potential hazard levels and expected damages from each of these scenarios. The committee also developed DRM strategies and a master plan

BOX 25.1

Principles for selecting large-scale earthquake scenarios and the actual earthquakes selected

- Repeated occurrence
- High probability of future occurrence
- Possibility of occurring within the next 100 years
- Not considered if an active fault earthquake has occurred in the last 500 years
- A significant number of occurrences can be identified in historical records
- Magnitude is between M 7 and M 8
- Consider the economic and social activities and central administrative functions to be protected

(Earthquakes meeting the above criteria)

1. Tokai earthquake (M 8.0)
2. Tonankai/Nankai earthquakes (M 8.6)
3. Japan and Chishima trenches earthquake (M 7.6–8.6)
4. Tokyo Metropolitan inland earthquake (M 6.9–7.5)
5. Chubu and Kinki inland earthquake (M 6.9–8.0)

for preventive actions as well as postdisaster response and recovery measures. DRM measures implemented at the national, prefectural, and municipal levels have traditionally been based on these strategies and plans.

The March 11, 2011, disaster occurred in the vicinity of the Japan and Chishima trenches—the region where the Central Disaster Management Council's committee had investigated trench-type earthquakes. From the list of past earthquakes in the region (map 25.2), eight were selected for consideration, based mainly on their intensity, frequency, and the possibility of recurrence in the same area. The selected historic earthquake scenarios included the Meiji Sanriku Tsunami of 1896, which generated a giant 20-meter-high tsunami, and Miyagi-ken-oki (Miyagi Prefecture) earthquakes that have been occurring at 40-year intervals. On the other hand, earthquakes such as those off the coast of Fukushima Prefecture were not selected because their probability of occurrence was estimated to be low, at 7 percent (map 25.3). Furthermore, the Jogan Earthquake of 869, believed to have caused massive tsunamis in the east Japan region, was excluded because the available modeling techniques were unable to replicate its seismic

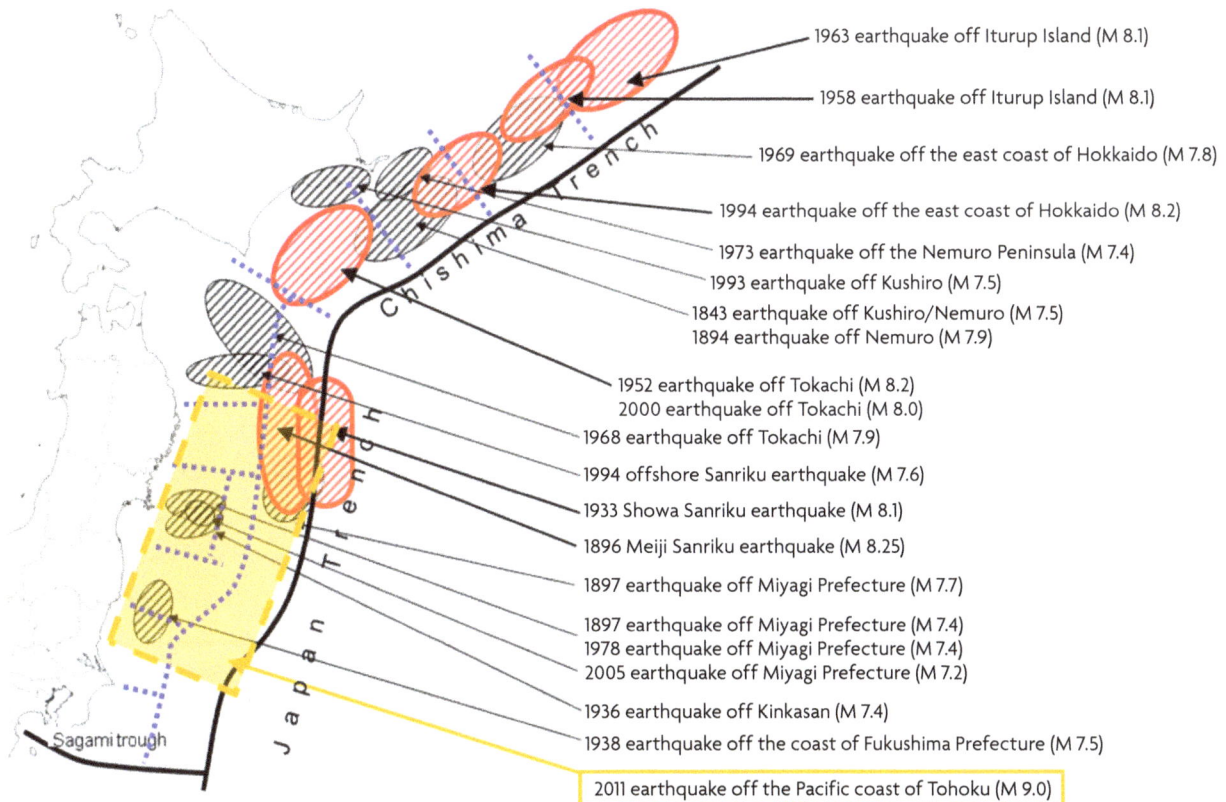

Map 25.2 Historical occurrence of trench-type earthquakes in the vicinity of the Japan and Chishima trenches.
Source: CAO.

The labels on the map read:

- 1963 earthquake off Iturup Island (M 8.1)
- 1958 earthquake off Iturup Island (M 8.1)
- 1969 earthquake off the east coast of Hokkaido (M 7.8)
- 1994 earthquake off the east coast of Hokkaido (M 8.2)
- 1973 earthquake off the Nemuro Peninsula (M 7.4)
- 1993 earthquake off Kushiro (M 7.5)
- 1843 earthquake off Kushiro/Nemuro (M 7.5)
- 1894 earthquake off Nemuro (M 7.9)
- 1952 earthquake off Tokachi (M 8.2)
- 2000 earthquake off Tokachi (M 8.0)
- 1968 earthquake off Tokachi (M 7.9)
- 1994 offshore Sanriku earthquake (M 7.6)
- 1933 Showa Sanriku earthquake (M 8.1)
- 1896 Meiji Sanriku earthquake (M 8.25)
- 1897 earthquake off Miyagi Prefecture (M 7.7)
- 1897 earthquake off Miyagi Prefecture (M 7.4)
- 1978 earthquake off Miyagi Prefecture (M 7.4)
- 2005 earthquake off Miyagi Prefecture (M 7.2)
- 1936 earthquake off Kinkasan (M 7.4)
- 1938 earthquake off the coast of Fukushima Prefecture (M 7.5)
- 2011 earthquake off the Pacific coast of Tohoku (M 9.0)

intensity and tsunami height, and the probability of recurrence in the same area was considered to be very low.

The magnitude of earthquake and tsunami hazards exceeded predisaster estimates

As illustrated in map 25.2, the March 11 earthquake had a very large epicentral and tsunami source area, larger than any earthquake recorded in Japan's history. Furthermore, its magnitude of M_w (moment magnitude) 9.0 exceeded the hazard level of any earthquake in the country ever considered for purposes of disaster management. Thus, the extent of the high seismic intensity area of the actual earthquake was much larger than expected, and the area that experienced Japanese seismic intensity of 5+ or larger was about 10 times the estimate (map 25.4). Furthermore, the actual tsunami height was twice the height used in the predisaster tsunami hazard predictions (map 25.5).

Because the magnitude of the earthquake and tsunami far exceeded the predisaster estimates, the Japanese government has been revising its methods of assessing earthquakes and tsunami hazards. The Basic Disaster Management Plan, revised after the Great East Japan Earthquake (GEJE), provides the following guidelines for estimating earthquakes and tsunamis:

- Earthquake and tsunami countermeasures should be based on scenarios that take into account the largest-possible earthquakes and tsunamis, which should be considered from every possible angle using all scientific means.

- Earthquake and tsunami scenarios should be based on the most accurate earthquake records available, going as far back in history as possible, and created in combination with an analysis of historical literature and topographical and geological studies, as well as other scientific findings.

As of October 2008

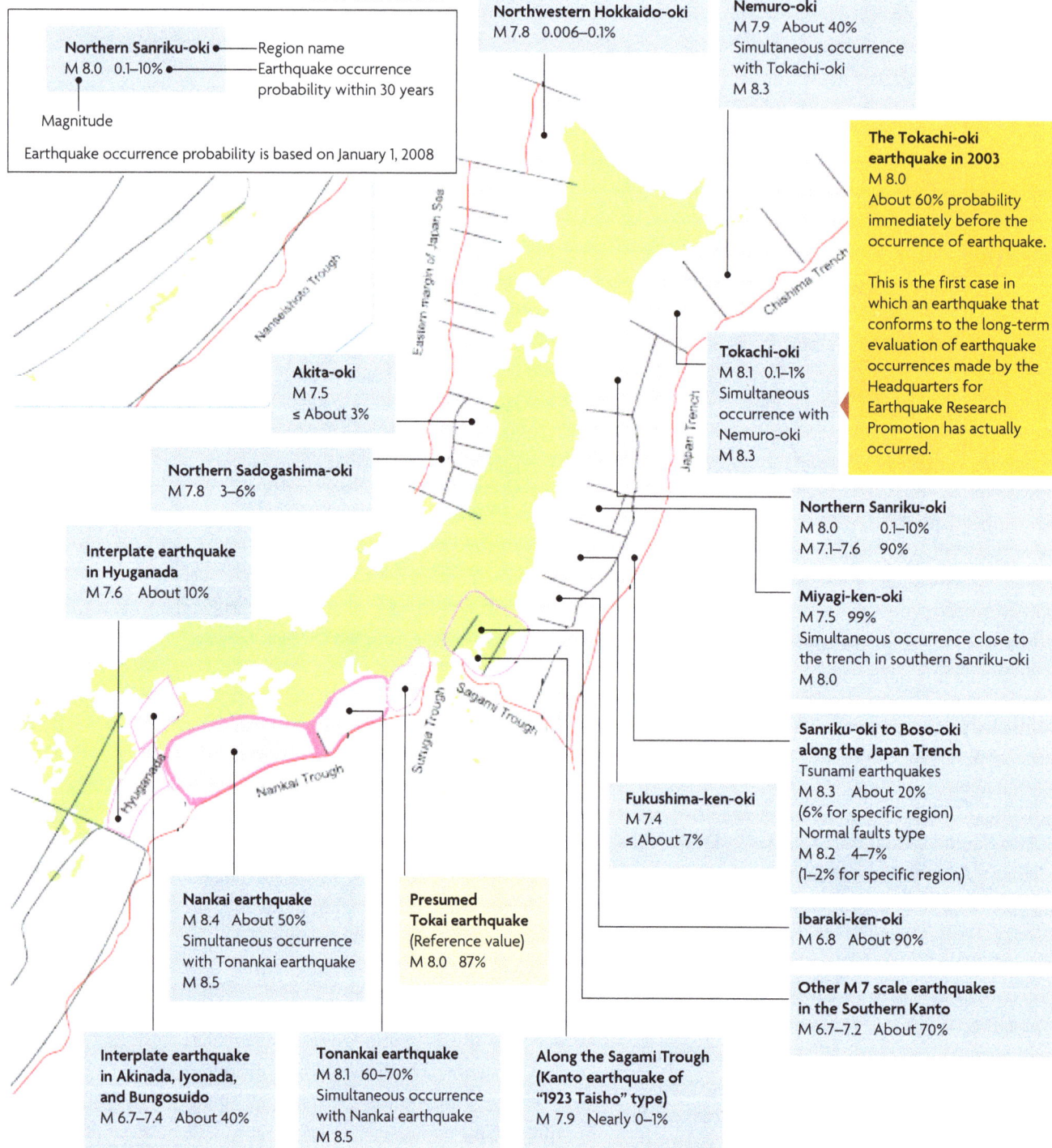

Northern Sanriku-oki ●——— Region name
M 8.0   0.1–10% ●——— Earthquake occurrence
                          probability within 30 years
Magnitude

Earthquake occurrence probability is based on January 1, 2008

Northwestern Hokkaido-oki
M 7.8   0.006–0.1%

Nemuro-oki
M 7.9   About 40%
Simultaneous occurrence
with Tokachi-oki
M 8.3

The Tokachi-oki
earthquake in 2003
M 8.0
About 60% probability
immediately before the
occurrence of earthquake.

This is the first case in
which an earthquake that
conforms to the long-term
evaluation of earthquake
occurrences made by the
Headquarters for
Earthquake Research
Promotion has actually
occurred.

Eastern margin of Japan Sea

Nanseishoto Trough

Chishima Trench

Japan Trench

Akita-oki
M 7.5
≤ About 3%

Tokachi-oki
M 8.1   0.1–1%
Simultaneous
occurrence with
Nemuro-oki
M 8.3

Northern Sadogashima-oki
M 7.8   3–6%

Northern Sanriku-oki
M 8.0   0.1–10%
M 7.1–7.6   90%

Interplate earthquake
in Hyuganada
M 7.6   About 10%

Miyagi-ken-oki
M 7.5   99%
Simultaneous occurrence close to
the trench in southern Sanriku-oki
M 8.0

Hyuganada

Nankai Trough

Suruga Trough

Sagami Trough

Sanriku-oki to Boso-oki
along the Japan Trench
Tsunami earthquakes
M 8.3   About 20%
(6% for specific region)
Normal faults type
M 8.2   4–7%
(1–2% for specific region)

Fukushima-ken-oki
M 7.4
≤ About 7%

Nankai earthquake
M 8.4   About 50%
Simultaneous occurrence
with Tonankai earthquake
M 8.5

Presumed
Tokai earthquake
(Reference value)
M 8.0   87%

Ibaraki-ken-oki
M 6.8   About 90%

Other M 7 scale earthquakes
in the Southern Kanto
M 6.7–7.2   About 70%

Interplate earthquake
in Akinada, Iyonada,
and Bungosuido
M 6.7–7.4   About 40%

Tonankai earthquake
M 8.1   60–70%
Simultaneous occurrence
with Nankai earthquake
M 8.5

Along the Sagami Trough
(Kanto earthquake of
"1923 Taisho" type)
M 7.9   Nearly 0–1%

Map 25.3 Potential earthquakes in Japan: their probability of occurrence, magnitude, and location

Source: Headquarters for Earthquake Research Promotion.

Note: -oki = offshore.

Estimating damage

The damage caused by the GEJE far exceeded any predisaster estimates. The number of completely destroyed buildings was about six times the estimated amount, and the number of human lives lost more than seven times (table 25.1). The conventional methodology

a. Tohoku earthquake off the Pacific coast (2011)

b. Estimation for trench-type earthquakes in the vicinity of the Japan and Chishima Trenches

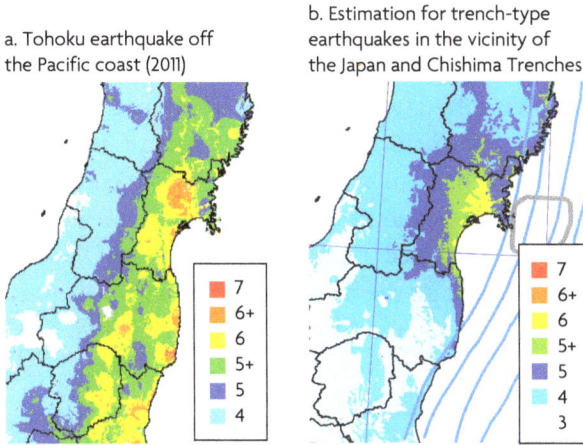

Map 25.4 Actual versus predicted seismic intensity distributions

Source: CAO.

Table 25.1 Comparison of estimated and actual damage

	ESTIMATION	GEJE	RATIO
Area with seismic intensity of 5+ or larger (km²)	3,540	34,843	9.8
Inundation area (km²)	270	561	2.1
Buildings completely destroyed	21,000	128,530	6.1
Disaster waste (tons)	1,400,000	24,900,000	17.8
Deaths (includes missing)	2,700	19,185	7.1

Source: CAO.

Note: The estimated figures reflect the larger of the damage estimates for the Miyagi offshore and Meiji-Sanriku earthquakes. Estimation of deaths uses the case of the Meiji-Sanriku earthquake with a low disaster awareness level. Deaths from the GEJE are as of January 31, 2012.

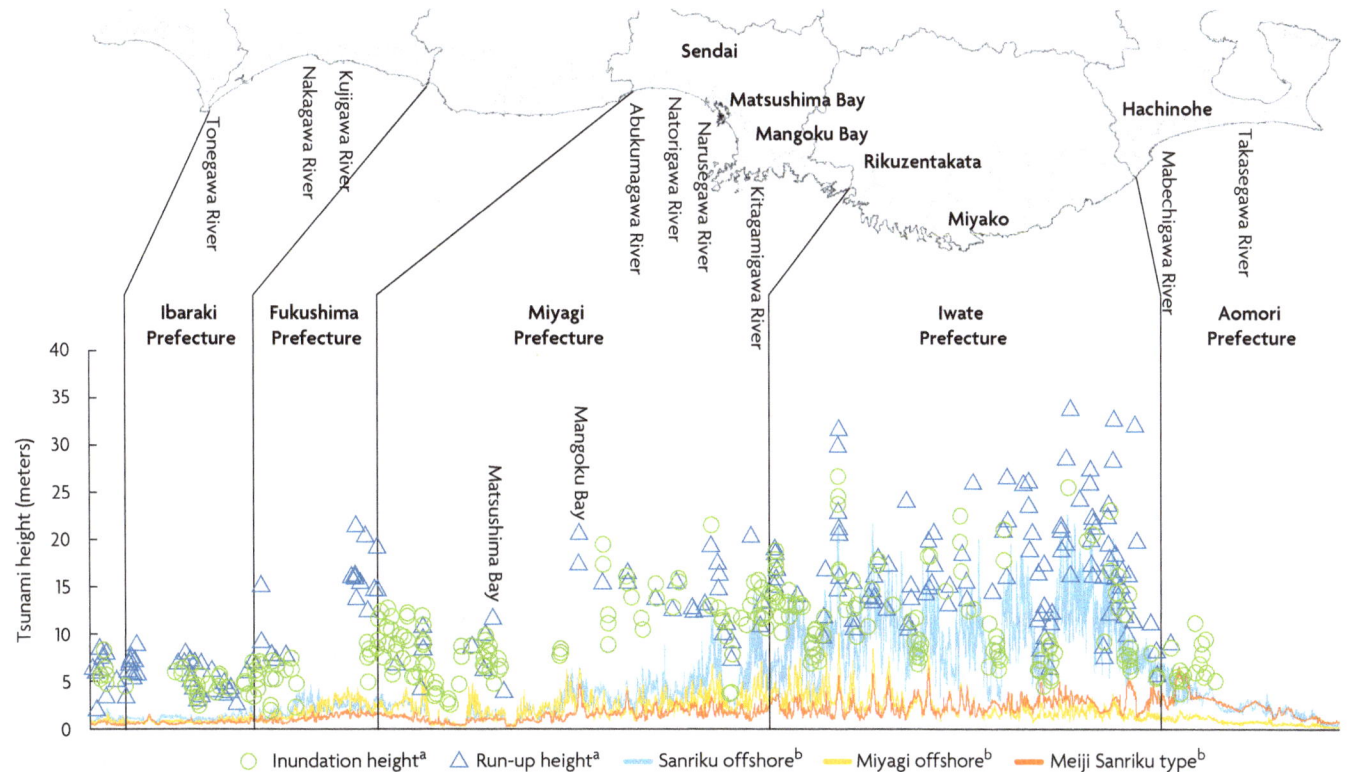

Map 25.5 Actual versus predicted tsunami height

Source: Ministry of Land, Infrastructure, Transport and Tourism (MLIT).

a. Actual records on March 11, 2011.

b. Simulation results before GEJE.

for estimating damages can be characterized as follows:

- Quantitative estimates include direct physical damage, human loss, damages to lifeline and transportation infrastructure, and economic losses (direct and indirect).

- Qualitative estimates include fires induced by tsunami; critical lifeline infrastructure facilities such as power plants, gas production plants, water and wastewater treatment plants, and so forth.

- Three scenarios were included, reflecting different seasons and times of day (winter 5 am, summer 12 pm, winter 6 pm), which are likely to affect fire scale and incidence.

- A facility is considered to have received no damage if it is equipped with enough mitigation measures against ground motion and fire.

A quantitative estimation of the impact was carried out using the relationship between the magnitude of the hazard (seismic intensity, maximum ground velocity, tsunami inundation depth, and so on) and the actual damage (number of destroyed houses, human loss, and so on), which was established based on historical earthquakes. For example, tsunami damage to buildings was estimated using the assumption that a building is completely destroyed if the inundation depth is 2.0 meters or more based on empirical evidence. Human losses caused by tsunamis were estimated based on the tsunami-affected population, historical records of death by tsunami inundation depth, and estimated evacuation rates (percentage of people who can obtain warning information and the time it takes for people to evacuate). These were calculated for 50-meter by 50-meter grid cells and overlaid on exposure data, such as spatial sociodemographic data available nationwide from the Geospatial Information Authority of Japan (GSI). Furthermore, infrastructure damage was estimated on the

basis of the estimated number of destroyed buildings, lifeline failure rates, and the number of days required for restoration, for which empirical relationships have been established based on previous disasters.

The underestimation of damage in the case of the GEJE was largely due to an underestimation of the magnitude of the hazards involved. Also, it has been pointed out that some factors—such as evacuation rates—used for damage estimation purposes were higher than actual rates, which could have further contributed to an underestimate of human losses. At the time of this writing, the damage estimation methodology is being revised.

Earthquake and tsunami simulation and hazard mapping

Hazard maps provide important information to help people understand the risks of natural hazards and to help mitigate disasters. Hazard maps indicate the extent of expected risk areas, and can be combined with disaster management information such as evacuation sites, evacuation routes, and so forth. In Japan, hazard maps are prepared and made available for various hazards such as earthquakes, tsunamis, floods, landslides, liquefaction, and volcanic eruption (chapters 26 and 27).

Japan's prefectural governments conduct hazard mapping, and the hazard data they prepare, for example, expected inundation depth and extent, is in turn used by the municipalities to prepare disaster management maps called hazard maps, which indicate not only the expected hazard but also information such as evacuation routes and evacuation sites (map 25.6). The Act on Special Measures for Earthquake Disaster Countermeasures, passed in 1995, mandates the prefectural governments and local municipalities to prepare these maps to promote awareness of earthquake and tsunami risks in their respective jurisdictions. As of 2010, more than 80 percent of the prefectures had prepared tsunami inundation maps and 50 percent of coastal municipalities were equipped with tsunami hazard maps.

The national government provides technical assistance and guidelines to promote hazard mapping by local governments. In 2004, the central government prepared *Tsunami and Storm Surge Hazard Map Guidelines* to help the municipalities in creating hazard maps and to promote the use of hazard maps throughout the country. The guidelines provide information on the basic concepts of tsunami and storm surge hazard maps, and the standard methodology for preparing them. The guidelines explain in depth the numerical simulation methodology for identifying inundation risk areas, which is the principal means of tsunami hazard mapping. Alternative methodologies, as shown in table 25.2, are also explained so that the best method can be selected according to the resources and data available. The numerical simulation of tsunamis generally requires the following steps:

- Development of a fault model

- Topographic data

- Setting of initial water level conditions (typically uses the vertical displacement calculated by the fault model)

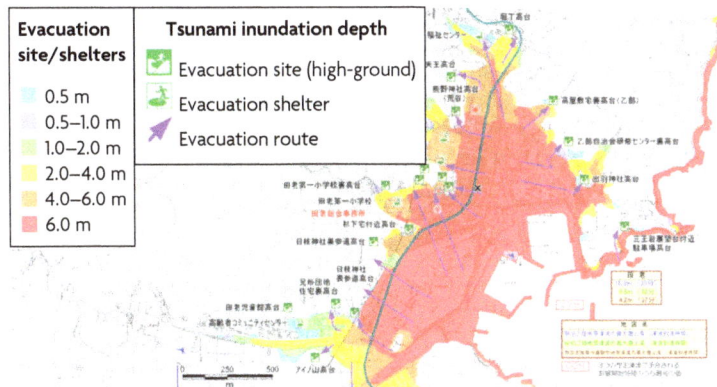

Map 25.6 An example of a tsunami hazard map, Miyako City, Iwate Prefecture
Source: Miyako City.

- Calibration and verification of the model

- Predictive simulation

Hazard maps in Japan have been used by the municipalities to design evacuation procedures. But they have not been utilized for land use or development planning. The lessons learned from the GEJE have prompted the Japanese government to implement a new act to create tsunami-resilient cities. The new legislative framework calls for the prefectural governments to prepare an inundation risk map, which is to be used for regulating land use and mitigating the effects of a tsunami (chapter 12).

Table 25.2 Methods for defining inundation risk areas

METHOD	PROCEDURE	ADVANTAGES/DISADVANTAGES
Numerical simulation in time series	Use numerical models to estimate inundation area as well as inundation depth and flow velocity, inundation time.	Precise assessment is possible and can take into account the effects of the disaster mitigation structures. Resource intensive.
Level-filling method	Calculate the inundation based on the height and width of the tsunami and estimate the extent of inundation based on the topographical data.	Not so resource intensive. Ignores the effects of structures and buildings and the momentum of water flow (tsunami run-up).
Prediction based on past inundation	Define the risk area based on the inundation area of historical tsunami events.	Simple and low cost. Cannot be used for areas with no historical records. Cannot reflect changes such as construction of disaster reduction facilities.
Estimation based on ground elevation	Define high-risk areas as those areas lying lower than the expected tsunami height.	Simple and low cost. Cannot take into account the effects of structures and buildings and the momentum of water flow (tsunami run-up).

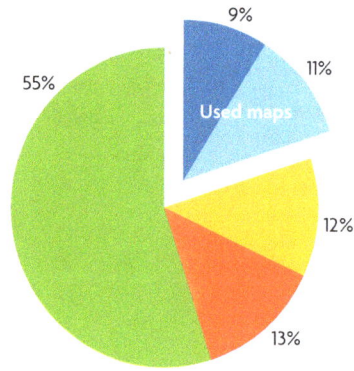

Figure 25.1 Hazard map usage patterns

Source: CAO.

Legend:
- Had map on wall at home
- Often referred to map at home
- Had map at home but didn't use at all
- Did not have map at home, but knew it was available
- Never saw nor heard of map

Pie chart values: 9%, 11%, 12%, 13%, 55%. "Used maps" labeled.

Hazard maps in the disaster-affected areas

All municipalities hit by tsunamis during the GEJE had prepared hazard maps before the earthquake and tsunami. But surveys show that only 20 percent of the people used these hazard maps (figure 25.1), and the extent of flooding indicated on the hazard maps was in many cases underestimated compared to the actual inundation area (map 25.7). It is likely that these maps provided residents with a false sense of safety, and prevented people from evacuating, resulting in greater human losses.

LESSONS

- *Hazard assessment is critical* since it serves as the basis for DRM policies. Earthquake and tsunami hazard assessment is conducted extensively in Japan to raise public awareness and to prepare for disasters.

- *Underestimation is frequent.* Predisaster damage estimation was low due to the underestimation of hazard levels. Past assessments did not adequately consider certain kinds of damage, including from long-period seismic waves, tsunami-induced fires, and nuclear accidents.

- *Recognizing the uncertainties associated with hazard assessment,* the largest-possible hazard scenario should have been used, drawing on all available information including not only seismological but also geological, archaeological, and historical studies looking at tsunami deposits, ancient documentation, and so on.

- *Hazard maps were developed by all municipalities in the disaster-hit areas* and served as important tools for designing evacuation procedures.

- *Hazard maps should facilitate and guide people's evacuation efforts and should not contribute to a false sense of safety.* Providing information on inundation risk zones for multiple levels of hazards including low-frequency events, or information directly linked with tsunami warnings would be effective. The meaning of the information provided on the maps needs to be clear and adequately explained to the users.

- *Risk information must be communicated to the public effectively.* In the GEJE, only 20 percent of the people made use of hazard maps.

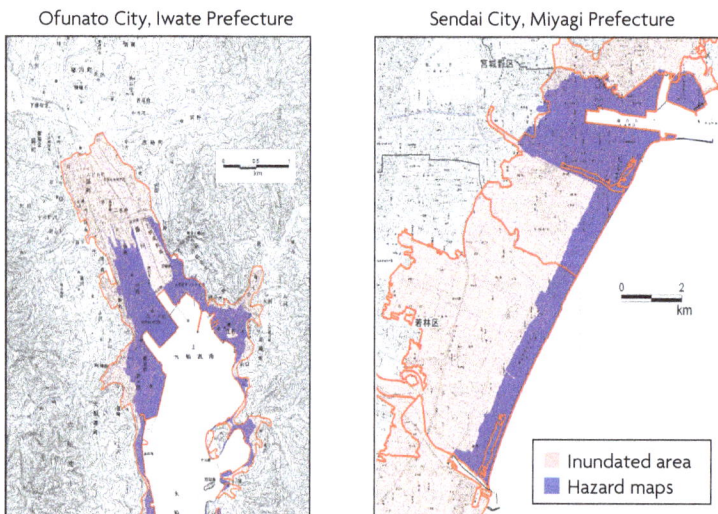

Ofunato City, Iwate Prefecture

Sendai City, Miyagi Prefecture

Legend:
- Inundated area
- Hazard maps

Map 25.7 Inundation area: Hazard map versus actual

Source: CAO.

RECOMMENDATIONS FOR DEVELOPING COUNTRIES

Assess risk. Understanding hazard and risk is a vital component of DRM. Quantitative estimation of potential damage is important as it informs the appropriate strategies and measures to be taken. Risk exposure data should be collected, mapped, and shared as they are vital components of risk assessment.

Prepare for the worst case. While bearing in mind that the hazard assessment of earthquakes and tsunamis has limitations and uncertainties, the largest possible hazard should be investigated and considered in formulating DRM policies. Hazard assessment should not rely solely on statistical analysis based on historically recorded earthquakes and tsunamis, because historical records may not account for the maximum-possible hazard levels that may occur in the future. Also, disasters have occurred for which there are no records available. The level of hazard to be used in designing structural measures should be selected based on local conditions. Hazard and risk assessment should be revised and updated periodically with the latest findings and in light of more recently experienced disasters.

Prepare and promote hazard maps. Hazard maps are effective tools for promoting risk awareness, for designing evacuation procedures, and for deciding the locations of evacuation facilities and shelters. Hazard maps should be easy to understand and easy to use for purposes of prompt evacuation, and users should be aware of the limitations and uncertainties of the information they contain. Considering budget and technical constraints, risk estimation methods can be selected as explained in table 25.2.

Share hazard and risk data and information. Data can be shared through central depositories that are open to the public, among other means (see chapter 26).

NOTE

Prepared by Junko Sagara, CTI Engineering, and Keiko Saito, World Bank.

BIBLIOGRAPHY

Cabinet Office. 2012. "Current Status of Efforts Toward Revision of Disaster Management Measures in Light of Lessons Learned from the Great East Japan Earthquake" [in Japanese]. Cabinet Office, Tokyo.

Central Disaster Management Council. 2011. *Report of the Committee for Technical Investigation on Countermeasures for Earthquakes and Tsunamis Based on the Lessons Learned from the 2011 Earthquake off the Pacific Coast of Tohoku* [in Japanese]. Cabinet Office, Tokyo.

Study Committee on Tsunami and Storm Surge Hazard Maps. 2004. *Tsunami and Storm Surge Hazard Map Manual* [in Japanese]. http://www.icharm.pwri.go.jp/publication/pdf/2004/tsunami_and_storm_surge_hazard_map_manual.pdf.

CHAPTER 26

Risk and Damage Information Management

In Japan, municipalities are responsible for producing hazard maps for floods, storm surges, volcanic eruptions, tsunamis, stagnant water, and landslides to which the municipality may be exposed. By combining exposure data with satellite images and aerial photographs, postevent damage assessments can be carried out with reasonable accuracy. Japan's experience with the disaster of March 2011 demonstrates that having exhaustive data on exposure expedites the damage assessment process, thereby reducing the time required for compensation payments and insurance payouts.

Japan is known for its disaster preparedness. Less well known but no less important for disaster response is the country's "data preparedness."

Communities need to understand the risks they face and to have access to early warnings. In Japan, maps that illustrate the likely extent of hazards and the location of evacuation centers and routes are distributed to households and public institutions, such as schools and hospitals, in an effort to raise public awareness of disaster risk. Immediately after the Great East Japan Earthquake (GEJE) and tsunami, information on the damage caused by the disaster was collected rapidly and shared among responding agencies using a variety of top-down and bottom-up tools, including remotely sensed data, public and private data sets, and online tools (such as the Ushahidi-based website www.sinsai.info). The data-collection and dissemination effort underpinned assistance to the affected population, timely allocation of resources to areas in need, and effective reconstruction planning.

FINDINGS

Pre-event dissemination of information on risks

In Japan, municipalities are responsible for producing maps related to the following hazards: floods, storm surges, volcanic eruptions, tsunamis, stagnant water, and geological hazards (landslides). These hazard maps include not only information on the expected intensity and extent of the hazard but also the location of evacuation centers and designated evacuation routes (chapter 25). The hazard map Web portal prepared by the Ministry of Land, Infrastructure, Transport and Tourism (MLIT) includes a link to all available hazard maps, providing a one-stop shop where information on risks from natural hazards can be accessed (map 26.1).

Post-event collection of damage data

Learning from their experiences with past events, Japan Self-Defense Forces (JSDF) have been upgrading their emergency response plans. One of the JSDF's tasks is to capture video footage of the affected region immediately following a major disaster event. In the case of the GEJE, helicopters were dispatched immediately after the main shock. They transmitted footage of the approaching tsunami live on national and global news channels, contributing to the quick mobilization of resources.

In the immediate aftermath of a natural disaster, the collection of information on the damage allows appropriate resources to be allocated for response activities. Traditionally, data have been collected by people sent to the affected areas. During the past decade, however, the use of remotely sensed data has become viable for damage data collection thanks to improvements in the spatial resolution of such data (less than 1 meter with optical satellite images) and reductions in acquisition costs.

Following a disaster, satellite data are the first to become available, followed by aerial photographs, which provide more detailed images. Aerial surveys are subject to logistical delays, whereas satellites are already in orbit and can generally deliver data within 24 hours to a few days, depending on the satellite. With aerial surveys, by contrast, weather conditions

Map 26.1 Interface of the MLIT hazard map Web portal

Source: Ministry of Land, Infrastructure, Transport and Tourism (MLIT).

Note: The interface shows (in green) the municipalities for which tsunami hazard maps have been published. Clicking on the municipality takes the user to the municipality's website, where the actual hazard map can be accessed.

must be good, and the area that a single image can cover is smaller than the area covered by a satellite image, prolonging the time required to photograph a given area.

The International Charter organization provides member states with a unified system of space data acquisition and delivery. Member states can request satellite data at no cost in the event of emergencies following natural or manmade disasters. Remotely sensed data are analyzed by predesignated value-adding vendors to derive and deliver the information requested by the affected country. After the GEJE, the International Charter was activated through the Cabinet Office of Japan, the designated authorized user in Japan. Products produced through the charter ranged from maps of the extent of inundation from the tsunami to areas of liquefaction, spot checks in areas of interest, and estimates of the volume of debris (table 26.1).

Public-private partnership between aerial survey firms and the Geospatial Information Authority of Japan

Japan has been using remotely sensed data following major natural hazard events for some time. In 1995, following the Great Hanshin-Awaji (Kobe) Earthquake, the National Broadcasting Corporation (NHK) flew helicopters with high-definition video cameras over Kobe city to capture the damage. Private aerial survey firms deploy aircraft to take aerial photographs and other types of remotely

Table 26.1 Excerpts from survey of end users on the use of satellite-based remotely sensed data carried out by JAXA, 2011

END USER	USE OF DATA
Cabinet Secretariat	Spot checks of areas of interest, for example, Sendai airport, Fukushima Nuclear Power Station. Pre- and postevent images. Maps of maximum inundation.
Cabinet Office	Overview map using ALOS postearthquake images. International Charter products. Imagery related to Fukushima Nuclear Power Station.
Ministry of Land, Infrastructure, Transport and Tourism	Maps of maximum extent of inundation. Data based on interpretation of PALSAR and AVNIR-2 taken on March 21, 25, and 30, 2011. Information on areas with stagnant water also continuously provided. Request to monitor 40,000 areas designated as being at high risk from landslide. Wildfire monitoring.
Ministry of Agriculture, Forestry and Fisheries (MAFF)	Request for information on inundation and presence of stagnant water in agricultural areas. MAFF estimates inundated agricultural area to be 24,000 hectares in six prefectures. Information on inundation in the northern parts of Chiba and Ibaraki prefectures also requested. Data to be used by MAFF to validate ground surveys and for recovery planning.
Fishery Agency	Collaboration sought to assist in offshore search for lost ships.
Ministry of Environment	Request to assist in mapping debris floating off the coast of Sanriku. 560,000 m^2 of debris already identified in vicinity of Rikuzentakata alone.
Ministry of Education, Culture, Sports, Science and Technology	Images of Fukushima Nuclear Power Station.
Geospatial Information Authority of Japan (GSI)	Providing all available imagery. Using electronic control points provided by GSI and InSAR data analyzed by JAXA; crustal deformation of 3.5 meters was identified in Oshika Peninsula.
Miyagi Prefecture	Sighting of an SOS sign in a park in Miyagi Prefecture was reported by the International Charter.
Iwate Prefecture/University	Monitoring of road accessibility.
Kanto Regional Development Bureau	Mapping of liquefaction areas provided through International Charter.

Source: JAXA 2011.

Note: ALOS = Advanced Land Observation Satellite; PALSAR = phased array type L-band synthetic aperture radar; AVNIR-2 = advanced visible and near infrared radiometer type 2; InSAR = interferometric synthetic aperture radar; JAXA = Japan Aerospace Exploration Agency.

sensed data (for example, LiDAR data, in the case of landslides or volcanic eruptions) following every natural disaster event in Japan. Currently the major aerial survey companies have a public-private partnership with the Geospatial Information Authority of Japan (GSI) under which they jointly capture damage information, thus avoiding duplication of effort. The agreement has been in effect for some years, resulting in an archive of records documenting the changes caused by natural disasters in Japan.

Following the GEJE and tsunami, the partnership spent a month taking aerial photographs of the entire Tohoku region coastline (approximately 500 kilometers, [km]).

Tsunami inundation mapping using remotely sensed data

As early as five days after the tsunami, the GSI announced the first estimate of the total inundation area as 400 square kilometers (km^2), based on manual interpretation of aerial photographs taken on March 12 and 13. One month after the event, on April 18, the government officially announced the total inundation extent to be 561 km^2. The increase reflected the availability of additional aerial photographs and high-resolution optical satellite images of areas previously not covered.

Although GSI's inundation mapping was considered the official information, other organizations used various methodologies and data sources to map the extent of inundation (a list of these can be found in Earthquake Engineering Field Investigation Team [EEFIT] 2011).

For 30 municipalities the Statistics Bureau of Japan compared the difference between the estimate of the population affected by inundation derived using GSI's aerial photographs with that produced by a private company. Some of the differences are shown in table 26.2. In most cases, the differences between the two estimates are negligible in relation to the total population in the respective municipalities. In a few cases, however, the difference amounted to more than 20 percent of the total population of that municipality. In Shiogama the difference between the estimates was more than 30 percent of the total population. The full comparison results can be found on the Statistics Bureau's website.

In an independent validation of the mapping performed using Japan Aerospace Exploration Agency's (JAXA) ALOS satellite image and GSI's aerial photographs, Sawada (2011) found a substantial difference in the area shown as inundated: interpretations based on aerial photographs reported twice as much inundated area as interpretations based on satellite images.

Table 26.2 Examples of the difference between estimates of the affected population in municipalities in Miyagi Prefecture using two different estimates of the extent of inundation

MUNICIPALITY	POPULATION TOTAL (2007 CENSUS)	POPULATION WITHIN INUNDATED AREA			DIFFERENCE AS PERCENTAGE OF TOTAL POPULATION
		AS ESTIMATED BY GSI	AS ESTIMATED BY PRIVATE COMPANY	DIFFERENCE BETWEEN GSI AND PRIVATE COMPANY	
Miyagino-ku	182,678	17,375	11,858	5,517	3.0
Wakabayashi-ku	129,942	9,386	8,700	686	0.5
Taihaku-ku	222,447	3,201	2,519	682	0.3
Ishinomaki	167,324	112,276	102,670	9,606	5.7
Shiogama	59,357	18,718	173	18,545	31.2

Source: Ministry of Internal Affairs and Communications.

Spatial data preparedness in Japan

Decision makers need spatial data to make informed decisions about disaster preparedness, post-event responses, and recovery planning. Spatial data provide information on the location of key infrastructure, populations, agriculture, industrial facilities, education and health facilities, and so on. In Japan these data sets are freely available from the GSI website in both raster and vector formats. Building-specific data on exposure levels are also commercially available for the entire country. Overlaying these data sets with the mapped hazard (for example, the extent of tsunami inundation) permits a rapid damage assessment. Commercial building-specific data sets were made available at no cost to enable response agencies to assist in the relief and recovery activities (figure 26.1, box 26.1).

Quick determination of government compensation and insurance payments through the use of aerial photographs

Aerial photographs were used in an innovative way to determine compensation payments from local governments and payouts of earthquake insurance. Because the area of inundation was clearly visible from aerial photographs, and because the tsunami was so powerful, it was deemed that structures located within the coastal inundation zones were completely destroyed. The owners, therefore, were eligible for full compensation.

The innovation in these cases lies in the fact that payouts were made without sending an inspector or a loss adjuster to the address—that is, the aerial photographs were the sole source of claim verification. This system expedited the claim-payment process, resulting in an average payout by the earthquake insurance schemes of $250 million per day during the last week of April 2011—1.5 months after the earthquake (see chapter 29).

Although data preparedness is advanced in Japan, some of the information is available only in Japanese, and navigating the web-

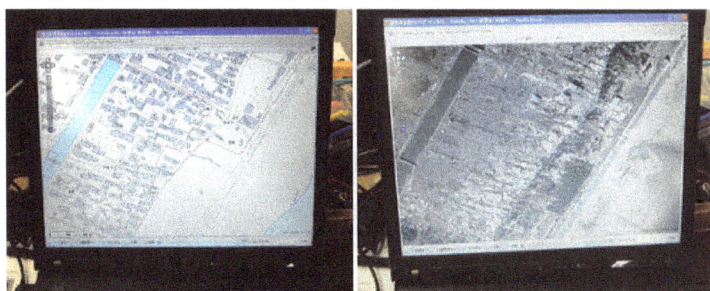

Figure 26.1 Commercial per-building exposure data set (left) and post-3.11 aerial photograph of the same area (right)

Source: All311 website.

Note: The open source platform (http://all311.ecom-plat.jp/) provides the ability to overlay the two. These maps can be accessed at volunteer centers in the Tohoku region. Local governments can apply to have the system installed at no cost in their areas.

BOX 26.1

Crowd-sourced damage assessment using remotely sensed data in Haiti and New Zealand

When hazard information is combined with geocoded data on key infrastructure and mechanisms to analyze "big data" (for example, crowdsourcing), it has the potential to provide damage information rapidly and with reasonable accuracy. In the case of the tsunami damage assessment following the GEJE, a binary damage-assessment system was used, in which building-level data on structures that had been geocoded before the event was overlaid on data on the extent of the disaster, permitting a high-confidence assessment of whether a building was destroyed.

Similar methodologies have been used and continue to be tested for earthquake damage assessment in Haiti and in Christchurch, New Zealand. Large-scale crowd-sourced earthquake damage assessments have been carried out with a view to operationalizing the methodology. Accuracy assessments are being performed to ascertain the level of accuracy that is achievable using these tools. Remotely sensed data has also been used for flood damage assessment. In all cases, it is clear that the accuracy of the damage assessment increases where pertinent data on key infrastructure are available, making a strong case for data preparedness.

---

sites where data are available can be difficult. OpenStreetMap (OSM) is an international volunteer technical community dedicated to producing freely available, detailed topographic data for the entire globe. Local volunteers donate their time to trace satellite images made available for the purpose. To accommodate the international community's need for topographic maps and English annotation,

Figure 26.2 Online interface of Geospatial Disaster Management Mash-up Service Study (GDMS)

Source: GDMS website, http://gdms.jp.

OSM volunteers created detailed maps of the entire Tohoku coastal region and began publishing the resulting topographic maps online just a few hours after the main shock.

The OSM maps are open; that is, the data can be used across different platforms and without any restrictions. Another characteristic of the maps is that all annotations are available in the local language as well as in English. Moreover, the styles used in the maps are standardized, providing a consistent feel. In some countries, the OSM platform is being used as a tool to raise awareness in

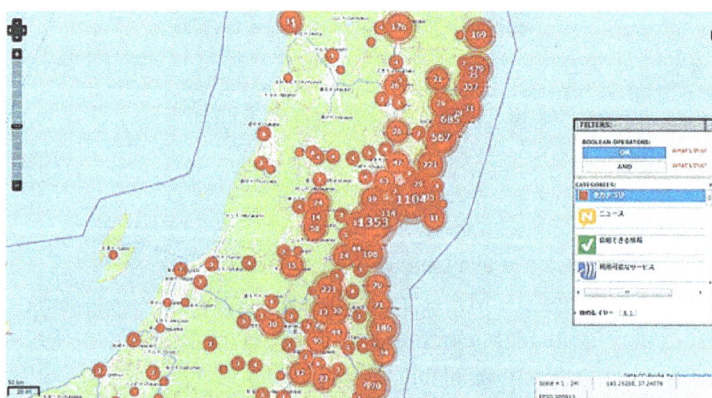

Map 26.2 The interface of Sinsai.info (based on the Ushahidi platform)

Source: http://www.sinsai.info/.

Note: The red circles show the number and locations of the requests from local communities. The diameters of the circles are proportional to the number of requests logged at that location. OSM Japan, prepared following the event by local and international OSM volunteers, is used as the backdrop.

communities at risk from natural disasters by involving them in collecting data on their own communities.

Online platforms to store and distribute spatial data following the earthquake and tsunami

Much of the spatial data created following the GEJE is open data. Several online platforms have been created to host and distribute these open datasets to assist in damage assessment, to facilitate response and relief activities on the ground, and to help local communities. Two such platforms are the Emergency Mapping Team (EMT) and the Geospatial Disaster Management Mash-up Service Study (GDMS, figure 26.2). Most of the platforms use a map interface, against which the data hosted on the system are visualized spatially.

Use of social media for bottom-up information sharing

In recent years, the use of social media in post-disaster settings has spread around the world. Even after the tsunami, when the entire phone network and Internet were down, information from the affected areas came through on social media such as Twitter and Facebook (chapter 15). Many families stayed in touch using these media in the immediate aftermath. Japanese mobile networks and telecommunication companies have well-established systems that allow subscribers to leave messages for their loved ones. Google set up an online person finder after the GEJE.

Twitter, Facebook, and new types of social media such as Ushahidi are establishing themselves as a global standard for collecting information on needs in local communities. Ushahidi is an open source online interface that allows bottom-up information sharing. Developed to ensure a fair election in Kenya in 2008, the platform is designed to allow anyone to upload information or requests for help, using Twitter or e-mail, which are visualized on a map interface (map 26.2), thus making them actionable items. Sinsai.info, a combination of Ushahidi and OSM Japan, was launched in the immediate aftermath of the GEJE, when OSM data was being used as the base map to display requests for help coming in from communities in the Tohoku region.

All311 is another site that was launched immediately after the event. Hosted by the National Research Institute for Earth Science and Disaster Prevention (NIED) and built using an e-community platform developed by NIED, the site is a one-stop shop for information on ongoing activities, both top-down and bottom-up, in the recovery process. Information is provided in Japanese only. Its e-community is an open source tool for developing information-sharing platforms with spatial content.

LESSONS

- *Satellite images are available before aerial photographs, but they do not reveal as much detail.* After the GEJE, a standing public-private partnership between the major aerial survey companies and GSI captured aerial photographs of the areas affected by the GEJE. GSI published an estimate of the inundated area five days after the event, based on manual interpretation of the aerial photographs then available.

- *The limits of technology for response activities should be recognized.* In the GEJE, the inundation area mapped from aerial photographs was much larger than that mapped from satellite images.

- *Aerial photographs expedited the claim-payment process.* By overlaying the tsunami inundation estimates with commercially available building-level data sets, it was possible, for insurance purposes, to designate structures that had been completely destroyed by the tsunami.

- *Crowd-sourced methods for collecting damage information have great potential.* After the GEJE, OSM volunteers were mobilized to create topographical maps of the region with annotations in English and Japanese.

- *Online platforms* were created to host and distribute spatial data useful for response and recovery. Sinsai.info and All311 are two examples.

RECOMMENDATIONS FOR DEVELOPING COUNTRIES

A one-stop online portal is a good way of disseminating hazard maps for a given country. However, in countries where Internet access is not readily available, an online portal may not necessarily be optimal. Conventional methods, such as paper maps and booklets, should be utilized as well.

Data preparedness is a key ingredient for both pre-event disaster risk management and post-event damage assessment and reconstruction planning. Data collection on key infrastructure should be carried out during normal times and kept up to date. The data can be used for other purposes such as town planning.

Satellite images and aerial photographs are now routinely used for postevent damage assessment. Damage assessment can be carried out with reasonable accuracy by combining data on infrastructure with exposure data. Collected data should have a specific, well-managed repository and be paired with appropriate tools to analyze the data for risk-assessment purposes.

New information and communications technology tools are increasingly being used in emergency situations. Open source portals, such as the Ushahidi-based sinsai.info, are important tools that allow requests for help from local people to be logged and acted upon. Creating protocols for how these volunteer-based communities can work with official government entities is increasingly important.

NOTE

Prepared by Keiko Saito, World Bank. Special thanks to the Earthquake Engineering Field Investigation Team (EEFIT), United Kingdom.

BIBLIOGRAPHY

ALL311. 2011. http://all311.ecom-plat.jp/ [in Japanese].

Corbane, C., K. Saito, E. Bjorgo, L. Dell'Oro, R. Eguchi, G. Evans, S. Ghosh, B. Adams, R. Gartley, F. Ghesquiere, S. Gill, T. Kemper, R. S. G. Krishnan, G. Lemoine, B. Piard, O. Senegas, R. Spence, W. Svekla, and J. Toro. 2011. "A Comprehensive Analysis of Building Damage in the January 12, 2010, M7 Haiti Earthquake Using High-Resolution Satellite and Aerial Imagery." *Photogrammetric Engineering and Remote Sensing,* Special issue on the 2010 Haiti Earthquake.

EEFIT (Earthquake Engineering Field Investigation Team). 2011. *The Mw9.0 Tohoku Earthquake and Tsunami of 11th March 2011—A Field Report by EEFIT.* London, U.K.: EEFIT. http://www.istructe.org/resources-centre/technical-topic-areas/eefit/eefit-reports.

Fire and Disaster Management Agency. 2011. "Disaster Management Measures of Municipalities and their Response to the Great East Japan Earthquake," slide 2 [in Japanese]. Paper presented at the first meeting of the Enhancement and Strengthening of Earthquake and Tsunami Countermeasures in the Regional Disaster Management Plan. http://www.fdma.go.jp/disaster/chiikibousai_kento/01/shiryo_05.pdf.

GFDRR (Global Facility for Disaster Reduction and Recovery). 2011. *Volunteer Technology Communities Open Development.* Washington, DC: World Bank Group. http://www.gfdrr.org/gfdrr/sites/gfdrr.org/files/documents/Volunteer%20Technology%20Communities%20-%20Open%20Development.pdf.

GSI (Geospatial Information Authority of Japan). 2011. "Estimate of the Inundation Area" [in Japanese]. http://www.gsi.go.jp/kikaku/kikaku60001.html (version 1); http://www.gsi.go.jp/common/000059939.pdf (version 5).

JAXA (Japan Aerospace Exploration Agency). 2011. "Use of Satellite Based Data for the Great East Japan Earthquake and Tsunami" [in Japanese]. http://www.jaxa.jp/press/2011/04/20110406_sac_earthquakes.pdf.

MLIT (Ministry of Land, Infrastructure, Transport and Tourism). 2011. "Disaster Portal" [in Japanese]. http://disapotal.gsi.go.jp/.

Sawada, H. 2011. "Remote Sensing for Emergency Mapping." PowerPoint presentation [in Japanese]. http://stlab.iis.u-tokyo.ac.jp/~sawada/files/GreatEarthquakePresentatio0425.pdf.

Statistics Bureau. 2011. http://www.stat.go.jp/info/shinsai/zuhyou/sai.xls [in Japanese].

Ushahidi. 2011. http://www.ushahidi.com/.

ZENRIN. 2011. http://www.zenrin.co.jp/news/110415.html [in Japanese].

CHAPTER 27

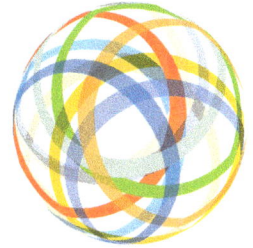

Risk Communication

Risk communication is an important component of disaster risk management because it shapes people's perceptions of risk and influences their actions with respect to disaster preparedness and disaster response. It also influences the intervention decisions that are made throughout the disaster management cycle. The credibility of the information source takes a long time to build and needs to be well established before a disaster strikes. In Japan, the level of trust in government and other official communications was sorely tested following the nuclear accident at the Fukushima Daiichi Nuclear Power Station.

Disaster preparedness is often perceived as being mainly a governmental responsibility, with information and directives traveling from the top down. That is the case to some extent, since local communities generally lack the tools and skills needed to conduct scientific risk assessments and fully understand the underlying risk in their localities without expert assistance. The problem with the top-down approach is that policies may be imposed on communities without taking local conditions into account, and communities may become overly dependent on information coming from the government. Recent experiences from the Great East Japan Earthquake (GEJE) showed that when the local community was involved in planning for disaster preparedness, and people took ownership of their own safety plans, they were better prepared and better able to take the necessary actions to protect themselves.

Successful risk communication occurs when there is holistic learning, facilitation, and trust. In holistic learning, the gap in knowledge between the information sender and receiver is minimal (figure 27.1). Hazard maps, booklets, and videos can all help narrow that gap when it comes to disaster education and risk communication.

Normally, the information generators or senders are government agencies, universities, or research institutions that have the capacity to assess risk and the political mandate to

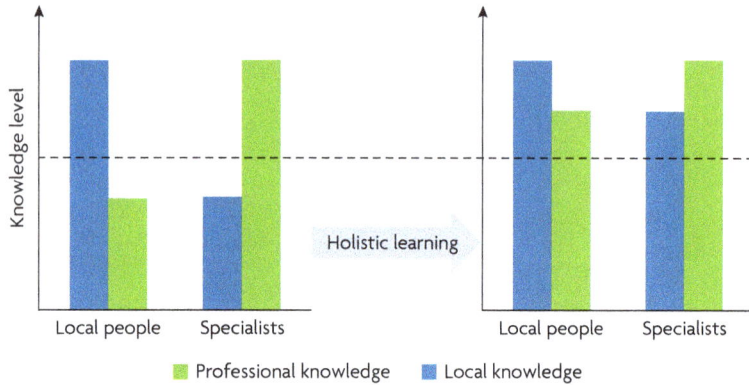

Figure 27.1 The concept of holistic learning: Narrowing the gap between local and specialist knowledge

Legend:
- ■ Professional knowledge
- ■ Local knowledge

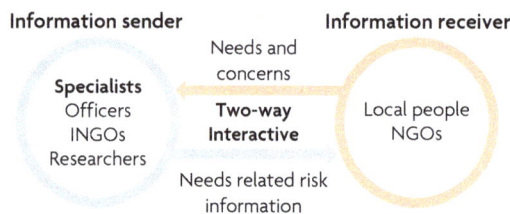

Figure 27.2 The risk communication framework

Source: Kikkawa 1999.

implement disaster risk management (DRM) measures. The information receivers are the communities, businesses, and individuals who have knowledge of the local area and are the ultimate users of the risk information (figure 27.2).

FINDINGS

The importance of trusting the information provider

Early warnings greatly influence how people perceive and evaluate the risks from imminent hazard and their subsequent decision to evacuate. In this respect, the level of trust in and the credibility of the person, institution, or medium issuing the warning is of crucial importance. Furthermore, factors such as fatalism can affect evacuation decisions. People who have responded to too many false alarms may not take the warnings seriously.

In some cases, the underestimation of the height of the tsunami in the warnings that went out on March 11 likely delayed evacuation and possibly increased fatalities (chapter 7). Japan's proposed new early warning scheme will not include any numerical values for tsunami height in the first warning but will use more descriptive expressions, such as "massive" or "very high" waves, in the event of earthquakes larger than magnitude 8. These terms will be further qualified by expressions such as a "tsunami height equivalent to the GEJE is expected."

Official risk communication tools: Hazard maps

In Japan, hazard maps indicate expected hazard levels and locations as well as the location of evacuation centers and routes (chapter 25). Map 27.1 was prepared by the village of Toni (Kamaishi City, Iwate Prefecture) in a local

Map 27.1 Hazard map produced by the village of Toni in Kamaishi City, Iwate Prefecture

Source: Kamaishi City.

workshop with community members. It includes predicted inundation depths indicated by colors, historical records of inundated areas, lead times, evacuation shelters, and telephone numbers for warnings. The hazard map was printed and distributed to all families in Toni before the GEJE.

Developing this type of disaster map through a participatory process is an effective way of communicating risk to the community at large. A postdisaster survey in the Toni area identified citizens' motivations for participating in the mapmaking process (figure 27.3).

Problems with the hazard maps in use

Mapping schemes differ in the colors and symbols used to convey hazard information. In the United States, efforts are being made to ensure the consistency of the content of hazard maps, as well as their design.

While hazard maps are useful tools to help communities understand the risks they face, there are, nevertheless, uncertainties associated with the assessment of the hazard risk itself—future disasters may exceed the levels indicated on the maps. In addition to producing and delivering the maps, their content should be presented to local communities, as was done in Toni Village. In the course of such presentations, governments and experts must explain the limitations of prediction technology. In the GEJE, the maps provided residents with a false sense of safety. Only 20 percent of residents utilized hazard maps for their evacuation in the GEJE (chapter 25).

Another way of raising awareness of risk is through evacuation drills carried out under as many different scenarios as possible, for example, at night or in rainy weather (chapter 11). Education at school is also effective to prepare for disasters (chapter 8).

Although risks from tsunamis are now well understood in the wake of the March 11 event, communities must also become aware of the risks from other possible disasters, such

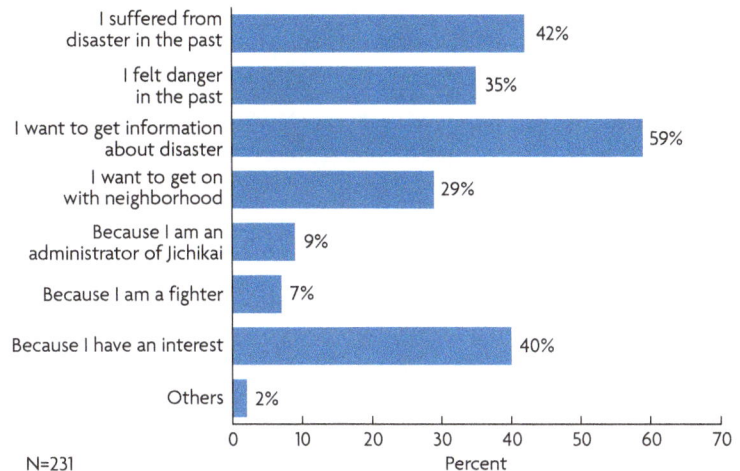

Figure 27.3 Reasons given by people in Toni Village for participating in the hazard mapping exercise before the GEJE

Source: Kyoto University.

as landslides or cyclones. A Web portal maintained by the Ministry of Land, Infrastructure, Transport and Tourism provides access to all hazard maps created throughout the country (see chapter 26 for details).

Informal tool: Local knowledge along the Sanriku Coast

The Tohoku region has two contrasting topographic features. One is the Sendai plain, south of Sendai City, which is relatively flat and offers little access to higher ground close to the coast. The other is the Sanriku-ria coast north of Sendai, where the mountains are near the coast. These topographical characteristics influence the kinds of informal evacuation strategies used in the respective areas.

Tendenko is a term used in the Sanriku coastal area, referring to self-evacuation without stopping to look for family members, neighbors, or relatives. The assumption is that everyone will self-evacuate, and therefore there is no need to be concerned about others. Depending on the location of an earthquake's epicenter, the lead time between the main shock and the arrival of the tsunami can be short. In these cases it is imperative that people self-evacuate without

Figure 27.4 Designated evacuation building (left) and evacuation road (right) in Kamaishi

Source: © Kyoto University. Used with permission. Further permission required for reuse.

delay. This is practical in the coastal area of Sanriku because of the proximity of higher ground (figure 27.4).

But the *tendenko* concept does not apply in the Sendai plain because there is no higher ground nearby (figure 27.5). There, public buildings such as schools or community centers are used as evacuation centers.

Risk communication following the accident at the Fukushima Daiichi Nuclear Power Station

The accident at the Fukushima Daiichi Nuclear Power Station highlighted the issue of risk communication in nuclear emergencies. The Investigation Committee on the Accident at the Fukushima Nuclear Power Stations (2011) reported:

- Communication from the government had been far from ideal. The government delayed providing urgent information, withheld press releases, and was unclear in its explanations. . . . Neither those directly

Figure 27.5 Flat area in Sendai Plain offering no possibility of evacuating to higher ground

Source: © Kyoto University. Used with permission. Further permission required for reuse.

affected by the accident at the Fukushima station nor the public at large believed that the government was providing truthful and accurate information in a timely manner. Examples include the government's information about the status of the reactor cores—core meltdowns in particular—and the critical condition of unit 3, as well as the unclear statement, repeated several times, that the radiation "will not immediately affect human bodies."

Nuclear and Industrial Safety Agency (2012) reported that "Seventy-four percent of people at the affected areas were dissatisfied with the information provided because:

- The background and the reasoning behind the reports and recommendations coming from the official sources were not well explained and therefore could not be trusted.

- The briefings did not include enough detail."

Also, the government committee pointed out that "water contaminated by radiation was discharged into the ocean without notifying neighboring countries. Although this did not violate any relevant international conventions, it may have led the international community to question Japan's competence in responding to nuclear disasters."

LESSONS

Earthquake and tsunami risk communication

Risk communication is meant to help people save their own lives. For communication to be effective, people must be able to trust the information and its source, and it takes a long time to build that trust.

There are formal and informal tools for communicating risk. Hazard maps and early warnings systems are the formal tools that Japan has used, both of which are being revised in light of the GEJE, since both underestimated the actual risk. Hazard risk information should be continuously updated.

Informal communication tools include local knowledge such as *tendenko,* practiced on the Sanriku coast, where self-evacuation without waiting for family members and others is encouraged as soon as a large ground shaking is felt. These types of approaches and local knowledge based on experiences with large tsunamis should be preserved and passed from generation to generation.

Participatory DRM planning by the local community is an effective way of communicating risk. Different forms of communication may have to be used for different age groups. The local social structure can be leveraged to facilitate emergency planning, for example, by enlisting local leaders in their various roles and functions.

Regular drills and education also have an important role in shaping the perception of risk in local communities.

Complacency is a constant problem. Even people who have already experienced disasters need to be reminded of the importance of being prepared. People can also become overly reliant on early warning systems.

Nuclear accident

Japan's Nuclear and Industrial Safety Agency, a government regulatory body, has proposed the following actions to improve risk communication in the event of nuclear accidents.

Develop technical capacity. The technical capacity of staff to analyze information on accidents and to implement countermeasures should be enhanced through specialist training programs.

Develop communication capacity. Communication officers should be trained in disaster risk communications. Preparing manuals, communication materials, and answers to frequently asked questions is also necessary. Communication channels should be established with the mass media, the public, embassies, and local agencies.

Develop coordination capacity. Mechanisms for information sharing should be established among relevant agencies such as the Office of the Prime Minister and the Ministry of Foreign Affairs. Communication equipment and manuals are also necessary.

RECOMMENDATIONS FOR DEVELOPING COUNTRIES

Establish trust between information senders (for example, the government) and receivers (local communities). Trust is a big part of effective risk communication. If the information source cannot be trusted, real communication is impossible—and it takes a long time to establish trust. Complacency is also an issue. Overreliance on early warnings, hazard maps, and incoming information should be discouraged.

Use a variety of tools to communicate risk. Risk communication tools range from sophisticated communication systems to participatory emergency planning, including community hazard mapping, disaster evacuation drills, neighborhood watches, instruction in schools, and the passing of experience from generation to generation based on previous events.

The way in which risk is communicated in the early warning system is also important. Although sophisticated early warning systems and technologies are important during a disaster, the public should understand limitations of prediction technology.

Leverage the interest that local leaders may have in community preparedness and be aware of social structures, which vary from country to country and place to place. Work with local change agents to provide training and to develop an appropriate risk communication strategy.

Take a multihazard approach. The difference in Japan's preparedness for the earthquake and tsunami versus its preparedness for the nuclear accident following the GEJE demonstrates the importance of considering all hazards, not just those that are most likely to happen (chapter 36). A good communication strategy is one piece of an overall response plan, which was lacking for the nuclear accident at Fukushima Daiichi.

Update and monitor. Risks are dynamic and change over time depending on population increases or decreases, the development of new industrial facilities and commercial properties, the availability of new hazard information, and scientific innovations. Risk information should be updated regularly and reflected in risk communication strategies.

NOTE

Prepared by Rajib Shaw, Yukiko Takeuchi, and Shohei Matsuura, Kyoto University; and Keiko Saito, World Bank.

BIBLIOGRAPHY

Committee on the Issuance Standards and Information Statement of Tsunami Warnings. 2011. *Recommendations for Issuance Standards and Information Statement of Tsunami Warnings* [in Japanese]. http://www.jma.go.jp/jma/press /1112/16a/teigenan.pdf.

EEFIT (Earthquake Engineering Field Investigation Team). 2011. *The Mw9.0 Tohoku Earthquake and Tsunami of 11 March 2011: A Field Report by EEFIT.* London: EEFIT. http://www.istructe.org /resources-centre/technical-topic-areas/eefit /eefit-reports.

Investigation Committee on the Accident at the Fukushima Nuclear Power Stations of Tokyo Electric Power Company. 2011. *Interim Report.* Tokyo.

Kikkawa, T. 1999. *Risk Communication: Aiming at Mutual Understanding and Better Decision Making* [in Japanese]. Tokyo: Fukumura Shuppan.

Nuclear and Industrial Safety Agency. 2012. *Issues Regarding Public Hearings and Public Relations Activities On the Accident at Fukushima Daiichi Nuclear Power Station of Tokyo Electric Power Company and Future Effort* [in Japanese]. Tokyo.

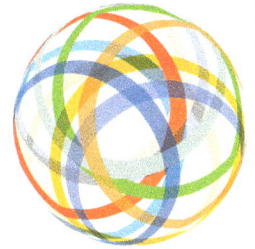

PART VI

The Economics of Disaster Risk, Risk Management, and Risk Financing

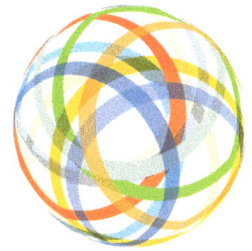

CHAPTER 28

Measuring the Cost-Effectiveness of Various Disaster Risk Management Measures

The Japanese experience shows that—if done right—preventive investments pay. The Japanese government invested about 7–8 percent of the total budget for disaster risk management (DRM) in the 1960s, a move that most probably decreased disaster deaths. Cost-effectiveness analysis and cost-benefit analysis of DRM projects have been widely implemented both at national and local levels in Japan. Different procedures for such analysis have been followed according to the type of project, the funds, and the governing entity responsible. The Japanese experience shows that cost-benefit analysis is applicable to DRM-related projects and is a useful tool in choosing among different options and understanding the effectiveness of a project.

The Great East Japan Earthquake (GEJE) and other recent disasters remind us of the importance of early actions to implement adequate prevention measures, mitigate risks, and establish sound postdisaster financing mechanisms to reduce human, economic, and financial impacts. Even if documented evidence is still lacking, there is a growing consensus that investing in disaster risk management (DRM) is cost-effective, though measuring cost savings is difficult. Several lessons can be derived from the cost-benefit analysis (CBA) and cost-effectiveness analysis (CEA) conducted in Japan.

FINDINGS

National budget for disaster risk management

Every year many people lose their lives and property in Japan due to natural disasters. Up until the 1950s, numerous large-scale typhoons and earthquakes caused extensive damage and thousands of casualties (figure 28.1). In the 1960s, DRM spending represented 7–8 percent of the national budget (figure 28.2). As mechanisms to cope with disasters and mitigate vulnerability to them have progressed (by developing DRM systems, promoting national

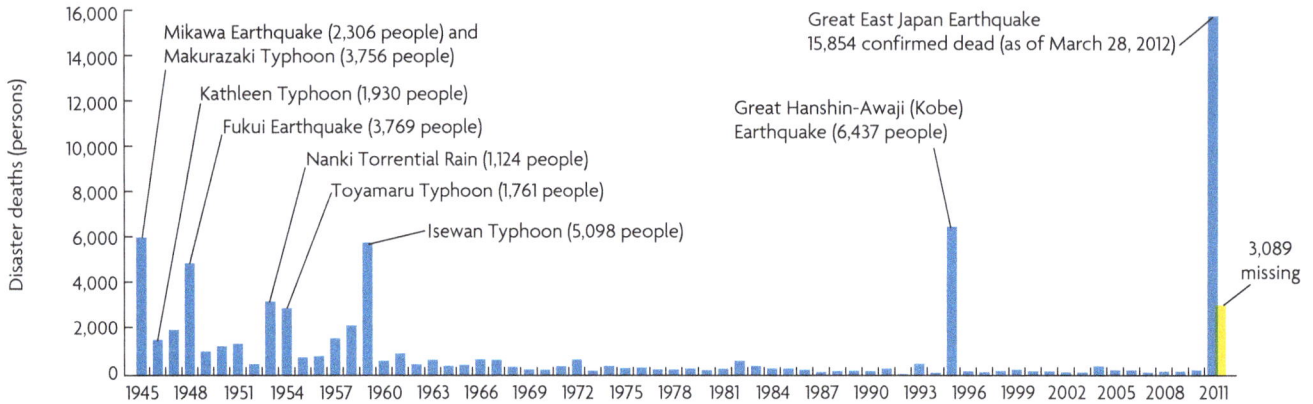

Figure 28.1 Disaster deaths in Japan, 1945–2011

Source: Ministry of Land, Infrastructure, Transport and Tourism (MLIT).

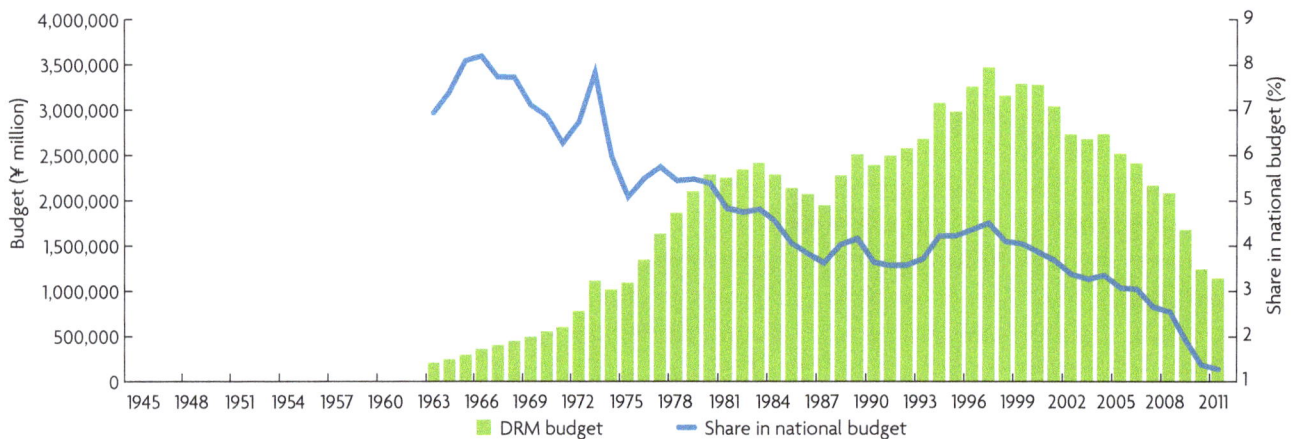

Figure 28.2 Change in DRM spending, 1963–2011

Source: MLIT.

land conservation, improving weather forecasting technologies, and upgrading disaster information communications systems), the number of disaster-related casualties, especially from floods, has been decreasing over the years with the exception of a few outliers.

Comparison of damage with other tsunami disasters

The GEJE is the strongest earthquake ever recorded in Japan; the destruction it caused is staggering. But it is clear that if Japan were not so well prepared, things could have been much worse.

A longstanding tradition of effective disaster prevention paid off. While almost 20,000 people lost their lives on March 11, the mortality ratio of the GEJE—which hit during the daytime—was considerably lower compared to the Meiji Sanriku Tsunami of 1896 (nighttime) or the Indian Ocean tsunami of 2004 (which also hit during the day) (figure 28.3).

Over the years, the Japanese government has invested in structural and nonstructural measures to prevent disasters and reduce their impacts. Around ¥1 trillion was invested in coastal dikes and breakwaters just in the areas affected by the GEJE, and yearly investments in earthquake monitoring and warning systems amounted to about ¥2 billion. Furthermore, a number of nonstructural measures—including community-based DRM (chapter 6), DRM

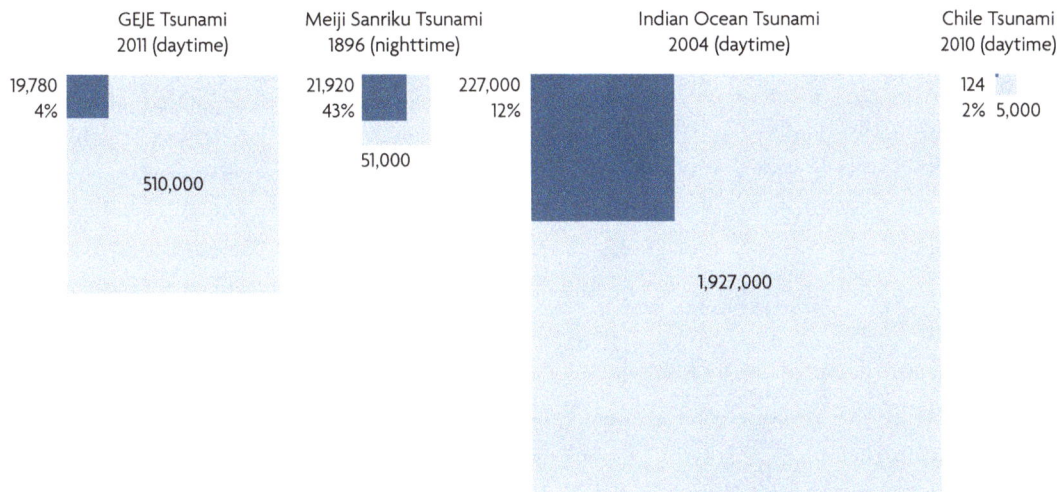

| GEJE Tsunami 2011 (daytime) | Meiji Sanriku Tsunami 1896 (nighttime) | Indian Ocean Tsunami 2004 (daytime) | Chile Tsunami 2010 (daytime) |

Tsunami (year)	Dead + missing (a)	Damaged houses	Population in affected area (b)	a/b (%)
GEJE (2011)	19,780	259,415	510,000	4
Meiji Sanriku (1896)	21,920	7,957	51,000[a]	43
Indian Ocean (2004)	227,000	1,700,000 (in population)	1,927,000[b]	12
Chile (2010)	124	1,500	5,000[c]	2

a. Number of damaged houses × average number of household members in Iwate (6.38).
b. Dead + population lost houses.
c. Number of damaged houses × average number of household members (3.5).

**Figure 28.3
Comparison of damage from four tsunami events**

Source: Cabinet Office and World Bank data.

education (chapter 8), and business continuity plans (chapter 9)—have been further developed over the years.

Measuring cost-effectiveness

It is essential to make sure that limited financial resources are used in a cost-effective way. Effective spending has high rates of return but is difficult in practice. There are varieties of criteria being used for evaluating the cost-effectiveness of projects, such as CBA, CEA, multicriteria analysis (MCA), and so on. CBA is a well-known tool, particularly useful for governments seeking to compare alternatives. CBA is used to organize and present costs and benefits of measures and projects and to evaluate cost efficiency. CBA was originally developed as a rate-of-return assessment and financial appraisal method to assess business investments. The main purpose was to compare all the costs and benefits of an investment (even if accruing across different sectors, in different locations, and in different

time periods) from the perspective of society. But for most DRM projects there is a lack of information, especially regarding benefits and profits, making it difficult to accurately estimate the cost-effectiveness of measures (Mechler 2005).

Cost-benefit analysis in Japan

In Japan project appraisals, including CBA, are conducted for public works projects before they are adopted, and every three to five years after adoption to evaluate project efficiency (figure 28.4). Committees for project appraisal (consisting of academic, business, or legal experts) are established for national and local entities responsible for project implementation, who evaluate the project efficiency of adopted projects. The committees assess the need, cost benefits, progress, possibilities for cost reduction, and the continuity of projects. The appraisal results and associated documents are made open to the public to ensure the transparency of decision making.

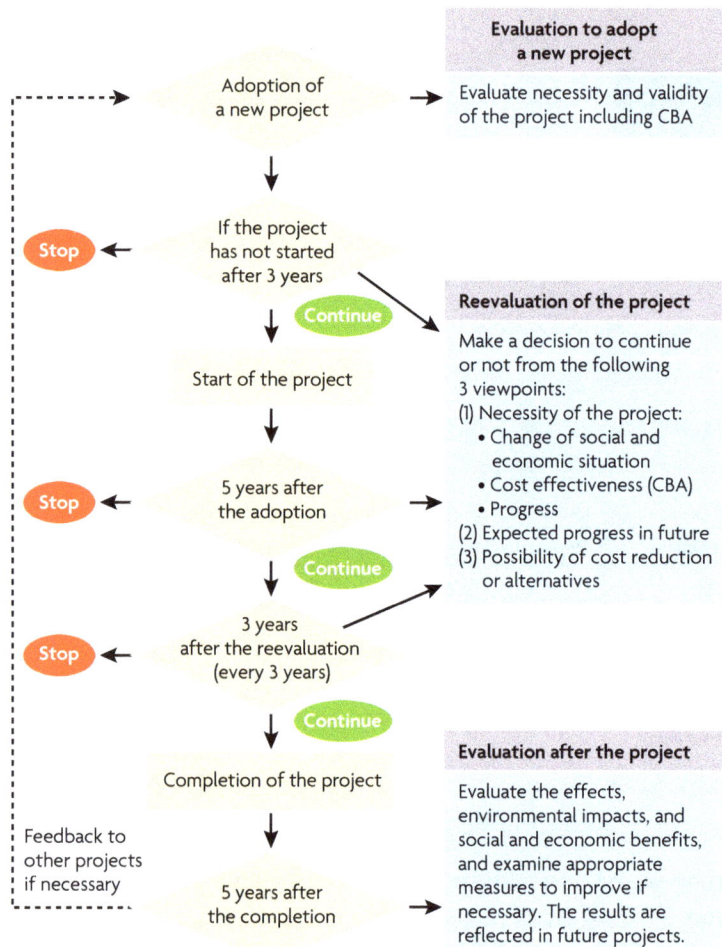

Figure 28.4 MLIT public works project evaluation process, based on Government Policy Evaluation Act (2002)

Source: MLIT.

Note: CBA = cost-benefit analysis.

A system for evaluating government policies was first introduced in Japan at the prefectural government level to reassess or conduct interim evaluations of ongoing projects. The first attempt at such evaluation was done by the Hokkaido prefectural government in 1997.

The central government, recognizing the importance of such a system, established the Government Policy Evaluations Act (GPEA) in 2001, to provide a legal framework for evaluating government policies. The GPEA aims to promote accountability; provide efficient, high-quality government services and projects; and ensure that the outcomes of these services and projects meet the needs of the nation.

The GPEA calls for all government policies, programs, and projects to be assessed before their inception, to be evaluated after their completion, and to be reassessed or subjected to interim evaluation when necessary.

Cost-benefit analysis for coastal projects

Under the Government Policy Evaluation Act in 2001, the Ministry of Land, Infrastructure, Transport and Tourism (MLIT) conducts CBA on every project based on the Technical Guidelines of CBA for Public Works Projects (2004). These guidelines set out the overarching principles to be followed by each individual department (such as those overseeing rivers, roads, or urban development) of the MLIT. Maintenance and management of existing infrastructure and disaster-rehabilitation works are excluded. The Reconstruction Agency has confirmed that post-GEJE rehabilitation efforts are not subject to CBA evaluation.

In 1987 the MLIT and Ministry of Agriculture, Forestry and Fisheries published the "Guidelines for Cost Benefit Analysis in Coastal Works." The guidelines were revised in 2004 following the inclusion of disaster prevention, environmental conservation, and seacoast utilization considerations into the objectives of the Seacoast Act (figure 28.5). The guidelines recommend that benefits from seacoast works projects should be quantified into monetary values as much as possible based on probabilities and risks relevant to the following issues:

• Protection of inland properties from flooding by tsunamis and storm surge (expected losses are estimated by multiplying the damage ratio to the value of properties such as buildings, crops, public infrastructure, and so on)

• Prevention or mitigation of damage to land and properties from erosion (the same method of protection of properties from flooding)

• Prevention or mitigation of damage by blown sands and sea spray on inland properties and crops, and negative effects on daily life such as through additional labor (expected losses are estimated by evaluating the depreciated value of buildings, damaged crops, and labor loads for cleaning)

- Protection of natural environments such as ecosystems and water quality, and the development of better landscape planning (the values of natural landscapes and ecosystems along the seacoast are estimated, as are the benefits of implementing projects; the seawater purification function of the beach is also valued)

- Utilization of the seacoast for activities such as recreation and sea bathing (the values of the expansion of recreation activities, fatigue recovery effects, land development, and so on are estimated)

Specific costs to implement a project—including major initial outlays for the investment effort and maintenance expenses—are estimated. The costs and benefits identified have to be discounted to ensure that current and future effects are comparable. Finally, costs and benefits are compared under the economic efficiency decision criteria, such as net present value, benefit-cost ratio (B/C), or the economic internal rate of return (EIRR).

The breakwater construction project in Kuji Port, Iwate Prefecture—started in 1990 and scheduled to be completed in 2028—is a good example of a CBA applied to a DRM project. The efficiency of the project was last reevaluated in 2010, when the costs were estimated at ¥108.5 billion and the benefits at ¥136.5 billion. The EIRR was calculated at 4.8 percent, and B/C at 1.3. In this evaluation, prevention of inland flooding and sea disasters were considered as monetized benefits, while a decrease in the affected population, improvement of moored vessels security, and stability and development of local industry were considered as qualitative benefits. The project is estimated to reduce the potentially inundated area from 377 to 50 hectares, and reduce the damage to

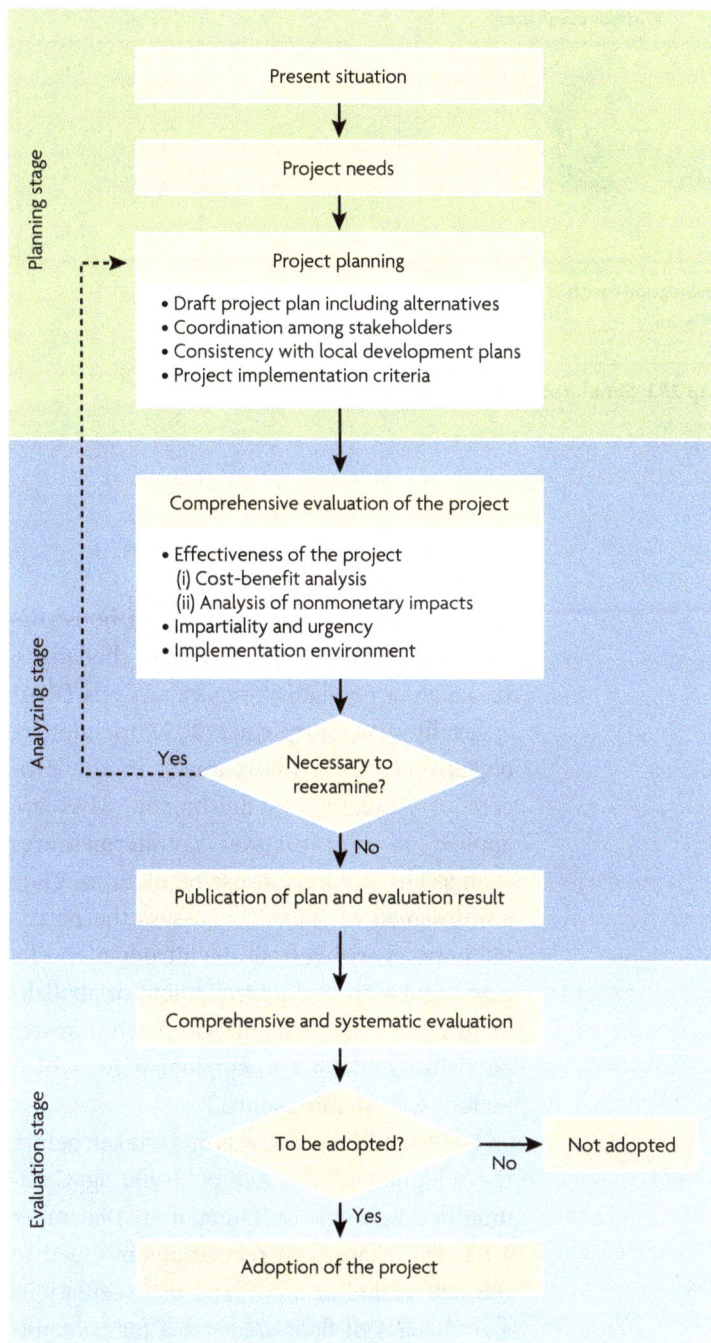

Figure 28.5 Seacoast works: From planning to implementation
Source: MLIT.

housing from 2,618 to 330 houses (map 28.1). Annual estimated benefits are

- Protection from inundation: ¥4.2 billion

- Protection from marine accident by storm: ¥5.6 billion

- Residual value: ¥11.4 billion

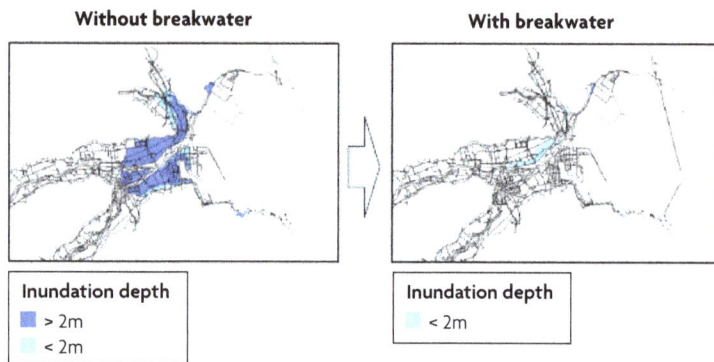

Without breakwater **With breakwater**

Inundation depth
■ > 2m
■ < 2m

Inundation depth
■ < 2m

Map 28.1 Simulated inundation areas

Source: MLIT.

Regulatory impact analysis assessing nonstructural measures in Japan

Assessing the cost-effectiveness of nonstructural measures presents specific challenges. In Japan, a regulatory impact analysis (RIA) is legally mandatory since 2007 to improve objectiveness and transparency in the process of regulatory establishment. RIAs are applied to nonstructural countermeasures such as changes in land-use regulations. They are designed to objectively assess the potential impacts arising from the introduction of a new regulation or the amendment or abolishment of an existing regulation. Each ministry publishes guidelines to conduct RIAs, which include CBA requirements.

For example, an RIA was undertaken before the adoption of the Act on Building Communities Resilient to Tsunami in December 2011. The changes in regulations outlined in the act—including new land-use regulations and changes of floor-area-ratios for tsunami-evacuation buildings in the designated zone—were assessed through the RIA. It was estimated that the benefits from these changes could outweigh the costs of implementation, as they develop more resilient urban areas through increased safety of housing and public facilities in tsunami-exposed areas and construction restrictions for potentially dangerous buildings. (For more information on the act, please consult chapter 12.)

The costs considered in the RIA include the costs associated with the approval processes for structures that contribute to tsunami evacuation; the costs of preparing evacuation plans or evacuation drills; and various administrative costs for approval, inspection, or monitoring of buildings or land use. The benefits, on the other hand, include prevention of inappropriate development, facilitation of prompt evacuation in case of tsunami disasters, and promotion of adequate maintenance of tsunami-disaster-mitigation facilities—all of which contribute to the protection of lives and the mitigation of damage in tsunami-risk areas. These costs and benefits were considered qualitatively in the RIA.

The MLIT has conducted approximately 50 RIAs since 2007. One was conducted, for example, when the Act on Promotion of Seismic Retrofitting of Buildings was revised in 2005 to add schools, welfare facilities, and buildings for storage or treatment of hazardous objects to those facilities under the guidance of administrative offices, and to establish "retrofitting support centers" nominated by the government.

New approach to evaluating the effectiveness of dual-purpose infrastructure

The Sanriku Expressway being constructed along the seashore in the tsunami-affected Iwate and Miyagi prefectures contributed to the recovery of this area (chapter 4). But the evaluation of the cost-effectiveness of such redundant infrastructure (that is, a road used as part of a DRM facility) has never been taken into account before in Japan. The Japanese government is now trying to modify its evaluation methodology to include the potential benefits of road projects from the perspective of disaster management and DRM.

Evaluation methodology is used when the MLIT selects a new road construction project that is expected to be a key route for rescue and relief supplies, materials, and resources for emergency response, and to form a wide range of road networks for DRM. The evaluation of the disaster mitigation function encompasses

- *Necessity.* Clarify why the project is needed based on DRM considerations (for example, for transportation of rescue and relief supplies, transportation to emergency medical facilities, and reaching core cities in and around the stricken area).

- *Efficiency.* Numerically estimate the level of improvement and evaluate its priority (for example, improvement of the disaster management function by securing transportation between core cities or within the regional network, like shortening of travel time, dissolution of isolated areas, and so on).

- *Effectiveness.* Compare the effectiveness of several alternative plans and similar projects.

LESSONS

CEA and, more in particular, CBA, has several limitations, including the difficulty of accounting for nonmarket values, the lack of accounting for the distribution of benefits and costs, and the issue of choosing the correct discount rate. In addition, CBA of DRM presents additional challenges related to the fact that the planning horizon of DRM measures is typically longer than that of policy makers, and that the occurrence of natural hazards needs to be captured with stochastic methods (Mechler 2005). Conducting probabilistic CBA often proves difficult because of the absence of reliable hazard and vulnerability data. This is perhaps the greatest challenge faced by the DRM community in conducting comprehensive economic studies of proposed DRM measures in developing countries. Despite limitations, CBA remains the most commonly used tool to analyze the benefits and costs of DRM measures. In a review of the existing literature on CBA of DRM measures in developing countries,

a Global Facility for Disaster Reduction and Recovery (GFDRR) study finds a wide variation in methodologies, assumptions, discount rates, and sensitivity analyses, suggesting that DRM analyses are highly context sensitive (GFDRR 2007).

CBA on infrastructure projects has been widely implemented both at national and local levels in Japan. Different procedures have been identified according to the type of project, the funds, and the governing entity responsible. Different types of costs are included in the analysis, such as operational, maintenance, and fiscal costs; also, different types of benefits are accounted for, such as the protection of inland properties and the natural environment or recreational utilization.

The Japanese experience shows that CBA is applicable to DRM structural projects and is a useful tool to help choose among different options (higher B/C is one of the variables to be taken into account when making decisions) and to understand the effectiveness of a project/measure. Nonstructural measures, such as land-use regulations and building codes, can be evaluated as well. For example, administration costs and other necessary costs can be compared when deciding among alternative measures.

The use of CBA must be adapted to the type of measure that is being evaluated. Infrastructure and soft measures require different approaches—not only different procedures and calculations, but also different objectives and bottom-line evaluations. It is also important to introduce clear guidelines about how, when, and where to implement CBA. The Japanese experience also proves that sectoral guidelines released by specific ministries are very helpful, as they describe in practical terms each step to be taken when implementing CBA.

While saving lives is the top priority, valuing such lives when assessing the potential benefits of different measures is extremely challenging and poses complex ethical and political questions. But ignoring the value of life implicitly considers people "useless"—and it would be unethical if property is protected but lives are not. For example, background work done for the joint United Nations–World Bank (UN-WB) report *Natural Hazards, UnNatural Disasters* shows how, if the value of lives saved were ignored, retrofitting buildings in the Turkish district of Atakoy would not be cost-effective, with a B/C lower than 1. Background work done for the report finds that including a value of life of $750,000 in the benefits, however, tips the scale toward retrofitting. And only by including the value of lives saved (at $400,000 each) did earthquake-strengthening measures for apartment buildings and schools in Turkey pass the cost-benefit test (UN-WB 2010). This example shows the limitations of CBA. Other techniques such as MCA have been explored and could be more acceptable from an ethical perspective. MCAs do not at present offer much help for practical decision making in Japan.

RECOMMENDATIONS FOR DEVELOPING COUNTRIES

Despite its limitations the CBA can be a powerful tool when deciding on and prioritizing DRM measures. It is useful when the issues are complex and there are several competing proposals, and particularly so when comparing alternatives.

It is important to set clear rules about when, how, and on what CBA should be performed. Regulatory frameworks, policy procedures, and specific guidelines (possibly at sectoral levels), overseen by specific ministries, can certainly improve the implementation of CBA for DRM.

Connections between decision making and CBA must be clear. CBA can be one informative input, or one of the main variables in decision making. Any decisions should be transparent and reviewed regularly. In the Japanese context, project appraisal committees consisting of external experts and academics evaluate the projects before their adoption, and then reassess their effectiveness to secure transparency and accountability in decision making.

NOTE

Prepared by Masato Toyama and Junko Sagara, CTI Engineering.

BIBLIOGRAPHY

Cabinet Office, Government of Japan. 2011. "White Paper on Disaster Management 2011" (in Japanese).

Central Disaster Management Council, Japan. 2011. *Report of the Committee for Technical Investigation on Countermeasures for Earthquakes and Tsunamis Based on the Lessons Learned from the "2011 Earthquake off the Pacific Coast of Tohoku."*

Fire and Disaster Management Agency, Japan. 2011. *Report of the Study on Fulfillment and Improvement of Earthquake and Tsunami Disaster Reduction in Regional Disaster Prevention Plans* (in Japanese).

GFDRR (Global Facility for Disaster Reduction and Recovery). 2007. "A Reference Paper on Benefit-Cost Studies on Disaster Risk Reduction in Developing Countries."

Mechler, R. 2005. "Cost-Benefit Analysis of Natural Disaster Risk Management in Developing Countries." Deutsche Gesellschaft für Technische Zusammenarbeit (GTZ).

MLIT (Ministry of Land, Infrastructure, Transport and Tourism, Japan). 2009. "Guidelines for Assessment of MLIT Policy on Regulatory Impact Analysis" (in Japanese).

MLIT and Ministry of Agriculture, Forestry and Fisheries, Japan. 2004. "Guidelines for Cost-Benefit Analysis in Seacoast Works" (in Japanese).

Onishi, T. 2012. "What Lessons We Should Learn from the Great East Japan Disaster?" Paper presented at the 5th GEOSS Asia-Pacific Symposium, Tokyo, April 2.

UN–WB (United Nations–World Bank). 2010. *Natural Hazards, UnNatural Disasters: The Economics of Effective Prevention.* Washington, DC: World Bank.

Yamamoto, C., and K. Sato. 2007. "Feasibility Studies under the System for Evaluating Government Policies in Japan." Paper presented at Improving Public Expenditure Management for Large-Scale Projects: Focusing on a Feasibility Study, Seoul, May 22–23.

CHAPTER 29

Earthquake Risk Insurance

The March 2011 earthquake that hit East Japan was the fourth-largest ever recorded. It was not only a human tragedy but an economic shock with losses estimated in excess of ¥16,900 billion, making it the costliest disaster in history. Despite this, the Japanese insurance industry is expected to emerge without significant financial impairment, thanks to a well-developed residential earthquake risk insurance dual program (with private nonlife insurers and cooperative mutual insurers) based on conservative control of insurers' liabilities (through insurance policy structures and reinsurance). Meanwhile, more than half of Japanese homeowners are still uninsured, creating a significant fiscal burden for the government.

FINDINGS

Residential earthquake insurance: A dual program with carefully controlled liabilities

Residential earthquake insurance coverage in Japan relies on two major actors: private nonlife insurers and cooperative mutual insurers. Despite major differences in their financial management of earthquake risk, these two insurance systems demonstrated their efficiency in claims settlements and their financial viability after the Great East Japan Earthquake (GEJE). Table 29.1 compares the residential earthquake insurance scheme offered by private nonlife insurance companies with the scheme offered by the largest cooperative mutual insurer, the National Mutual Insurance Federation of Agricultural Cooperatives (also known as JA Kyosai).[1] While the perils covered, assets covered, and extent of coverage are similar across the two programs, earthquake coverage is offered on a voluntary basis with risk-based premium rates by private insurers, and on an automatic basis with flat rates by cooperative mutual insurers.

Both programs are based on conservative control of insurers' liabilities. In both programs, the claims payments are not intended to provide complete coverage: the maximum coverage is limited to 50 percent of the fire insurance amount (subject to upper limits).

Table 29.1 The dual residential earthquake insurance system in Japan

	PRIVATE NONLIFE INSURERS	COOPERATIVE MUTUAL INSURER JA KYOSAI
Perils covered	Earthquake, volcanic eruption, tsunami	Earthquake, volcanic eruption, tsunami
Assets covered	Residential dwelling and content	Residential dwelling and content
Extent of coverage	30–50 percent of fire insurance amount with limits	Up to 50 percent of fire insurance amount with limits
Coverage purchase	Optional endorsement to residential fire insurance policy	Automatically included in building endowment policy
Premium rate	Risk-based rates (by risk zone and type of construction)	Flat rates (wooden/ nonwooden)
Reinsurance	Japan Earthquake Reinsurance Co. (JER) and Japanese government	International reinsurance and capital markets
Loss adjustment	Three-step system	Proportional system
Penetration of earthquake coverage (percent households)	25%	11%

Source: World Bank compilation.

Likewise, both programs rely on sophisticated reinsurance strategies. The reinsurance protection of the private insurance scheme relies on a catastrophe insurance pooling mechanism, the Japanese Earthquake Reinsurance Co. (JER), backed by the government of Japan. In contrast, reinsurance protection for cooperative mutual insurers is provided by international reinsurance and capital markets, with no government intervention. In both cases, the use of reinsurance serves to limit the liability of the private or cooperative risk carriers.

Penetration under the private nonlife insurance program is estimated at about 25 percent of Japanese households, with just under 13 million residential earthquake insurance policies in force: an estimated 48 percent of all fire insurance policies in force include earthquake coverage. Cooperative mutual insurance programs cover about 14 percent of Japanese households, so that total penetration is estimated at 39 percent.[2] JA Kyosai holds a very

large share of the cooperative mutual insurer market, with 5.4 million households holding building endowment policies covering residential earthquake risk (11 percent of total Japanese households). The cooperative mutual insurer Zenrosai has an additional 1.7 million natural disaster policies covering residential earthquake risk, accounting for a further 3 percent of total Japanese households.

Private nonlife insurance companies and the Japanese earthquake reinsurance company

Earthquake insurance offered by private nonlife insurance companies is available as an optional endorsement to fire insurance policies. Earthquake coverage is available at policy limits of 30–50 percent of the fire insurance limit, with maximum limits of ¥50 million per dwelling and ¥10 million for personal property.

A three-step claims settlement allows for rapid damage assessment and claims settlement. Payouts are not proportional to damage, but based on a three-step system: total loss, half loss, and partial loss—which allow for 100 percent, 50 percent, and 5 percent of the earthquake insurance policy limit, respectively.

The premium rates are risk based and vary according to the prefecture where the dwelling is located (divided into eight risk zones) and type of construction (wooden or nonwooden). For an insured amount of ¥10 million, the annual premium varies between ¥5,000 for a nonwooden structure in Nagazaki Prefecture, and ¥31,300 for a wooden structure in Tokyo. Discount rates of up to 30 percent apply when the building is earthquake resistant, according to the Japanese Housing Performance Designation Standards, including a 10 percent discount for buildings constructed after 1981. The premium rates, calculated by the Non-Life Insurance Rating Organization, consist of the pure premium rate and a loading factor. It should be noted that the rates do not include any loading for profit since the program is not for profit. Despite this rating and because of

Japan's considerable earthquake exposure, rates are still considered high.

The 1966 Earthquake Insurance Law (enacted after the Niigata earthquake of 1964) established the JER, to whom private nonlife insurers were obliged to offer earthquake insurance and cede 100 percent of the earthquake premium and liabilities. The JER thus acts as the sole earthquake reinsurer for the private insurance market. The JER can be seen as an earthquake reinsurance pool, retaining a portion of the liability and ceding the rest back to private insurers (based on their market share) and to the Japanese government through reinsurance treaties. The reinsurance program is designed such that the liability of private insurers and the JER itself does not exceed the accumulated reserves from earthquake insurance premiums. Figure 29.1 describes the Japanese earthquake reinsurance program as revised in May 2011 after the GEJE. The total claims-paying capacity of the program is currently ¥5,500 billion, which is estimated to correspond to the scenario of the 1923 Great Kanto earthquake with a return period of 220 years.[3] Should insured earthquake losses exceed this amount, claims would be prorated.

The role of the Japanese government is central to the program. The maximum liability of the government of Japan, JER, and private insurers is 87 percent, 10 percent, and 3 percent, respectively. It should be noted that under the previous reinsurance program (before May 2011), the government's liability was only 78 percent, and the rest was shared equally between the JER and private insurers. The revision of the reinsurance program, leading to an increase of the government's liability share, is the direct consequence of a depletion of the earthquake reserves of both the JER and private insurers after the GEJE.

Japanese accounting standards allow the insurers to build up pre-event catastrophe reserves (by accumulating the earthquake insurance premiums received, less expenses and any underwriting gains and investment

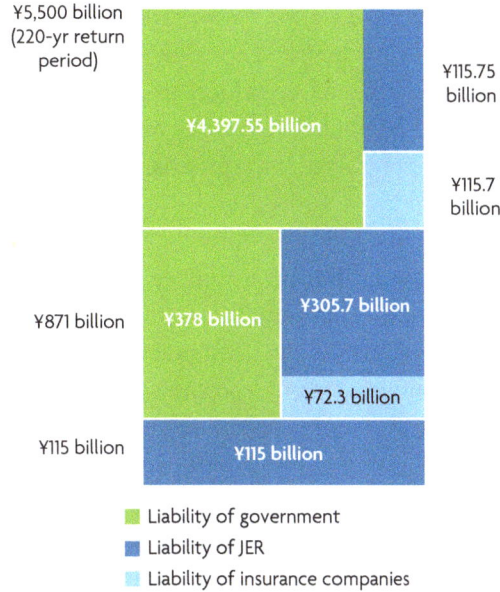

Figure 29.1 Japanese earthquake reinsurance program (as of May 2011)

Source: JER 2011a.

income) over time with separate resources to pay claims, the size of which is based on the probable maximum loss of the insurer's portfolio. Likewise, the government of Japan has set up a special account to accumulate its reserves. Table 29.2 shows the amount of reserves at the end of fiscal year 2010—that is, before the GEJE. The GEJE wiped out about half of the program's earthquake reserves.

It is noteworthy that the total reserves supporting the Japanese Earthquake Reinsurance Program, even before the GEJE, represent only a fraction of the liability of all stakeholders. The size of this potential gap is largely due to the government's reserve-to-liability ratio

Table 29.2 Reserves under the earthquake insurance program

	RESERVE AT END OF FISCAL YEAR 2010 (¥ BILLION)
Government	1,343
JER	424
Private insurers	489
Total	2,256

Source: JER 2011a.

under the program, which appears low. In case of a major earthquake exceeding the reserves available, it would be critical to immediately mobilize additional resources to ensure the financial solvability of the program.

Cooperative mutual insurers

Residential earthquake insurance is also available through cooperative mutual insurers. These insurers conduct insurance operations on behalf of Japan's cooperative societies. The largest of these cooperatives is JA Kyosai, which holds an estimated 85 percent market share of all the homeowners insurance written through cooperative mutual insurers. Like any cooperative, JA Kyosai operates on a nonprofit basis. Its insurance products are different from those of private insurers. Cooperative mutual

insurers offer building endowment policies: these policies offer more comprehensive coverage than the policies available through the private insurers and can therefore be seen as a savings mechanism that provides funding for home repairs, whether caused by natural disasters or other adverse events. The five-year (or longer) term policy automatically covers residential dwellings and personal property from damage caused by fire, flood, earthquakes, and other natural disasters. If the policy expires and the policyholder has not claimed a total loss, he or she is entitled to a partial refund of the premium. At the start of 2011, JA Kyosai's client base comprised more than 11 million building endowment policies.

Earthquake insurance is automatically included in the building endowment policies offered by JA Kyosai. The policy limit is 50 percent of the fire insurance limit, up to ¥250 million. The average fire insurance amount is ¥30 million; hence the average earthquake insurance limit is ¥15 million.

Under the building endowment policy available through JA Kyosai, the claims settlement process in case of an earthquake is proportional: a loss assessor estimates the damage percentage of the house, and this rate is applied to the earthquake policy limit.

The premium rate is flat, that is, the same wherever the dwelling is located. It only differs according to whether the building is a wooden or nonwooden structure.

Cooperative mutual insurers are not subject to the Earthquake Insurance Law and do not participate in the JER. They work outside the nonlife insurance regulatory framework and are instead accountable to their respective ministries; for example, JA Kyosai reports to the Ministry of Agriculture, Forestry and Fisheries. In contrast to private nonlife insurers, cooperative mutual insurers cede a significant portion of their liabilities to the international reinsurance market. JA Kyosai is known to have one of the largest reinsurance programs in the world, with reinsurance capacity in excess of

---

BOX 29.1

Innovative catastrophe risk financing: Capital markets protect Japanese farmers against earthquake

In 2008, Munich Re, a reinsurance company based in Germany, issued JA Kyosai's second catastrophe (Cat) bond, a $300 million issue, through the special-purpose vehicle (SPV), Muteki Ltd.

Cat bonds are index-linked securities that secure financial resources on the capital markets, to be disbursed in case of the occurrence of a predefined natural disaster. Cat bonds generally cover the highest level of risk and are mainly issued for specific perils with an annual probability of occurrence of 2 percent or less (that is, a return period of 50 years or more). Unlike traditional reinsurance, Cat bonds are fully collateralized and offer multiyear coverage (usually three to five years).

The three-year Muteki Cat bond provided fully collateralized protection for Japanese earthquake exposure indirectly to JA Kyosai/Zenkyoren, through a reinsurance agreement with Munich Re, which served as counterparty on the transaction. Like other Cat bonds in Japan, the Muteki Cat bond was parametric, triggered by the location and magnitude of an earthquake rather than the actual losses. Following the GEJE disaster, the Muteki Cat Bond became the first Cat Bond to pay out on the occurrence of an earthquake event. The instrument released the full coverage limit of $300 million in response to the event.

In February 2012, Guy Carpenter and Company announced the placement of a $300 million Cat bond, through the SPV Kibou Ltd., which would ultimately benefit JA Kyosai. It provided protection on a parametric basis, using earthquake data gathered from various recording stations from the Kyoshin-Net network of seismographs.

¥75 billion. Its large and well-diversified asset base also allows it to retain a significant portion of its liability. In addition to traditional reinsurance, JA Kyosai has issued catastrophe (Cat) bonds to better spread its risk. See box 29.1.

Industrial and commercial earthquake insurance

Traditionally, industrial and commercial earthquake insurance has been issued as a reduced indemnity policy, which provides limited coverage on a proportional basis. The extent of the coverage depends on the location of the asset, for which the country has been divided into 12 risk zones. The indemnity limit varies from 15 percent in Tokyo up to 100 percent in Niigata. Following the enactment of the Insurance Business Law in 1996, which largely deregulated the insurance market in Japan, insurance policies on a first-loss basis were also offered, which generated a significant increase in the sum insured (the maximum amount that could be paid out). Loss of revenue and business interruptions caused by earthquakes have not traditionally been marketed and have low penetration rates.

Other classes include earthquake fire expense insurance. This is a limited amount for fire following an earthquake, which is provided automatically with some insurance policies, such as the storekeepers' comprehensive policy. The coverage is limited to 5 percent of the fire sum insured, up to certain fixed limits. Other insurance policies that generally include earthquake coverage are cargo insurance, motor insurance, and engineering insurance.

Economic and insured losses

The GEJE caused major direct economic losses, with current estimates of ¥16,900 billion (chapter 28; box 29.2). Private (residential, commercial, and industrial) buildings represented 62 percent and public infrastructure represented 13 percent of the (direct) economic losses (see figure 29.2). Insured losses were estimated at ¥2,750 billion, or 16 percent

of total economic losses. Residential assets represented 78 percent of insured losses. Fifty-six percent of the residential insured losses were covered by private insurers and the JER, and 44 percent were covered by cooperative mutual insurers (see figure 29.2).

Despite significant differences, both private and mutual residential earthquake insurance

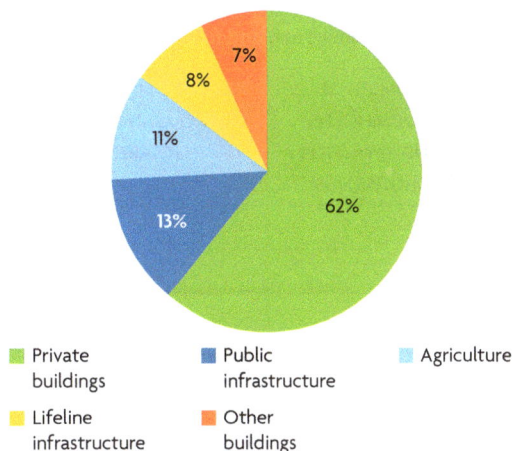

Figure 29.2 Economic and insured losses of the Great East Japan Earthquake

Sources: Cabinet Office (CAO); JER 2011a and 2011b; JA Kyosai 2011.

Legend:
- Private buildings
- Public infrastructure
- Agriculture
- Lifeline infrastructure
- Other buildings

a. Economic losses by sector, as percentage of total loss (¥16,900 billion)

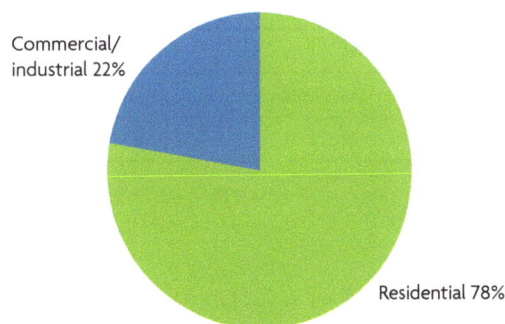

Commercial/industrial 22%
Residential 78%

b. Insured losses by sector, as percentage of total insured losses (¥2,750 billion)

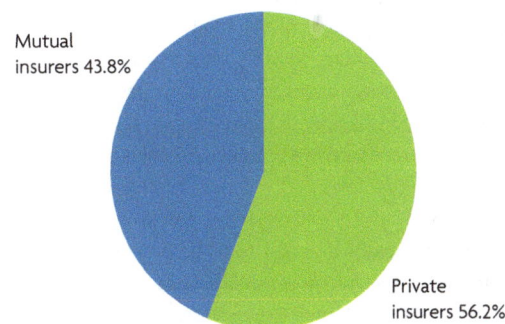

Mutual insurers 43.8%
Private insurers 56.2%

c. Insured residential losses by scheme, as percentage of total insured residential losses (¥2,137 billion)

Agriculture and fishery insurance

Insurance schemes in agriculture and fishing helped farmers and fishermen stabilize their businesses by compensating them for losses and damages caused by the GEJE. Insurance paid for some level of damage sustained by almost all fishing boats. In Japan these schemes began as cooperative activities by local farmers and fishermen. They were subsequently turned into voluntary mutual aid programs established by the government, which subsidizes the premiums paid by farmers and fishermen, covers part of the administrative costs, and reinsures the insurance associations.

Policies in force for agricultural, fishing boat, and fisheries insurance in 2009

	NUMBER OF HOUSEHOLDS UNDERWRITTEN (THOUSANDS)	AREA UNDERWRITTEN (THOUSANDS OF HECTARES)	VALUE COVERED (¥ MILLION)	PENETRATION
Farm products				
Paddy rice	1,752	1,479	1,223,157	91% (area)
Field rice	0.4	0.2	46	5% (area)
Wheat and barley	49	252	83,277	95% (area)
Fruit trees				
Harvest mutual relief	76	45	107,200	26% (area)
Tree mutual relief	4	1	7,000	2% (area)
Livestock	89	6.665 (number of livestock)	724,585	42% (number of livestock)
Field crops	82	259	140,400	62%
Fishing boats	192 (boats)	n.a.	1,028,517	>100% (number of boats)
Fisheries	61	n.a.	394,155	52% (number of households)

Source: Ministry of Agriculture, Forestry and Fisheries (MAFF).
Note: n.a. = Not applicable.

Fishery insurance

The earthquake and tsunami damaged some 25,000 fishing vessels, at a cost of ¥170 billion. Ninety percent of the vessels in Iwate, Miyagi, and Fukushima prefectures were damaged, which had an enormous effect on the fishing industry since these vessels were used for aquaculture as well as fishing. Before the tsunami, the three prefectures accounted for 10 percent of the total catch in Japan (excluding aquaculture). Aquaculture industries were also severely damaged, particularly in the Iwate and Miyagi prefectures, where production of oysters and *wakame*, or seaweed, is widespread. Damage to aquaculture amounted to ¥131 billion: ¥57 billion for production and ¥74 billion for facilities.

The fisheries insurance system in Japan is well organized, providing essential insurance services at a reasonable cost to all fishermen including small-scale producers. The fishing vessel insurance system, which was established in 1952 under the Fishing Vessel Damage Compensation Law, aims at stabilizing fishing businesses by covering the loss of and damages to their fishing vessels. The system includes the following insurances:

- *Fishing vessel insurance* covering basic damage caused by accidents and disasters, and including special insurance for damage caused by war and seizure

- *Protection and indemnity insurance* covering compensation for the crew and damages incurred during navigation

- *Owner-operator insurance* covering the death of owner-operators

- *Cargo insurance* covering the loss of catches or cargo

- *Pleasure boat insurance* covering compensation, rescue costs, and damages

- *Transshipped catches insurance*

- *Crew salary insurance* covering crew salaries if vessels are seized

The fisheries mutual insurance scheme, which was established in 1964 under the Fisheries Disaster Compensation Law, aims at stabilizing small- and medium-size fishing and aquaculture operations by covering losses from poor catches caused by natural disasters. The system insures fish harvests, aquaculture, special aquaculture, and fishing gear.

The government subsidizes one-third to one-half of the premium. While fishing vessel insurance enjoyed a surplus of ¥16.5 billion in 2010, the fisheries mutual insurance scheme suffered a deficit of ¥28.9 billion.

The Ministry of Agriculture, Forestry and Fisheries estimates that total claims would amount to ¥120.4 billion, of which the central government will cover ¥94 billion (or 78 percent) for the GEJE as figure shows. As of March 13, 2012, ¥63.4 billion in claims have been paid out: ¥47.5 billion under the fishing vessel insurance system, and ¥15.9 billion under the fisheries mutual insurance scheme. Sixty percent of vessels were insured under the vessel insurance scheme, of which some 80 percent of boats were over 20 tonnes. Some 80 percent of the insured vessels were more than 15 years old. Since the schemes cover the residual value of the vessels, the claims paid out may not cover the replacement costs.

Billion ¥

	FISHING VESSEL INSURANCE SYSTEM	FISHERIES MUTUAL INSURANCE SCHEME	TOTAL
Government	72.7 (78%)	21.3 (77%)	94.0 (78%)
Reserve of government special account	11.0 (12%)	—	11.0 (9%)
Associations at national level	1.4 (2%)	3.0 (11%)	4.4 (4%)
Associations	7.8 (8%)	3.2 (12%)	11.0 (9%)
Total	92.9 (100%)	27.5 (100%)	120.4 (100%)

Source: MAFF.
Note: — = not available.

Agriculture Insurance

Damage to agricultural production and facilities from the GEJE event amounted to ¥63 billion. Rice is an important crop in Japan, but because the GEJE happened before the rice-growing season, insurance almost did not cover rice production losses. Since compensation related to the accident at the Fukushima Nuclear Power Station has not yet been decided, the total payout on agricultural insurance is uncertain. In Miyagi Prefecture the agricultural insurance scheme has covered damages to greenhouses in the amount of ¥1 billion.

The Farm Losses Compensation Law introduced the agricultural insurance scheme in 1947 to help farmers stabilize their businesses by covering damages caused by natural disasters; the scheme offers insurance coverage for almost all major agricultural products. It was started by local farmers as a cooperative initiative to set up a reserve fund to pay for insurance premiums, which evolved into agricultural mutual relief associations. The insurance scheme includes rice, wheat, and barley insurance (mandatory for paddy fields of more than 20 hectares); livestock insurance; fruit and fruit tree insurance; field crop and horticultural insurance; greenhouse insurance; and insurance for houses and properties. The government subsidizes half of the farmers' premiums.

Note: Prepared by Mikio Ishiwatari, World Bank.

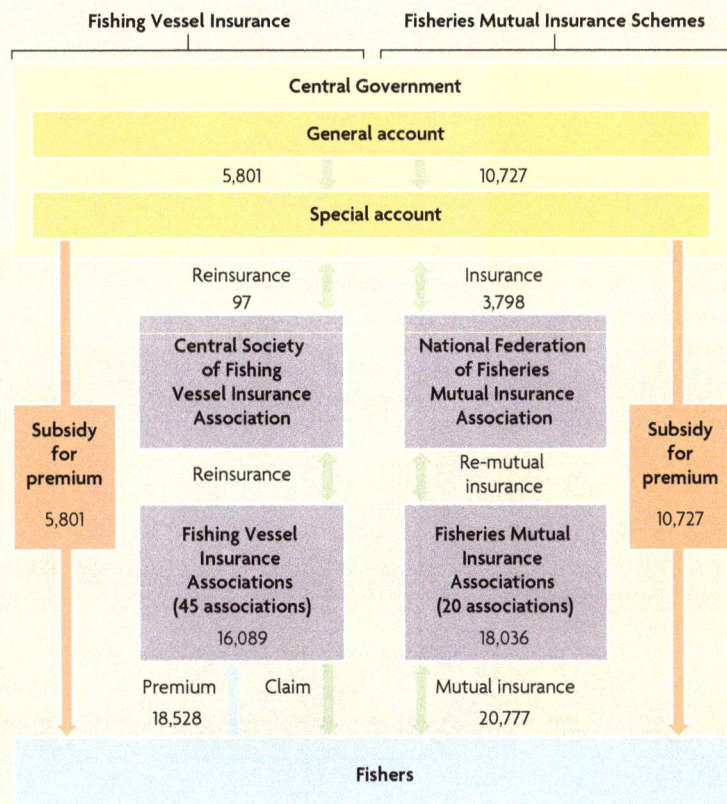

Source: MAFF.
Note: Numbers are millions of JPY in FY 2011 budget.

**Figure 29.3
Estimated GEJE
insured residential
losses, by earthquake
insurance program**

Sources: JER 2011a and 2011b;
JA Kyosai 2011.

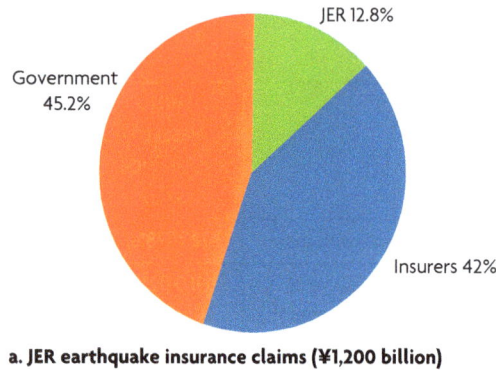

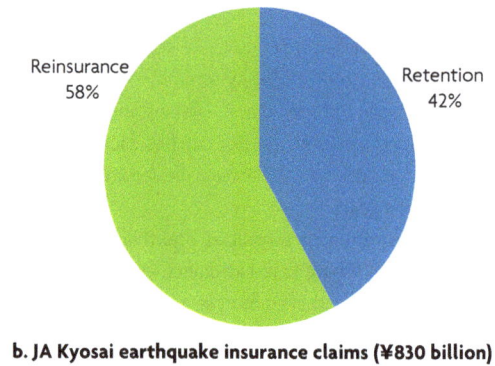

a. JER earthquake insurance claims (¥1,200 billion)

b. JA Kyosai earthquake insurance claims (¥830 billion)

programs had adequate capacity to meet their claims obligations, thanks to efficient management of exposure to losses through a combination of policy limits and reinsurance protection. The earthquake insurance program managed by the private nonlife insurance companies faced an estimated total loss of ¥1,200 billion, with 42 percent retained by private insurers, 13 percent retained by the JER, and 45 percent retained by the government (see figure 29.3). This event, however, severely depleted the earthquake reserves of both the private insurers and JER, leading to an increase in

government liability in the revised reinsurance program of 2012. Earthquake losses incurred by JA Kyosai were estimated at ¥830 billion, 90 percent of which were residential losses. It is estimated that about 58 percent of those losses were reinsured.

The three-step earthquake claims settlement system implemented by the private insurance companies allowed claims to be settled rapidly. Satellite images were also used to identify total losses on buildings, which further helped speed up claims settlements. In the aftermath of the disaster, the General Insurance Association of Japan designated specific total loss zones, based on satellite imagery (chapter 26). Any total loss claims filed within these areas did not require additional confirmation of incurred losses, thereby speeding up the payout process. Out of ¥1,200 billion generated by the 741,000 claim payments made after the GEJE, 60 percent was paid within two months and 90 percent within five months.

Comparative analysis of the GEJE with other recent earthquakes

It is interesting to compare the economic and fiscal impact of the GEJE with the impact of other recent earthquakes: the 2010 earthquake in Chile and the 2011 earthquake in Canterbury, New Zealand. All three earthquakes were very large in magnitude and caused severe economic losses in their countries. Table 29.3 summarizes this comparative analysis. While the GEJE caused the largest economic losses

Table 29.3 Comparative analysis of the Tohoku (GEJE), Canterbury, and Maule earthquakes

	TOHOKU, JAPAN	CANTERBURY, NEW ZEALAND	MAULE, CHILE
Year	2011	2011	2010
Magnitude	9.0	6.3	8.8
Estimated direct economic losses ($ billion)	225	15	20
Estimated direct economic losses (% GDP)	4	9	9
Estimated direct losses borne by government (as % of government expenditures)	8	11	n.a.
Estimated insured losses (% of direct economic losses)	16	80	40
Estimated insured losses covered by international reinsurance (%)	23	73	95

Sources: Aon Benfield 2011; Swiss Re 2012; New Zealand Treasury 2011; RMS 2011; Ministry of Finance, Japan.
Note: Direct economic losses are defined as damage to physical assets (including infrastructure); n.a. = not applicable.

in absolute terms, losses as a percentage of gross domestic product (GDP) are lower than those in Chile and New Zealand given the size of the Japanese economy. The government's portion of direct losses (that is, additional expenditures), expressed as a percentage of total government expenditures, were estimated at 8 percent for the GEJE and 11 percent for the Canterbury earthquake in New Zealand. Finally, the fraction of the insured losses covered by international reinsurance was estimated at 95 percent in Chile, 29 percent in New Zealand (where the Earthquake Commission retained a large fraction of the losses), and 23 percent in Japan. This last figure hides a large difference between the JER, which relies on public reinsurance and cooperative mutual insurers (such JA Kyosai) that purchase most of their reinsurance capacity abroad.

LESSONS

Some key lessons can be drawn from the review of Japan's earthquake insurance programs in light of the GEJE:

- *There is no one-size-fits-all insurance program.* The dual earthquake insurance programs in Japan illustrate that there is no one-size-fits-all catastrophe insurance program. Two very different schemes can coexist successfully within a country significantly exposed to earthquakes, offering earthquake coverage to about four households out of 10 in Japan.

- *Resilience is critical for earthquake insurance programs.* Both programs managed to fulfill their obligations after the GEJE without difficulties, because of the sound management of policy limits and conservative reinsurance coverage. The apparent resilience of the current setup does not mean, however, that there is no room for these schemes to improve without compromising sustainability. For example, the earthquake

insurance limit offered by JA Kyosai started at 10 percent and has increased progressively to 50 percent currently.

- *Rapid claims settlement can be achieved, even after a major disaster.* The three-step claims adjustment system implemented by the private insurers allows for rapid damage assessment and claims settlement. It also takes into account that, immediately after a major disaster, large numbers of loss assessors have to be deployed at the same time. The simplicity of the three-step system allows this to happen.

- *Insurance penetration in Japan is high, but there is still considerable room for expansion.* About 40 percent of Japanese households have earthquake insurance coverage, leaving 60 percent of households without coverage. International experience shows that it is very difficult, if not impossible, to increase the penetration rate beyond a certain level on a voluntary basis. Compulsory earthquake insurance could therefore be considered.

The GEJE also highlighted certain challenges of earthquake insurance programs run by private insurance companies:

- *The JER claims-paying capacity is limited in the aggregate.* The aggregate limit was initially set at ¥5,500 billion (increased to ¥6,200 billion in 2012), which would be sufficient for a major earthquake such as the Great Kanto earthquake in 1923. But this does not take into account the occurrence of consecutive major earthquakes, which could jeopardize the solvency of the program.

- *The government's liability under the JER exceeds its ex ante financing arrangements.* The government's maximum liability is adjusted based on the balance of earthquake reserves of the private insurers and the JER, and the maximum defined liability under the program. The government currently holds 87 percent of the total liability of the program. Its current special account would not be sufficient to cover this level of liability and would require an immediate budget appropriation or reallocation in case of a major disaster.

- *Limited policy coverage may not meet the needs of the insured.* The program is designed to provide partial coverage (up to 50 percent of the fire insurance coverage limit) to "stabilize the livelihood of the earthquake victims" (article 1 of the 1966 Earthquake Insurance Law). There seems to be a growing demand for higher coverage, but such an increase in coverage should be carefully evaluated to maintain the financial sustainability of the system.

- *The claims settlement process introduces significant basis risk and could be revised.* Although the three-step claims adjustment process allows for rapid settlement of claims, there is a big gap between payouts for partial loss (5 percent) and half loss (50 percent). This increases the risk that payments will not match the needs of the insured party following the occurrence of damage (basis risk). A fourth intermediate step could be introduced to reduce this risk.

- *Catastrophe risk modeling for Japan is sophisticated, but could be improved.* State-of-the-art catastrophe risk models have been developed for Japan, but need to be further refined as secondary loss perils (such as tsunamis, which caused about 30 percent of the total losses from the GEJE) and liquefaction are not included as standard in all models. These models could also be used to further assess the catastrophe risk exposure of public buildings and infrastructures.

RECOMMENDATIONS FOR DEVELOPING COUNTRIES

Japanese earthquake insurance programs demonstrated considerable resilience after the GEJE. From this experience, recommendations can be made to disaster-prone developing countries willing to promote catastrophe risk insurance to help them promote viable and affordable programs and clearly define the role of the government in public-private partnerships (PPPs).

Structure policies to allow for sustainable and affordable programs. Catastrophe risk insurance policies should be designed to enable insurance companies control their liabilities and offer affordable coverage. The policy structure can be revised over time to better respond to the needs of the policyholders, while also ensuring the system's resilience to major disasters. The partial coverage produced by both Japanese earthquake insurance programs and the simplified loss adjustment process of the private insurer system helps to keep costs down.

Price insurance premiums based on the underlying risks. Insurance premiums should reflect the underlying risks with respect to the various risk zones and types of construction. Risk-based insurance premiums make policyholders aware of the underlying cost of risk, thereby providing financial incentives to engage in disaster risk mitigation. Even in cases where the full cost of cover is not passed onto the policyholder, it is still possible to signal the underlying cost of risk by making subsidies transparent.

Provide incentives to invest in disaster risk mitigation. Additional financial incentives, such as discounts on premium rates or lower deductibles, can be offered to the policyholders who invest in risk reduction.

Consider mechanisms for enforcing insurance purchase. Voluntary catastrophe risk insurance does not typically generate high penetration rates, even in highly developed insurance markets. Some type of compulsory mechanism, such as an automatic catastrophe guarantee in fire insurance policies, may be necessary to ensure that a large proportion of the population is insured against natural disasters.

Promote multiple-catastrophe risk insurance delivery channels. Catastrophe risk insurance should leverage existing nonlife insurance delivery channels, such as private insurers or mutual insurers. The Japanese system demonstrates that different segments of the population may be best served by different delivery channels, even for very similar products. Multiple distribution channels for catastrophe risk insurance should therefore be explored.

Develop detailed catastrophe risk models. Detailed catastrophe risk models and databases are essential for detailed risk assessment, premium rate calculation, and efficient management of catastrophe risk insurance liabilities. In addition to a strong hazard model, such assessments also require detailed exposure databases of at-risk assets (buildings and infrastructure) and detailed vulnerability functions to translate hazard values into dollar losses. These models are typically developed by private risk-modeling firms and licensed to the insurance industry. But for some less-developed insurance markets, governments and donors have funded or partially funded the development of such models as public goods to support market development.

Develop catastrophe risk insurance market infrastructure. Catastrophe risk insurance markets require major investments in basic infrastructure, such as catastrophe risk models, exposure databases, product design and pricing, and the like. Governments can play a major role in developing this kind of infrastructure to help the private insurance industry offer cost-effective and affordable insurance solutions.

Promote enabling legal and regulatory environments. Unlike traditional lines of insurance business such as automobile insurance, catastrophe risk insurance can generate large correlated losses for insurers. The legal and regulatory framework should enforce adequate pricing, reserving, and reinsurance buying to ensure that insurers will meet their claims in full in the event of a disaster.

Promote PPPs for catastrophe insurance programs. Governments can play an important role in building an affordable and sustainable earthquake insurance program. As the private insurance sector brings its technical expertise and financial capacity to the table, governments can support the development of public goods and risk-market infrastructure to foster sustainable market-based insurance solutions.

Governments can play a role as the financier of last resort. Governments may want to act as financiers of last resort when private reinsurance capacity is unavailable or too expensive to allow domestic insurers to offer cost-effective insurance solutions. Governments should not compete with the private reinsurance market but rather complement it. When needed, governments should make financial capacity available to domestic insurers through public reinsurance or (contingent) credit.

NOTES

Prepared by Olivier Mahul and Emily White, World Bank.

1. Also known as Zenkyoren.
2. The number of households is estimated at about 51 million (Government of Japan, Statistics Bureau). Policy-in-force data from the Japanese Non-Life Insurance Rating Organization (2010), JA Kyosai Business Operations (2011), and Zenrosai Annual Report (2010). Cooperative mutual insurer figures extrapolated based on an 85 percent estimate of the JA Kyosai market share.
3. The total claims-paying capacity of the program will increase to ¥6.2 billion in 2012 (Ministry of Finance 2012).

BIBLIOGRAPHY

Aon Benfield. 2011. "Earthquake Insurance Business in Japan." December 2011. Aon Benfield, London, England.

General Insurance Association of Japan. 2011. *Annual Report 2010–2011.* Tokyo.

JA Kyosai. 2011. *Annual Report 2010, Business Operations.* Tokyo. http://www.ja-kyosai.or.jp/ebook/2010annual/index.html.

Japan Credit Rating Agency Ltd. 2011. "JCR Affirmed AAp/Stable on Japan Earthquake Reinsurance." News Release, December 28. Tokyo.

JER (Japan Earthquake Reinsurance Co., Ltd). 2011a. *Annual Report 2011*. Osaka, Japan.

———. 2011b. "Response to the Great East Japan Earthquake by the General Insurance Industry." Paper presentation, World Forum, Jamaica, October 25–26.

McAllister, S., and E. Cohen. 2011. "Japanese Casualty Insurers Show Resilience." http://www .contingencies.org.

Muir-Wood, R. 2011. "Designing Optimal Risk Mitigation and Risk Transfer Mechanisms to Improve the Management of Earthquake Risk in Chile." OECD Working Papers on Finance, Insurance and Private Pensions No. 12, Organisation for Economic Co-operation and Development (OECD), Paris.

New Zealand Treasury. 2012. *Economic and Financial Overview 2012*. New Zealand Government, Wellington. http://www.treasury.govt.nz /economy/overview/2012/nzefo-12.pdf.

Non-Life Insurance Rating Organization of Japan websites. http://www.giroj.or.jp/english/index .html.

RMS (Risk Management Solutions Inc.). 2011. "The M9.0 Tohoku, Japan Earthquake: Short-Term Changes in Seismic Risk." Newark, CA. http://www.rms.com/resources/publications /natural-catastrophes.

SCOR Global P&C. 2011. Technical Newsletters, December and October 2011. Paris, France.

Swiss Re. 2012. "Lessons from Major Earthquakes." Economic Research and Consulting. Zurich, Switzerland.

Zenrosai. 2011. *Annual Report 2010*. https://www .zenrosai.coop/english/pdf/2010/annual_report_ 2010.pdf.

CHAPTER 30

Economic Impact

Following the Great East Japan Earthquake (GEJE), the government of Japan responded promptly to stabilize markets and ensure a swift recovery. Economic activity has since started picking up, thanks in part to domestic demand driven by the massive reconstruction effort. Uncertainties remain, however, surrounding the restructuring of power supply and both national and global economic prospects. The year 2011 will be remembered for the severe challenges to the global supply chain posed by the GEJE and the Thai flood. As an important part of the networked production system, developing countries must share responsibility in making the supply chain more resilient under international cooperation.

FINDINGS

Following the Great East Japan Earthquake (GEJE), the government of Japan initially estimated the direct damages between ¥16 trillion and ¥25 trillion (see box 30.1). The Cabinet Office (CAO) later put estimated damages at ¥16.9 trillion (approximately $210 billion), or about 4 percent of Japan's gross domestic product (GDP). Before the disasters, approximately two-thirds of nonfinancial assets were held by the private sector. This was in line with the breakdown of the direct damage figures released by the CAO (table 30.1).

Table 30.1 Direct economic impact of the GEJE

CATEGORIES	DAMAGE (¥ TRILLION)	SHARE OF TOTAL DAMAGE (%)
Buildings (housing, offices, plants, machinery, and so on)	10.4	62
Lifeline utilities (electricity, gas, water, communication, and so on)	1.3	8
Social infrastructure (waterways, roads, harbors, drainage, airports, and so on)	2.2	13
Others (including agriculture and fisheries, and so on)	3.0	17
Total	16.9	

Source: Cabinet Office (CAO).

BOX 30.1

Government of Japan's estimates of the economic impact of the GEJE

The CAO released two different sets of estimated economic damages (damage on capital stocks) of the GEJE (table B30.1.1).

Table B30.1.1. Estimated economic damages of the GEJE by the CAO1
¥ trillion

	DISASTER REDUCTION SECTION	ECONOMIC AND FINANCIAL ANALYSIS SECTION	
		CASE 1	CASE 2
Buildings and houses	10.4	11	20
Utilities	1.3	1	1
Infrastructure	2.2	2	2
Others		2	2
Agriculture	1.9		
Others	1.1		
Total	16.9	16	25

Source: CAO.

Note: Case 1 uses damage rates twice as high as the Kobe earthquake, while Case 2 employs even higher damage rates against buildings and houses for the tsunami-affected areas.

The economic impacts are estimated separately for damages (on capital stocks) and losses (on flow). The estimation results for damages in table B30.1.1 are calculated by multiplying the existing predisaster capital stock data (based on the CAO's macroeconomic database) by damage rates twice as high as the ones observed for the Great Hanshin-Awaji (Kobe) Earthquake for Case 1, and by even higher damage rates against buildings and houses for Case 2 to take into account the damages from the tsunami. In this estimation, the damaged areas include the prefectures of Iwate, Miyagi, and Fukushima (the above-mentioned damage rates are applied to the tsunami-affected areas in these prefectures, while damage rates equivalent to the Kobe earthquake's are used for the non-tsunami-affected areas) and the surrounding prefectures of Hokkaido, Aomori, Ibaraki, and Chiba, for which damages are calculated by multiplying the capital stock data by damage rates modified based on the seismic intensity of each prefecture (details unknown).

The estimation of the economic impact from the GEJE (not included in table B30.1.1) covers the same prefectures and is carried out for three fiscal years (table B30.1.2).

Table B30.1.2. Estimated economic impact of the GEJE
¥ trillion

	FY2011		FY2012	FY2013
	FIRST HALF	SECOND HALF		
Production loss due to damages	−1.25 to −0.5	−1.25 to −0.5	−2.25 to −1.25	−2.25 to −1.25
Production loss due to supply chain disruption	−0.25	—	—	—
Production loss due to limited power supply	—	—	—	—
Production gain from recovery and reconstruction	2 to 3	3 to 5	6 to 9.5	5 to 7.75
Total	0.5 to 2.25	2 to 4.25	3.75 to 8.25	2.75 to 6.5

Source: CAO.

Note: — = not available.

The estimated production losses due to damages (first-order loss) by the GEJE are calculated based on the damages listed in table B30.1.1 using the production function of each sector. The production loss due to supply chain disruption (roughly equivalent to a higher-order loss) is estimated with the calculated production loss (the above first-order loss) and an interregional input-output table (between Tohoku and the rest of Japan). While the production losses due to limited (electric) power supply were considered, they were not estimated due to the uncertainty of effects on production (resiliency, conservation, or use of other adaptive measures). The production gains from recovery and reconstruction activities are derived by distributing the amount of estimated damages in table B30.1.1 over three years (meaning it is assumed that all the damaged capital stocks will be restored).

Most of the damages were concentrated in three prefectures of the Tohoku region: Fukushima, Iwate, and Miyagi. The sparsely populated Pacific Coast of the Tohoku region, where agriculture and fishery are the main activities, accounts for only 2.5 percent of the total Japanese economy in terms of industrial production (figure 30.1).

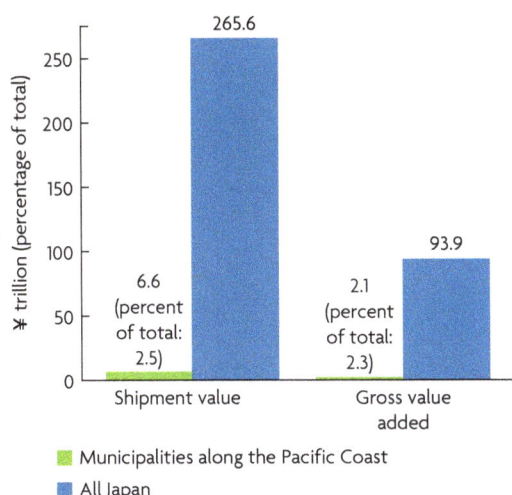

Figure 30.1 The extent of economic activity in the municipalities along the Pacific Coast

Source: Ministry of Economy, Trade and Industry (METI).

Table 30.2 Damage caused by the GEJE

SECTOR		DIRECT DAMAGE	MONETARY DAMAGE (¥100 MILLION)
Fisheries	Fishing vessels	25,014	1,701
	Fishery harbor facilities	319 harbors	8,230
	Aquaculture facilities	—	738
	Aquaculture products	—	575
	Common use facilities	1,725 facilities	1,249
SUBTOTAL, FISHERIES			**12,493**
Agricultural land, facilities	Damaged agricultural land	17,456 areas	4,012
	Damaged agricultural facilities	21,866 facilities	4,290
SUBTOTAL, AGRICULTURAL LAND AND FACILITIES		**39,322**	**8,302**
Agricultural crops, livestock, and production	Agricultural crops and livestock	—	140
	Agricultural livestock production facilities (mainly country elevators, agricultural warehouses, PVC greenhouses, livestock barns, compost depots, and so on)	—	487
SUBTOTAL, AGRICULTURAL CROPS			**626**
Forestry	Desolation of forest land	429 areas	238
	Damage of facilities for maintaining forest	255 facilities	1,167
	Damage of forest road	2,632 areas	42
	Damage of forests	(1,065 hectares)	10
	Processing and marketing facilities	112 facilities	508
	Cultivating facilities for forest products	473 facilities	25
SUBTOTAL, FORESTRY		**3,903 FACILITIES**	**1,989**
TOTAL			**23,410**

Source: Ministry of Agriculture, Forestry and Fisheries 2011.

Note: — = not available.

Despite the relatively small extent of economic activities in the affected region, the GEJE had severe and widespread economic impacts, partly due to the accident at the Fukushima Daiichi Nuclear Power Station and ensuing energy supply disruptions, and the supply chain disruptions (compounded by widespread flooding in Thailand a few months later).

In the first quarter of 2011, Japan's GDP contracted by 3.5 percent. According to the International Monetary Fund (IMF), GDP contracted by 0.7 percent in all of 2011, and the estimates for 2012 put GDP growth at 2 percent, stimulated by reconstruction work.

There are approximately 80,000 businesses in the tsunami-affected areas, 740,000 businesses in the earthquake-affected areas, 8,000 businesses in the evacuation zones of the Fukushima nuclear accident, and 1.45 million businesses in the prefectures covered by the Tokyo Electric Power Company.

Impacts on agriculture, forestry, and fisheries

The amount of damage to agriculture, forests, and fisheries by the GEJE was estimated as ¥2.34 trillion (table 30.2).

Around 24,000 hectares of agricultural land (approximately 80 percent of paddy fields and 20 percent of farmland) were flooded by the tsunami. Over 95 percent of the damaged agricultural land was located in the three prefectures most severely affected: Iwate, Miyagi, and Fukushima.

It is estimated that the area of agricultural land that will be restored and cultivated again by 2012 could be less than 50 percent in Iwate and Miyagi prefectures, and only up to 20 percent in Fukushima Prefecture as a result of the nuclear accident.

Many plywood-processing factories in Iwate and Miyagi prefectures, where about one-third of plywood products are produced, were damaged.

The Fukushima nuclear accident further impacted the agriculture, forestry, and fisheries sectors. Based on the provisional regulation on radiation (instated on March 17, 2011), shipping of food products containing radioactive iodine above a certain threshold has been restricted. In addition to the national regulation, some prefectures and local associations set additional restrictions on the shipping of food products.

The accident also affected trade flows of food products with other countries. Import controls for Japanese food products were intensified in 43 countries, and Japanese exports declined.

Impacts on the tourism industry

The GEJE has severely affected the tourism industry in Japan, but according to a report by the World Travel and Tourism Council (WTCC), recovery has been more rapid than previously expected for both domestic and international tourism.

Foreign visitor arrivals in the month immediately following the GEJE were 62 percent lower than the previous year. Recovery was swift and, by the fall of 2011, arrivals were only 15 percent down compared to the previous year. Inbound international travel was

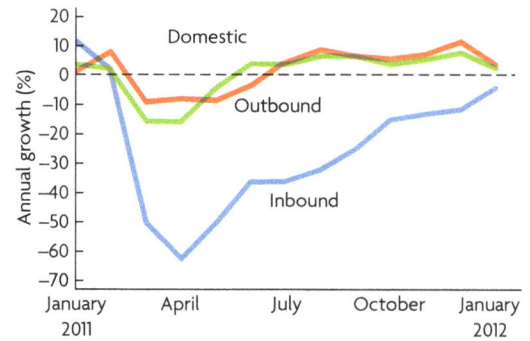

Figure 30.2 Japanese tourism demand, 2011–12
Source: WTTC 2012.

more severely affected compared to outbound international travel and domestic travel (figure 30.2). This trend reflects fears generated by the nuclear accident and loss of competitiveness as a result of the appreciation of the yen in the months following the disaster.

The WTCC estimates that the negative impact of the GEJE on the tourism industry amounts to approximately ¥0.7 trillion.

Impacts on financial and currency markets

Financial and currency markets stabilized quickly after the earthquake. Equity markets fell by over 15 percent in the first weeks after the earthquake, but recouped roughly one-third of their losses by mid-June 2011.

Figure 30.3 shows the Nikkei Index from January 2011 to June 2012. The Nikkei Index is a stock market index for the Tokyo Stock Exchange (TSE). It is a price-weighted average (the unit is yen), which indexes 225 companies in the TSE (components are reviewed once a year).

The figure clearly shows the fall after March 11 and the recovery until summer 2011. High volatility followed, but those values cannot be strictly connected to the recovery process, as the international financial crisis impacted the TSE.

In the immediate aftermath of the earthquake, the yen appreciated sharply because of speculation around sizeable repatriation flows by insurance companies, corporations, and households. The value of the yen touched a record ¥76.25 per dollar on March 17, before retreating to the 80-yen level. After concerted intervention in coordination with the Group of Seven (G-7), the yen/dollar rate has traded in a band of 80–84. Approximately a quarter of developing East Asia's long-term debt is denominated in yen. For China, 8 percent of its external government debt is in yen; the figure for Thailand is about 60 percent; for Vietnam about 35 percent; for the Philippines about 32 percent; and for Indonesia about 30 percent. A 1 percent appreciation in the value of the yen translates into a $250 million increase in annual debt servicing on yen-denominated securities by East Asia's developing countries.

After the disaster, the Bank of Japan injected liquidity to ensure that there would be no shortage of cash or funds to lend and no spikes in Japan's interest rates. Massive liquidity injections flattened the Japan Government Bond yield curve, with the 10-year rate moving in a narrow range between 1.1 and 1.2 percent.

One of the critical challenges for the Japanese economy remains overcoming deflation to return to a sustainable growth path with price stability. The Bank of Japan and the government are working together to prevent the economy from falling into a vicious cycle between yen appreciation and deflation.

Impacts on energy supply

The damage resulting from the earthquake and tsunami is being compounded by the resulting shortages in energy supply. Energy supply disruptions have caused rolling blackouts that have disrupted Japan's production capacity in its industrial heartland in the Kanto region, which accounts for about 40 percent of national GDP.

The Fukushima nuclear accident has pushed the government to explore alternative energy sources. The Ministry of Economy, Trade and Industry (METI) established the Fundamental Issues Subcommittee under the Advisory Committee for Natural Resources and Energy to advise a new long-term energy plan. In the

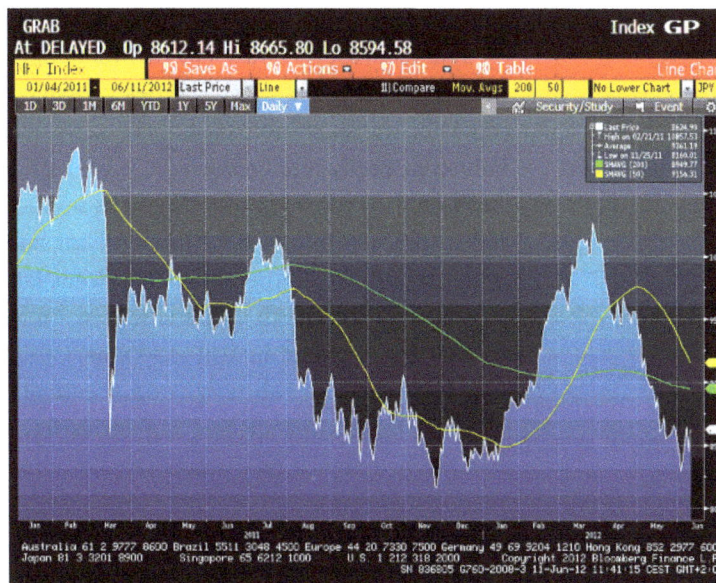

Figure 30.3 Nikkei Index, January 2011–June 2012

Source: © Bloomberg.com 2012. Used with permission. Further permission required for reuse.

interim report, the committee emphasized the need to reform the demand structure, including energy conservation measures and controls on peak-time electricity demand.

In the short term, the shift toward other energy sources will boost imports from oil- and petroleum-exporting countries in the East Asia region, in particular Indonesia, Malaysia, and Australia.

Impacts on industrial production

The main economic activities in the affected region are agriculture (mainly rice paddy fields) and fisheries, but manufacturing accounts for about a quarter of production in the region, and plants in the most severely damaged areas supply parts and products used in manufacturing elsewhere in Japan and Asia.

Damage to Japan's industrial facilities caused a sharp drop in production following the GEJE, but swift reconstruction has minimized the long-term impact on production.

Japan's METI reported that, as of August 2011, restoration works had been completed for 93 percent of the 91 production bases directly affecting Japan's major manufacturing industries, including machinery, automotive, and consumer electronics. The automotive industry recorded the greatest fall in production, but recovered rapidly as facilities reopened and vital transport networks were repaired. Industrial production rebounded from April onwards with a growth of 6.2 percent in May and 3.8 percent in June. But this is still not sufficient to fully offset the initial 15 percent fall experienced in March. Production in June remained lower than in 2010 and was 5 percent lower than in February, on a seasonally adjusted basis. Most affected industries have now reached almost predisaster levels of production (figure 30.4).

Double debt

The "double debt problem" generally refers to the financial difficulties facing individuals

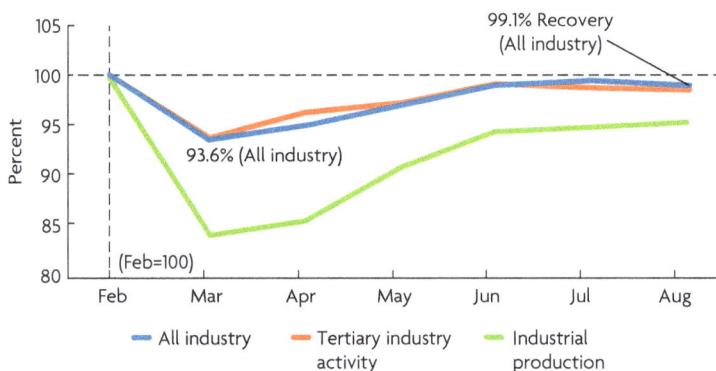

Figure 30.4 Indices of all industry activity (seasonally adjusted), 2011

Source: METI 2012.

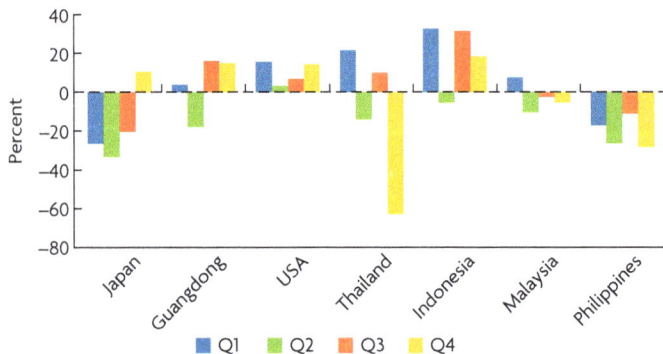

Figure 30.5 Impact of the GEJE and Thai flood on the global automobile industry

Source: Kobe University.

and business owners stricken by the GEJE who need to borrow to rebuild their destroyed houses and offices. But as they have existing loans on such premises, borrowing additional money results in two debts on the same property. The Japanese government as a whole worked on policy responses and formulated the Policy for the Double Debt Problem, which was released on June 17, 2011 (as explained in box 30.2).

GLOBAL SUPPLY CHAINS

It's a small (networked) world after all

With the rapid progress of information and transport technology together with the promotion of free trade, humans have developed an extensive network of production, trade, and investment throughout the world. Moreover, we have intensive agglomeration of production and consumption in major cities throughout the world, which are mutually connected through a dense supply chain network. Today's global production system is a complex, networked system that has operated efficiently under normal conditions. Nevertheless, recent megadisasters in Japan and Thailand have revealed the networked world's vulnerability to major disasters.

The magnitude of the Japanese economic impact is partially attributable to supply chain network disruptions. The disaster-affected areas served as major sources of the supply chain of goods (from procurement of parts to the delivery of finished products) for Japan's manufacturing industry. Failures of parts and material deliveries from these areas have forced many manufacturers nationwide to suspend their operations. The automobile industry, the electronic equipment industry, and the metal industry were affected most severely because they particularly depended on key parts and basic materials produced in the disaster-affected areas. Figure 30.5 shows that Japanese automobile production in the first

Measures to address the double-debt problem

Individual debtor guidelines for out-of-court workouts

Guidelines for individuals on out-of-court debt restructuring, that is, the Individual Debtor Guidelines for Out-of-Court Workouts were released on July 15, 2011, and took effect on August 22, 2011.

The guidelines are aimed at individual debtors who are unable, or deemed certain to soon become unable, to repay their existing loans—in other words, those who would in principle qualify to initiate bankruptcy or civil rehabilitation procedures. The creditors subject to the guidelines consist primarily of private sector banks, cooperative financial institutions, government-affiliated financial institutions, money lenders, and leasing companies.

As of March 30, 2012, the accumulated total number of cases consulted was 1,850, of which 538 cases were in the process of restructuring loans. This system is the first of its kind in Japan and is unprecedented even in the world.

Clearer application of financial inspection manuals

In the case of a company resuming or continuing its operations while repairing damage sustained from the earthquake and tsunami, there is a risk that its capital has been impaired due to the impact of the disaster. Capital augmentation is therefore urgently needed.

The Financial Services Agency introduced measures to apply its financial inspection manuals in a clearer manner, aiming to promote more active use of capital-eligible debt and thereby enable undercapitalized companies to improve their balance sheets and management.

These measures are expected to yield a number of positive effects. For example, even if a company's capital has been impaired due to the impact of the disaster, the company is able to exchange its existing loans for new ones that satisfy the requirements for capital-eligible debt (that is, a debt-debt swap). As a result, its balance sheet will become healthier, which will in turn lead to greater chances of obtaining new loans from financial institutions.

Measures for financial institutions

Some financial institutions located in the disaster-stricken area sustained significant damage; some institutions' operational bases were almost entirely destroyed by the disaster. It is imperative to maintain and strengthen the financial functions of banks and other institutions to revitalize the regional economy. To that end, special provisions concerning the disaster have been added to the Act on Special Measures for Strengthening Financial Functions. First, special provisions for disaster-affected financial institutions in need of the government's capital injection to strengthen its financial functions have been added. For instance, when such a financial institution draws up a management enhancement plan, its top executives are not held responsible or required to set profitability and efficiency targets, on the grounds that the impact of the earthquake and tsunami is beyond their control. Furthermore, the costs the financial institution bears for receiving capital injection are substantially lower than the costs needed under normal conditions. In addition, a much longer period is allowed for securing the repayment funds. In return for receiving this capital injection under very favorable conditions, the financial institution is expected to play its financial intermediary functions in an even more active way. Second, special provisions have been incorporated for *shinkin* banks, credit cooperatives, and other cooperative financial institutions to further ease the requirements for capital injection. Under the amended law, the government and the central organization of a financial institution jointly inject capital, and the financial institution is required to conclude a management guidance agreement with the central organization. In the event that the injected capital is highly unlikely to be repaid by the set date, said capital will be liquidated and the financial institution's business restructured. The Deposit Insurance Corporation's funds are used as the source of capital injection. The amendments also include a five-year extension to the end of March 2017 of the time limit for applications for the government's capital injection.

As of March 30, 2012, the government has decided to inject capital (¥191.0 billion in total) into 10 financial institutions—three banks, four *shinkin* banks, and three credit cooperatives—operating in the disaster-stricken areas in accordance with the Act on Special Measures for Strengthening Financial Functions.

---

and second quarter of 2011 were 25 percent and 33.8 percent less, respectively, than those in the same period the prior year.

Eastern Asia today, often called the "world factory," is based on a supply chain network centering around dozens of major cities and industrial agglomerations. Consequently, the impact of the GEJE and tsunami disaster could not remain limited to Japan. Figure 30.5 shows that automobile output in China's Guangdong Province and Thailand declined by 17.3 percent and 11.5 percent, respectively, in the second quarter. Other Asian countries such as Indonesia, Malaysia, and the Philippines were also affected. The impact extended beyond Asia. In the United States—where automakers, including those of Japanese origin, depend on the supply of some crucial parts from Japan—production growth plunged from 15.6 percent in the first quarter to 2.3 percent in the second. These results reaffirm that disruption in a specific region affects the world through the supply chain network.

In the fourth quarter of 2011, when Japanese manufacturing industries had almost recovered from the impact of the disaster, the East Asian supply chain was challenged again by the great flood in Thailand—the worst in 50 years. Automobile output in Thailand dropped by 61.5 percent in the fourth quarter compared to the same period of the prior year. Affected by the shortage of parts supplies from Thailand, Japanese automobile production was limited to a 4.5 percent year-on-year growth in November after recording 20.3 percent growth in October (although the impact was short-lived and growth returned to 13.4 percent in December). Being the local hub of the automobile supply chain in the Association of Southeast Asian Nations (ASEAN), the Thai effect was felt more severely in Malaysia and the Philippines, while the impact on Indonesia was sharp and short (year-on-year growth rates dropped to 0.7 percent in November but showed greater than 20 percent growth in October and December).

Thailand is also known as the global center of hard disk drive production—accounting for almost 20 percent of world exports, on par with China. According to a market survey conducted by Kakaku.com, compared to the beginning of October 2011, retail prices of popular hard disk drives (1 terabyte capacity and 7,200 rpm spin speed) in the Japanese market shot up 150–200 percent by mid-November before settling down, but remained about two times as high as the preflood level at the beginning of February 2012.

Disaster strikes when you least expect it

Recent experiences remind us of the vulnerability of supply chain networks, which contain some critical nodes wherein the production of particular parts and components is concentrated among a few suppliers. Importantly, such concentrations do not result from planning failures. Rather, they are self-organized through market interactions. Because of scale economies, production concentration is preferred by both suppliers and customers. Although a trade-off relation exists between scale economies and transport costs to deliver products to distant customers, lower transport costs make the concentration of production more profitable, as shown in figure 30.6. Consequently, globalization (decline of broadly defined international transport/transaction costs) tends to enhance the formation of agglomeration within a global supply chain. Because of self-organization, it is not feasible to eliminate potential risks by agglomeration in highly complex supply chains. To complicate matters further, when a disruption occurs, it is impossible to find replacements from other suppliers, at least in the short run, because of a high degree of customization. An example from the 2011 disaster was the Renesas Electronics Corporation's Naka plant, located in Ibaraki Prefecture. It produces a micro control unit (MCU) for high-quality motor vehicles that makes extensive use of electronic control technology. Over the years, Renesas has become a supplier of customized MCUs for major automobile companies throughout the world.

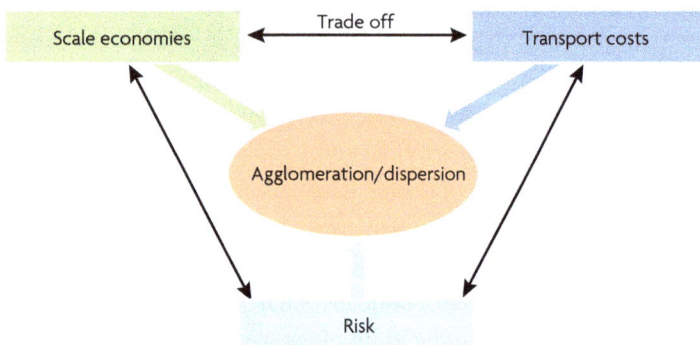

Figure 30.6 Trilateral trade-offs in global resilience

We might find other cases of dispersion forces if concentration increases the potential risk of disruption for the entire supply chain. Dispersions in this case might involve building sufficient safety stocks (dispersion of products), use of multiple suppliers, and duplication of production facilities. These actions, which are components of so-called business continuity plans (BCPs), are aimed at increasing redundancy and resiliency. They garner great attention in the supply chain management literature.

Individual firms are rarely capable of taking sufficient actions to mitigate the potential loss from supply chain disruptions because they are generally reluctant to assume the loss of efficiency derived through scale economies. High-impact/low-probability events, such as huge earthquakes and tsunamis, make our predictions more diverse and imprecise. Generally, although people's awareness of risk is tuned to a high level soon after experiencing an important natural disaster, heterogeneity in beliefs will increase with the passage of time. Moreover, uncertainty will be high in the decision-making process because the valuation of risks is difficult. In such a case, the market equilibrium can only reflect the opinion of the more optimistic firms, which avoids the costs of risk management. Agency problems might also be an issue. A risk-conscious buyer might wish to enforce a BCP on its supplier in the business contract, but the supplier's implementation could be partial if monitoring costs are high.

Actually, the 2011 disaster was not the first supply chain crisis in East Asia, even in recent times. A strong earthquake in Taiwan in March 2000 shut down large liquid crystal display factories agglomerated around the Hsinchu Science Park. The outbreak of the SARS epidemic in southern China in 2002–03 sent ripples through the global supply chain. Japan itself also suffered disruptions after the Great Hanshin-Awaji (Kobe) Earthquake of 1995 and the Chuetsu Offshore Earthquake of 2007.

Those disasters and their effects notwithstanding, critical nodes still widely persist.

Better to be brisk and slapdash than slow and elaborate

Prompt measures to remove bottlenecks are undoubtedly necessary to avoid prolonged dysfunction of supply chain networks. Agglomeration has a lock-in effect: that is, firms take actions reflexively to restore the agglomeration after it is damaged by temporary shocks. Collaboration among firms and/or government support of such efforts hasten rehabilitation.

Auto production in Japan recovered nearly to a normal level in August, five months after the shock. We might consider that the rapid recovery showed the high resilience of supply chain networks in the Japanese automobile industry. This was in part due to emergency relief measures taken by the private sector, such as sending technical personnel from all rival customer firms collaborating to help rehabilitate damaged suppliers' factories. The rapid revival of transportation networks (highways, railways, airports, and seaports) was also of fundamental importance.

After the Thai flood, the government implemented some measures to support firms striving for continuing production. These measures included permission for temporary production relocation and outsourcing and the exemption of import tariffs on locally unavailable parts, components, and industrial equipment. Additional corporate tax exemption was also given to flood-hit companies. For the automobile industry, imports of assembled cars were allowed free of tax. Entry of foreign experts to engage in rehabilitation of factories was made flexible.

These measures were complemented by international cooperation. The Japanese government issued temporary work visas for six months to Thai workers employed by flood-hit factories of Japanese affiliates. By the end of 2011 about 3,700 workers had participated in the program. This program benefited Japanese firms who needed a quick start-up of backup production in Japan to mitigate the disruption of the supply chain; it benefited Thai workers who might have lost jobs otherwise. The Bank of Thailand and the Bank of Japan launched a cooperative effort to provide Thai baht loans to flood-hit Japanese affiliates backed by Japanese government bonds.

Providing is preventing: Finding opportunity in crisis

There is no time to lose in emergencies. At the same time, it is necessary to consider whether returning to the predisaster situation is truly desirable if potential risks latent in agglomerations become glaringly apparent. We now confront the urgent task of promoting global disaster risk management of highly networked supply chains while our memory of 2011 is still fresh.

Individual firm/industry level

The main issue will be to enhance the resiliency of the supply chain while maintaining its efficiency. To minimize supply disruption, each company can seek the best mix of the following strategies at the individual firm level:

1. Elaborate a workable BCP that includes remote backup production provisions. Although this does not mean actual dispersion of production under normal conditions, repeated simulation training is necessary.

2. Procure key parts and materials from multiple sources routinely, sharing the costs of dispersion between buyers and suppliers.

3. Divide production and locate productive facilities in different locations, whether interregionally or internationally, even under normal conditions. Innovative production technology must be promoted by which higher-scale economies

are obtainable with smaller production volumes.

4. Coordinate standardization and sharing of parts and materials among companies. Avoiding excessive company-specific customization, such coordination provides sufficient lot size to suppliers by which dividing production facilities becomes economically viable.

These strategies have already been put into practice to some degree. Regarding strategy (1), when the earthquake halted desktop computer production at the Fukushima plant of Fujitsu, the company was able to restart production 12 days later at a factory 740 kilometers away in Shimane Prefecture in western Japan, which usually produced notebook computers, as had been simulated many times. This operation enabled Fujitsu to minimize the disruption period. Regarding strategy (2), Nissan has pursued a strategy of standardizing and sharing parts and materials aggressively through its experience of partnership with Renault. In fact, Nissan was able to recover production from the impact of the Thai flood quickly because it was able to switch to other suppliers of its global procurement network. For strategy (3), high global market-share companies have recognized the importance of risk-averse dispersion to maintain their market positions. One such company, Nidec-Shimpo Corp., which supplies small motors used in various machine products, boasts an 80 percent share of the global hard disk drive motor market (according to the company's website). When its three plants in Thailand were damaged by the flood, Nidec-Shimpo reacted quickly by increasing production capacity in China by 50 percent and in the Philippines by 60 percent to compensate for the loss of operations in Thailand. This action avoided the collapse of hard disk drive production. The company announced that the proportion of the production in Thailand would be reduced from its original 60 percent even after the rehabilitation of the factories, thereby reducing the risk of concentration. As an example of strategy (4), companies are usually reluctant because they are concerned that the use of standardized parts would require compromises in product quality, leading to the loss of competitiveness. After the GEJE, however, METI took initiatives to coordinate parts sharing in the Japanese automobile industry, and it is expected that more concrete measures will be taken as well.

Local and national government level

As might be expected, local and national governments have roles in areas where private initiatives cannot suffice. Typically, public policies are expected to enhance the resilience of infrastructure of all kinds supporting industrial production and the daily life of people. For example, in Japan, earthquake-resistance standards for public facilities and infrastructure were revised based on analyses of the damage that occurred. Still, the 2011 disaster left us the lesson to not mythologize safety: provisions in land-use planning are necessary where there is a tsunami risk because tide walls can never be sufficiently high. Moreover, society must take a hard look at the benefits and shortcomings of dependence on nuclear power generation. Strengthening local infrastructure for prevention of urban flooding in developing countries should be greatly emphasized. In this aspect, international cooperation is necessary; for example, the Japan International Cooperation Agency (JICA) will aid the Thai government in presenting a new master plan for flood mitigation in the Chao Phraya Delta.

In broader perspectives, national spatial planning must be readdressed to decentralize the overconcentrated economic-political functions in capital cities (for example, Tokyo, Bangkok, Manila, and Jakarta), and to develop a more resilient nationwide system of regions.

There is a need for accelerating the integration of the private sector into existing platforms and activities. One effective example of partnership and cooperation among national and local governments, volunteers, and the private sector is the Global Compact Network Japan (GCNJ). GCNJ joins the top corporate management of leading Japanese companies in a platform for linking corporate social responsibilities with business activities. GCNJ was established in 2003 and currently has a membership of more than 160 leading companies. GCNJ has been providing a platform for the private sector to address issues such as climate change and water, and create an enabling environment for public–private partnerships. After the GEJE, GCNJ organized a collective action program in which companies provided volunteer assistance to several disaster-affected cities in Miyagi Prefecture by utilizing and combining the resources and strengths of each company.

International cooperation

As we noted above, firms' risk aversion functions to some degree as a dispersive force, but this necessarily involves additional transport costs. Because dispersion will be international, we must recognize transport costs in a broad sense including import tariffs and nontariff barriers, customs clearance procedures, communications costs, and even exchange rates. Countries must join forces to mitigate widely various costs related to cross-border transactions. Such cooperation will increase connectivity to the global supply chain and thus the chance of attracting investment.

The 2011 earthquake and tsunami disaster came as a further blow to the Japanese manufacturing sector, which had already been threatened by high factor costs and a strong yen. But when firms were inclined to transfer more production overseas, the Thai flood occurred, compelling firms to revise their risk assessments of excessive concentration of operations overseas. Given the existence of critical parts and material suppliers within Japan, Japanese firms will find it attractive to determine an appropriate mix of production in Japan and overseas. That will seem preferable to accelerating the hollowing out of the

business environment for the improvement of taxation and expansion of free trade agreement networks.

Recently, the Thai government proposed to Japanese local governments and industrial groups that small and medium-size firms in local industrial clusters invest as a group and establish *sister clusters* in Thailand. Sister clusters can operate with vertically linked specialization at normal times, thereby realizing cost reduction, while they can mutually back up production in cases of large natural disasters. Firms can thereby enjoy the same collective efficiency overseas through familiar face-to-face contacts as they do in Japan. This would promote locational diversification of small firms, for which related costs are unaffordable.

LESSONS

- *Measuring the full extent of the GEJE's economic impacts will take time.* All industrial sectors as well as services suffered significant direct and indirect impacts. A lot will depend on how the government will address the energy supply issues.

- *The Bank of Japan's swift intervention to ensure immediate liquidity* was instrumental in mitigating impacts related to yen appreciation and access to financing.

- *Quick release of supplementary budget and ad hoc regulations are key.* The government played an important role in alleviating the impacts on households and businesses thanks to the subsequent approvals of supplementary budgets and regulations such as the Policy for the Double Debt Problem (chapter 31).

- *Unplanned concentration in supply chains is self-organized because of agglomeration economies.* The network of agglomerations is efficient in normal times, but the global production system is thereby vulnerable to natural disasters.

- When agglomeration is locked in, firms promptly react to restore the original structure against the damage of disaster. *Cooperation among firms and supporting policies can accelerate the process.*

- Although quick restoration is necessary to avoid exacerbation of a crisis through prolonged dysfunction of supply chains, *structural changes must be provided to enhance the resiliency of a supply chain,* without mythologizing the safety of the status quo.

- *Resilience of supply chains demands a certain degree of geographical dispersion.* To mitigate the loss of efficiency by dispersion, the previously described individual firm strategies (1)–(4), government policies, and international cooperation are in order.

RECOMMENDATIONS FOR DEVELOPING COUNTRIES

Consider possible effects on supply chains. In today's networked world, most countries are involved in the global supply chain, of which developing countries are an important part. A major disaster occurring in one country can have a global impact. Consequently, it is expected that developing countries will share the burden of strengthening the global resilience of supply chains.

Vulnerability is particularly high in many developing countries because political and economic activities are excessively concentrated in capital cities. An urgent need exists for bold measures aimed at decentralization and establishing backup systems for emergencies. Furthermore, recent rapid urbanization during economic growth has led to the destruction of natural systems of disaster prevention such as the water retention capacity of forests, thereby increasing risks of flooding. Moreover, urban sprawl is occurring in marginal areas where the infrastructure is unprepared for severe natural events.

A pressing need exists to remedy such weaknesses under international cooperation. Coordination among neighboring countries is also necessary in such areas as cross-border transportation systems and water resource management. Policy makers should assess natural disaster risks in a new light—as a mainstream issue that must be addressed by a country to play a major role in global production networks.

Consider widespread impacts. It is important that the impacts of a large-scale disaster such as the GEJE are not assessed and addressed in isolation but also by taking into account potential regional and worldwide impacts. Many countries in developing East Asia have strong ties with Japan and would be affected by an appreciation of the yen. In the immediate aftermath of the earthquake, when the yen appreciated sharply because of speculation about sizeable repatriation flows by insurance companies, corporations, and households, the Japanese authorities and the G-7 undertook a concerted effort to stabilize the course of the yen to avoid repercussions for the rest of the world, and East Asia specifically. Coordination among countries is fundamental in mitigating potential impacts of large-scale disasters.

NOTE

Prepared by Masahisa Fujita, Research Institute of Economy, Trade and Industry; Nobuaki Hamaguchi, Kobe University (on global supply chain); Financial Service Agency (on the double debt problem); and Junko Sagara, CTI Engineering; and with contributions from Bianca Adam, World Bank.

BIBLIOGRAPHY

Fujita, M., and N. Hamaguchi. 2012. "Japan and Economic Integration in East Asia: Post-Disaster Scenario." *Annals of Regional Science* 42 (2): 485–500.

IMF (International Monetary Fund). 2011. *Japan—2011 Selected Issues.* IMF Country Report No. 11/182, Washington, DC.

———. 2012. *World Economic Outlook. April 2012.* Washington, DC: IMF.

METI (Ministry of Economy, Trade and Industry). 2012. *Japan's Challenges.* Tokyo: METI.

Ministry of Agriculture, Forestry and Fisheries. 2011. November 25. Tokyo. http://www.maff.go.jp/e/quake/press_111125-2.html.

Ministry of Economy, Trade and Industry, Agency for Natural Resources and Energy. 2012. "Electricity Supply-Demand Outlook & Measures in Summer 2012." Tokyo.

Nanto, D. K., W. H. Cooper, J. M. Donnelly, and R. Johnson. 2011. "Japan's 2011 Earthquake and Tsunami: Economic Effects and Implications for the U.S." Congressional Research Service, Washington, DC.

Schnell, M., E. David, and D. Weinstein. 2012. "Evaluating the Economic Response to Japan's Earthquake." Policy Discussion Paper Series 12-P-003, Research Institute of Economy, Trade and Industry (RIETI), Tokyo.

Sheffi, Y., and J. B. Rice Jr. 2005. "A Supply Chain View of the Resilient Enterprises." *MIT Sloan Management Review* 47 (1): 41–48.

World Bank. 2011. "The Recent Earthquake and Tsunami in Japan: Implications for East Asia." East Asia and Pacific Economic Update, World Bank, Washington, DC. http://siteresources.worldbank.org/INTEAPHALFYEARLYUPDATE/Resources/550192-1300567391916/EAP_Update_March2011_japan.pdf.

WTTC (World Travel and Tourism Council). 2011. "The Tohoku Pacific Earthquake and Tsunami." December 2011 Update. London, U.K.

———. 2012. "The Tohoku Pacific Earthquake and Tsunami." March 2012 Update. London, U.K.

CHAPTER 31

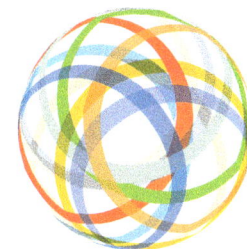

Financial and Fiscal Impact

The Great East Japan Earthquake occurred against the backdrop of a struggling economy and public finance system under stress, implying an exceptional fiscal cost and imposing a fiscal management challenge to the government of Japan (GoJ). In response, the government committed to a full-scale national initiative that has evinced its ability to quickly mobilize short-term liquidity but leaves in question its reliance on debt issuance and taxation measures to finance longer-term reconstruction. This note examines the fiscal costs of the event, the financial measures taken by the GoJ to fund these expenses, and the fiscal implications of these actions. Lessons learned and recommendations for developing countries are distilled from this discussion.

The Great East Japan Earthquake (GEJE) inflicted massive physical damage on private and public assets, destroyed livelihoods, and disrupted local and national economies. In the aftermath of the event, the government of Japan (GoJ) announced a full-scale national response in which the government would support (1) rebuilding disaster-resilient regions, (2) restoring the livelihoods of the disaster-affected population, and (3) reviving the local economy and industry. To finance this approach, the GoJ mobilized a portfolio of fiscal measures that minimized the financial burden on local governments, residents, and industry but significantly increased the financial burden of the central government, and thus, indirectly, of the current and future Japanese population and economy. According to the Cabinet Office (CAO), the GEJE was a "crisis in the midst of a crisis" for the Japanese economy and its public finance (CAO 2011c). The GoJ has had to balance financing and executing an effective postdisaster response against planning how to spread the costs of this response across generations.

FINDINGS

Understanding the GoJ's postdisaster roles and responsibilities, as stated in Japanese laws and as evidenced by past disasters, helps to explain the GoJ's expenditures and revenues related to the GEJE. Japanese law clearly defines the roles and responsibilities, including financial, of the local and central governments in disaster response. A number of laws lay out a broad scope for the GoJ's legal contingent liability[1] in the event of natural disasters, inclusive of responsibilities for disaster response, reconstruction of public and certain private assets, and social and economic restoration. At the center of these laws are the Disaster Relief Act and Disaster Countermeasures Basic Act (table 31.1).

Other laws—such as the Act on Special Financial Support to Deal with Extremely Severe Disasters (1962) and the Natural Disaster Victims Relief Law (1998)—further extend the scope of the government's financial responsibility. Additionally, a series of laws that provide for government support to provision certain lines of insurance (earthquake, agricultural, fisheries, fishing boat, and forest; see chapters 28 and 29) establish a contingent liability of the government to pay its portion of reinsurance payouts under these schemes.

Cost of the GEJE to the GoJ

The GoJ estimates that the GEJE caused direct economic damages to private and public capital and infrastructure in the amount of ¥16.9 trillion (approximately $210 billion), 4 percent of Japan's gross domestic product (GDP). The indirect costs of the event in the short, medium, and long term are difficult to quantify but are likely much greater.[2] Although originally forecasted to grow during 2011, Japan's GDP contracted by 3.5 percent during the first quarter and by 0.7 percent for the full year (IMF 2011).

While the public sector's share of the direct and indirect losses from the GEJE is difficult to determine, it is undoubtedly significant. More easily analyzed are the fiscal costs of the government's relief, recovery, and reconstruction measures after the GEJE. For short- to medium-term costs, government budgetary and cash-flow data (that is, disaster-related expenditures and revenues) can be used. For the assessment of longer-term fiscal impacts, projections are more difficult, as they embody a great deal of uncertainty due to possible variances in expected tax revenues, changes in the Japanese bond market, and/or changes in the GoJ's debt-management capacity. Furthermore, fat-tailed risks, such as the possibility of long-term impacts from the nuclear accident in Fukushima, could increase the fiscal costs of the disaster in the long run.

Central government spending on the GEJE

As of mid-2012, total central government funding allocated to the GEJE totals ¥19.17 trillion (table 31.2). This total includes spending

Table 31.1 Key laws framing the GoJ's contingent liability in the case of disaster

LAW(S)	RELEVANCE TO THE GOJ'S CONTINGENT LIABILITY IN NATURAL DISASTERS
Disaster Relief Act (1947)	• Provides for disaster relief and welfare support (including repair of private housing, cash transfers and/or loans, and so on) to affected populations. • Establishes subsidization of local governments' measures by the central government. • Mandates the establishment of a disaster relief fund for emergency relief activities by each prefecture.
Disaster Countermeasures Basic Act (1961)	• Is the cornerstone of Japan's disaster risk management (DRM) system. • Sets out local and central governments' responsibilities at all points in the DRM cycle, including levels and forms of the local and central governments' postdisaster responsibilities. • Embeds financial measures as one of the eight core components of Japan's DRM system; this section defines disaster-expense-sharing fiscal mechanisms that can be employed by the government post disaster (for example, subsidy, tax, and debt measures).

Table 31.2 Approved central government spending on the GEJE, FY10–FY12

DATE	FISCAL YEAR	FINANCING MECHANISM	AMOUNT (¥ BILLION)
14-Mar-11	10	FY10 General Contingency Budget	67.8
19-Apr-11	11	FY11 General Contingency Budget	50.3
2-May-11	11	1st Supplementary Budget	4,015.3
25-Jul-11	11	2nd Supplementary Budget	1,998.8
21-Nov-11	11	3rd Supplementary Budget	9,243.8
8-Feb-12	11	4th Supplementary Budget	6.7
1-Apr-12	12	FY12 Bridge Budget	9.3
6-Apr-12	12	FY12 Budget	3,775.4
TOTAL			**19,167.4**
Total FY11			15,314.9

Source: Based on data from the Ministry of Finance (MOF).

Note: The Third Supplementary Budget included a ¥2,489.3 billion allocation to repay the financing borrowed from FY11 pension funding. This repayment has been considered in this accounting of the GoJ spending on the GEJE.

Table 31.3 Estimated costs of the GEJE to the central Government of Japan

	PERCENTAGE OF FY10 GDP	PERCENTAGE OF FY11 INITIAL GENERAL ACCOUNT BUDGET
Total, FY10–12	4.0	20.7
Total, FY11	3.2	16.6

Source: Based on data from the MOF and Cabinet Office (CAO).

from the first contingency funding approved in Japan's fiscal year (FY) 2010,[3] through the approved funding for FY12. While earlier funding (that is, up to and including the second supplementary budget) was primarily for relief and recovery costs, the later budgets were primarily for reconstruction. Thus, a significant share of the later budgets may be disbursed for reconstruction projects over multiple FYs.

The GEJE imposed an exceptional cost on Japan's central government: total central government funding for the event through mid-2012 represented 4 percent of FY10 GDP and 20.7 percent of GoJ's initial FY11 general account budget (table 31.3).[4] Considering *only* the costs incurred during FY11 following the event, these represent 16.6 percent of the initial general account budget and 3.2 percent of FY10 GDP. In comparison, central government

spending on the Great Hanshin-Awaji (Kobe) Earthquake of 1995 totaled about 1 percent of Japan's GDP at the time (IMF 2011).

The GEJE reconstruction period is planned for 10 years, with the first 5 years as the concentrated reconstruction period. The latest GoJ figures for central and local government reconstruction expenditures (released July 29, 2011) estimate at least ¥19 trillion[5] until the end of FY15 and ¥23 trillion for the full 10 years (Reconstruction Agency 2011). As central government spending through FY12 had already exceeded ¥19 trillion, it is likely that total public expenditures on the GEJE will run fairly above these levels.

The central government is also responsible for its portion of insurance payouts under the public-private insurance programs for earthquakes, agriculture, fisheries, fishing boats, and forests (see chapter 29). Payments for the government's liability under the fisheries and fishing boat insurance, ¥93.9 billion, are included in the first supplementary budget. The central government's share of payouts for the GEJE under the agricultural and forest insurance programs is still undetermined.[6] Its payment under the earthquake insurance program, not financed by the supplementary budgets, totals ¥540 billion.

Allocation of central government expenditures on the GEJE

The most significant funding allocations by the central government on the GEJE from FY10 through FY12 are for economic and social support programs and miscellaneous

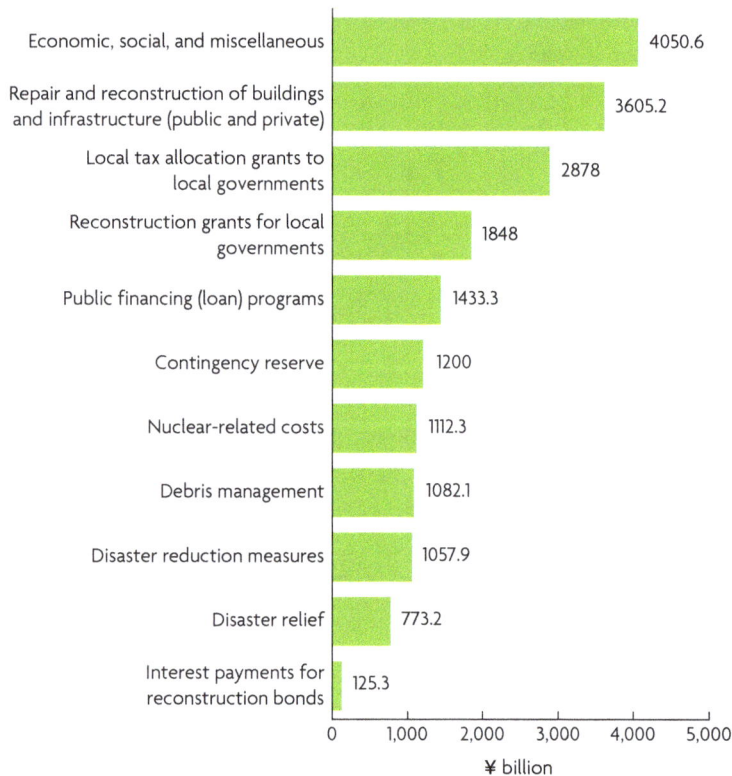

Figure 31.1 Central government funding allocation for the GEJE, FY10–FY12

Source: Based on data from the MOF.

Note: Due to rounding, the sum of these totals is not exactly equal to total central government expenditure, ¥19,167.4 billion. This categorization of allocations is based on that used by the GoJ. In some cases, two GoJ categories have been combined when funding is for similar activities.

expenditures, followed by repair and reconstruction costs for public and private buildings (figure 31.1 and table 31.4). If transfers to local governments under local tax allocation grants for discretionary spending and reconstruction grants are aggregated, however, these take the lead, being greater than ¥4.7 trillion.

While these figures are informative, they must be interpreted with care. Some categories provide estimates of close to final or final totals for allocations to the category; others, such as repair and reconstruction and interest payments for reconstruction bonds, will continue to grow. In addition, because the local tax allocation grants to local governments represent a discretionary spending category, the governments can allocate these funds across the remaining categories (that is, the total of

central and local government spending on disaster relief may be greater than what is captured here); similarly, the reconstruction grants for local governments increase the total amount spent on repair and reconstruction of buildings and infrastructure.

Costs to local governments

The fiscal impact of the GEJE on local governments (prefectural and municipal) is much more difficult to assess, in great part due to the very limited availability of information on disaster-related expenditures and revenues at local levels. The scale of the disaster—primarily in the three most-affected prefectures—Fukushima, Iwate, and Miyagi—suggests that it far exceeded the capacity of local public finance to fund a significant share of reconstruction costs.

From the designation of the GEJE as an "extremely severe disaster" the day after the event, the GoJ's decisions and policies have aimed to shift as much of the financial burden of the GEJE to the central government. For example, under the Natural Disaster Victims Relief Law, which provides subsidies up to ¥3 million to affected households, the central and local governments equally share the liability. Following the GEJE, however, the law was amended with the central government's share being increased to 80 percent for the GEJE.[7] The central government budgeted ¥352 billion between the first and second supplementary budgets to fund its additional liability under the program.

The role of the central government in funding reconstruction is emphasized in the central reconstruction policy—the *Basic Guidelines for Reconstruction in Response to the Great East Japan Earthquake*. The *Basic Guidelines* promote a full-scale national response that will "make use of all possible measures to support reconstruction efforts of the disaster-afflicted local governments," and establish a Special Zone for Reconstruction within which local governments, residents, and industries are

Table 31.4 Explanation of central government funding allocations for the GEJE

ALLOCATION CATEGORY	AMOUNT (¥ BILLION)	ADDITIONAL EXPLANATION
Repair and reconstruction of buildings and infrastructure (public and private)	3,605.2	Repair and reconstruction of public and private buildings (airports, facilities, housing, schools, and so on) and infrastructure (sanitation, roads, railroads, and so on)
Local allocation tax grants to local governments	2,878	Special tax allocation for discretionary spending
Reconstruction grants for local governments	1,848	Program for municipalities in the Special Zone for Reconstruction
Public financing (loan) programs	1,433.3	Loan programs for small and medium enterprises (SMEs), agriculture and education industries, homeowners, and so on
Economic and social restoration measures and miscellaneous expenses	4,050.6	Support to economic restoration such as employment measures, measures for SMEs, agriculture-related industries, and so on. Support to social restoration such as housing grants, health-care support, education assistance, and so on. Miscellaneous costs such as self-defense and police forces; food, fuel, electricity, and natural resource supplies; international information sharing; and so on.
Contingency reserve for recovery and reconstruction from the GEJE	1,200	
Debris management	1,082.1	
Disaster relief	773.2	Temporary housing, condolence money, and so on
Disaster reduction measures	1,057.9	Earthquake-resistant building of schools (national)
Reconstruction from nuclear damage	836.9	
Compensation for damage by nuclear accident	275.4	Security money, investment
Interest payments for reconstruction bonds	125.3	

Source: Based on data from the MOF.

eligible for tax reductions and incentives and budget and financial subsidies. One of the most significant supporting subsidies is the reconstruction grant program for local governments. Under this program, after having their reconstruction plans approved, municipalities receive grants worth 50 percent of project costs for infrastructure and asset reconstruction and 80 percent for supporting projects. The remainder of the project costs can be financed by the special local allocation tax provided by the central government, effectively eliminating any additional expenses to the municipal government (Reconstruction Agency 2012).

Through FY12 the central government provided ¥1.6 trillion in GEJE reconstruction grants and about ¥3 trillion in local allocation tax grants to local governments. Restrictions on the use of the special local allocation tax grants have been relaxed for the GEJE reconstruction, allowing for spending at the discretion of local governments.

Reduced tax revenues from special tax measures

The GoJ implemented a series of special tax measures designed to increase the cost sharing of disaster recovery and reconstruction by the Japanese population and private sector

Table 31.5 Special tax measures in response to the GEJE

TARGET	GOAL OF MEASURE(S)	MEASURE(S)
Japanese population and private sector	Encourage contributions to recovery and reconstruction efforts	• Increase of maximum deduction from income tax for contributions to the GEJE • Income tax deduction for investments in companies contributing to the regional recovery
Disaster-affected population and enterprises	Relieve financial and administrative burden	• Individuals: Special treatments for casualty losses, property damage (housing, household assets, motor vehicles), pension savings, and so on • Firms: Special treatments for inventory and asset losses, withholding taxes, and so on
	Promote investment and growth in reconstruction zones	• Tax incentives to promote investment, employment, and research and development in selected industries (for example, renewable energy, agriculture, and medical)

Source: Based on information from the National Tax Agency 2011.

(table 31.5). Many tax incentive measures also aimed to attract the development of priority industries in the reconstruction zone. These tax incentives were complemented by financial incentives through subsidies in some cases.

In the longer term, these tax measures would help to widen and deepen the government's tax base and raise tax revenue. In the short term, however, they reduced the tax revenues of the central and local governments. The central government, therefore, bore the full costs and compensated the local governments for their decrease in revenues (Reconstruction Agency 2012).

GoJ's short-, medium-, and long-term disaster financing methods and their fiscal impacts

Short-term financing mechanisms

The GoJ moved with remarkable speed to mobilize emergency relief funding following the GEJE. Within three days, the CAO was determined to draw down on Japan's FY10 general contingency budget[8] to procure and transport emergency relief supplies to the disaster-affected areas. A total of ¥67.8 billion was mobilized before the end of March; in April, another ¥50.3 billion was drawn down from the FY11 general contingency budget for transitional shelter. This funding was quickly mobilized because, unlike supplementary budgets, prior parliamentary approval was not required.[9] Thus, the general contingency budget provided immediate bridge financing till more substantial funding could be mobilized (figure 31.2).

Within two months the GoJ approved a ¥4,015.3 billion supplementary budget for

Figure 31.2 The GEJE financial allocations and timing, FY10–12

Source: Based on data from the MOF 2012.

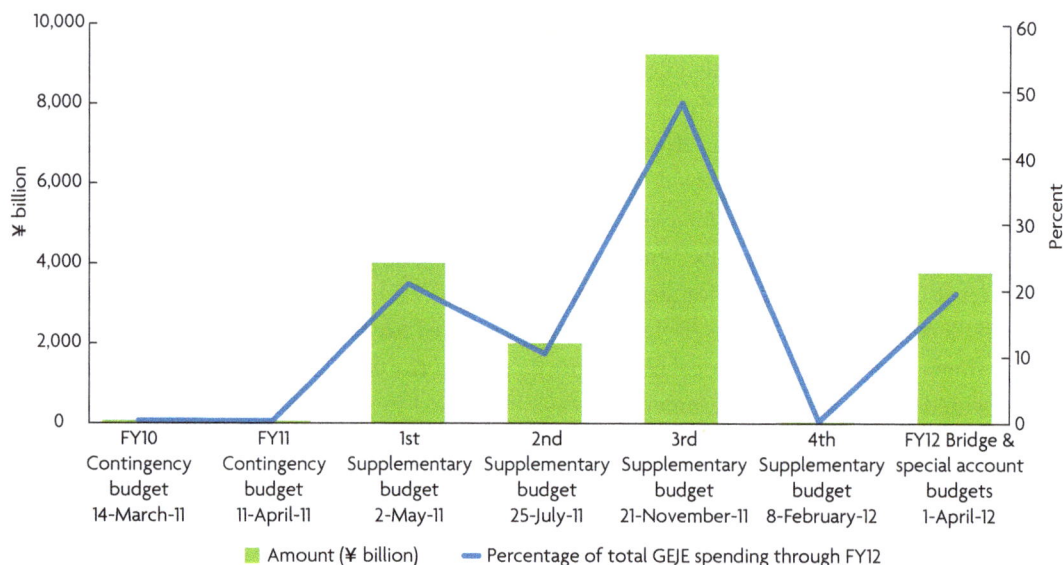

relief-and-recovery costs. For this First Supplementary Budget, the Ministry of Finance looked within the existing budget for funding sources. The approved budget relied on a combination of budget reallocation (¥660.6 billion), borrowing from the pension fund (¥2,489.7 billion), contribution from public works projects (¥55.1 billion), and liquidation of the full FY11 allocation to the Contingency Reserve for Economic Crisis Response and Regional Revitalization (¥810 billion).

This approach illustrates the GoJ's resourcefulness, but also demonstrates the limitations of ex-post budget adjustments to finance disasters. Budget reallocation was used for the first supplementary budget and again for the third (¥164.8 billion). In sum, though, less than 1 percent of the FY11 general account budget was reallocated to the GEJE recovery efforts, and budget reallocation contributed only 5.4 percent of current total central government spending on the event. Furthermore, more than half the funding for the first supplementary budget was borrowed from the pension fund, which allowed the government time to mobilize additional resources that have to be repaid at a later date. Finally, the government redirected the full FY11 Contingency Reserve for Economic Crisis Response and Regional Revitalization toward the disaster—the intent of this reserve, however, was not for natural disasters but for economic measures required to stabilize Japan's economic situation during times of financial crisis.[10]

In late July, the smaller second supplementary budget was passed. The GoJ was able to fund this budget with surplus from FY10, the result of higher-than-expected FY10 tax revenues and unused funds.

Medium- to long-term financing mechanisms

The government's short-term measures funded relief-and-recovery activities while it formulated its reconstruction policy. When the *Basic Guidelines* policy document was released at the end of July, it set a conceptual framework of sharing the costs of the GEJE

reconstruction within this generation and not passing them on to future generations of Japanese. Financial resources provisioned for use by the *Basic Guidelines* are as follows:

- Reduction of government expenditures
- Selling of state-owned properties
- Reviews of the special accounts and personnel salaries of public servants
- Increases in nontax revenues
- Temporary taxation measures

On November 30, 2011, the bill on special measures to secure financing for the GEJE reconstruction was passed. Its approval followed a great deal of debate about what debt and tax measures the government should take for the GEJE. Under the approved plan, issuance of Japanese government bonds (JGBs) financed the majority of the estimated reconstruction costs. The bulk of repayment costs for these bonds were secured through tax increases. Personal income tax, in the form of a surtax, was raised for 25 years starting in 2013. A 5 percent corporate income tax cut that was initially planned in 2011 was postponed, and a ¥1,000 increase in per capita local tax (currently ¥4,000 per year) was included. Table 31.6 provides details on the increases and their projected revenue generation.

While the tax measures will be phased in starting in FY12, reconstruction bond issuance

Table 31.6 Special reconstruction taxes: Schedule and projected revenues

TAX ITEM	INCREASE	PERIOD	PROJECTED REVENUE
Income tax	2.1% surtax	1/13–12/37 (25 years)	¥7.3 trillion
Corporation tax	Delay of a planned 5% cut (effectively, a 10% surtax)	4/12–3/15 (3 years)	¥2.4 trillion
Individual inhabitant tax (local tax)[a]	¥1,000 per person (annual)	6/14–6/24 (10 years)	¥0.8 trillion

Source: Based on data from MOF.

a. Revenue from the local tax increase is not directed toward reconstruction in the disaster zone, but to finance urgent disaster mitigation projects, such as retrofitting public buildings to reduce earthquake risks in individual localities.

commenced in early December 2011. In total, slightly more than ¥14.2 trillion of JGBs were issued: approximately ¥11.6 trillion for the third supplementary budget of 2011 and nearly ¥2.7 trillion for FY12. Interestingly, about 25–30 percent of reconstruction bonds are being sold to retail investors with 3-, 5-, and 10-year maturities. A portion of these bonds are reconstruction supporters bonds that facilitate financial support and solidarity from the Japanese public. These bonds offer the lowest possible interest rate for government bonds (0.05 percent) for three years, before converting to standard JGB rates. The GoJ has recruited Japanese celebrities to market the bonds and is offering gold and silver commemorative coins to purchasers (figure 31.3).

Figure 31.3 Design of commemorative coin by elementary school student

Source: MOF.

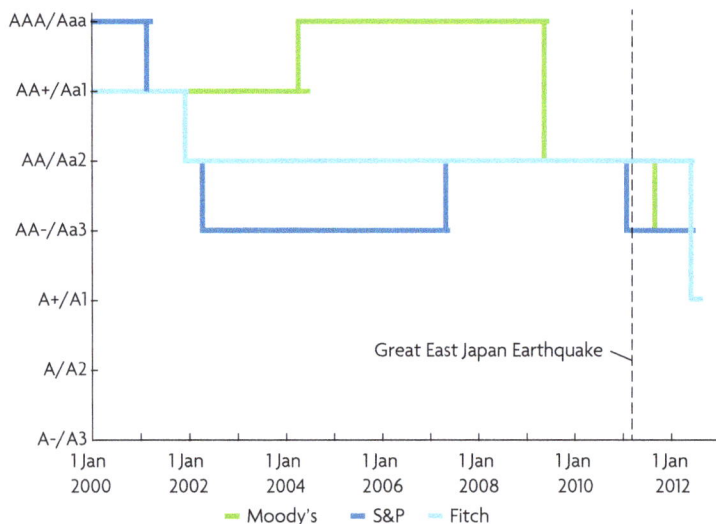

Figure 31.4 Sovereign credit rating of Japan by major rating agencies, 2000 to current

Source: Based on data from Moody's, S&P, and Fitch.

Fiscal impacts of the GoJ's financial measures

Although the GoJ is endeavoring to minimize debt costs and tax increases, the financial measures it has taken for reconstruction have had significant fiscal impacts. The GEJE was "a crisis in the midst of a crisis," and the financial burden of the event has placed significant additional strain on public finance.

Even before the GEJE, Japan's public finance was under stress, as budget deficits of the central and local governments grew. Credibility of the JGBs and its sovereign debt rating was, and still is, declining—it is now rated at the same level as China by each major rating agency (figure 31.4). Compared to its accumulated central government debt-to-GDP ratio at the time of the 1995 Kobe earthquake, which was lower than one-half of GDP, the GoJ's central government debt was about 140 percent and growing at the start of FY11 (debt ratios of Hyogo prefectural and municipal governments doubled and remain higher than prior to the event).

One of the factors driving the government's increasing dependence on debt has been Japan's aging population and decreasing tax revenue. The population share aged 65 and above is expected to increase from 21.5 percent in 2007 up to 40 percent in 2050. Such aging is already increasing the fiscal burden on the government, as it needs to spend more on social expenditure. In addition, in recent years, tax revenues have been declining due to the global financial crisis and tax cuts. While Japan still can increase some tax forms, others, such as the corporate income tax, are already high.[11]

In sum, at the time of the GEJE, the GoJ had little leeway in terms of either its ability to utilize debt financing or taxation measures. Debt issuance increases demand for fiscal reconstruction that further undermines confidence in the creditworthiness of the JGBs. Regarding tax increases, the government was relying on existing room for tax increases to finance rising social expenditures. The aging of the population means that the government is less able to spread the costs of the GEJE intergenerationally because there is already such a high burden placed on the young and future generations.

While initial policy goals following the GEJE were to minimize debt issuance and to keep taxation measures temporary, the plan finally agreed upon was somewhat different than that initially proposed. Issuance of reconstruction bonds was widely accepted as a short-term measure to finance the reconstruction costs. Opinions differed, however, regarding their redemption period. Standard construction bonds have a 60-year maturity, leaving the burden of repayment to future generations. For reconstruction bonds, though, the GoJ proposed that they be paid back within 10 years, with tax increases also within the redemption period to secure revenues to redeem them.

Ultimately, negotiation and compromise resulted in the final package of debt and tax measures for the GEJE. A much-discussed increase of consumption tax was left out of the package.[12] The marginal increase of personal income tax was low, but the surtax was put in place for 25 years, placing the public debt burden on the "shrinking," relatively speaking, younger generation. Furthermore, there is a risk that reconstruction tax revenues will not match with expenditures for servicing reconstruction debt, which is being aligned with the broader plan for government debt issuance.[13] In addition, long-term uncertainty about macroeconomic conditions increases the risk of mismatch between projected and actual tax revenues.

In the context of the government's gross outstanding debt, the additional reconstruction bonds issued in FY11 and FY12 make small contributions (figure 31.5). That said, they

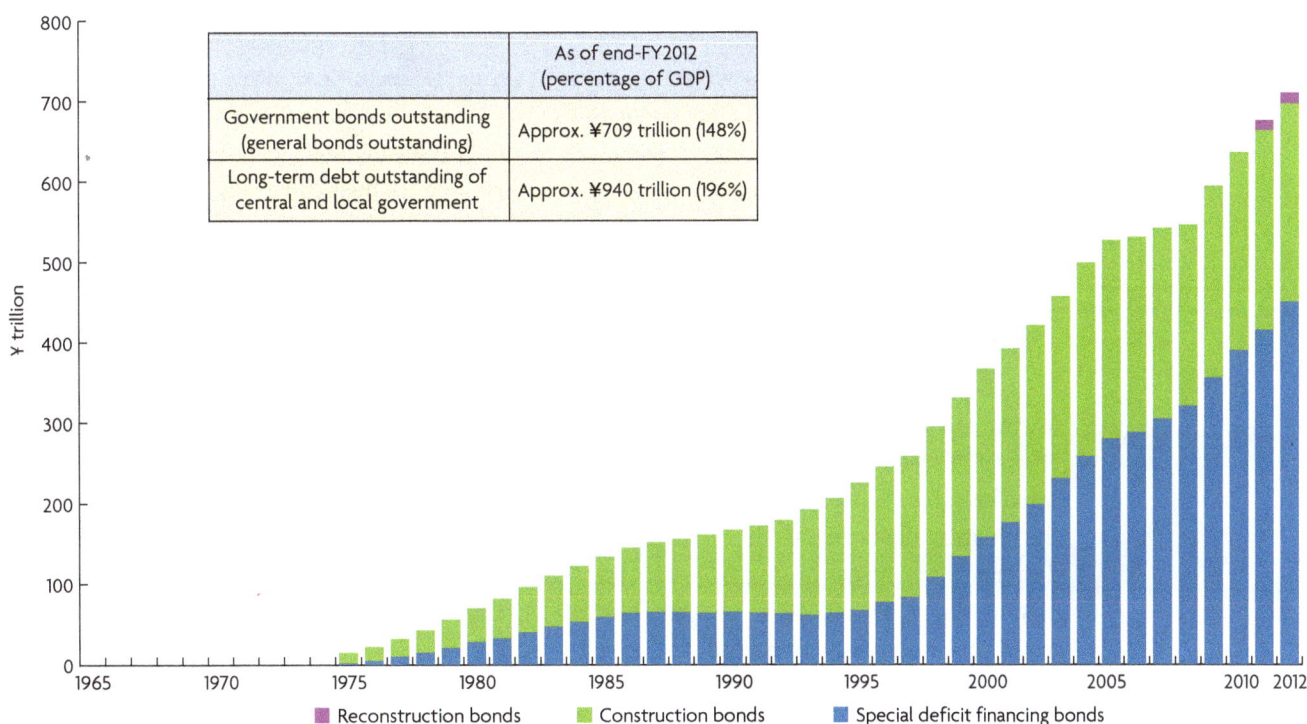

	As of end-FY2012 (percentage of GDP)
Government bonds outstanding (general bonds outstanding)	Approx. ¥709 trillion (148%)
Long-term debt outstanding of central and local government	Approx. ¥940 trillion (196%)

■ Reconstruction bonds   ■ Construction bonds   ■ Special deficit financing bonds

Figure 31.5 Accumulated GoJ bonds outstanding, FY1965–FY2012

Source: MOF 2011.

force a change in the government's medium-term fiscal policy to reduce debt issuance year on year, as the total amounts issued in FY11 and FY12 are greater than the reduction in non-reconstruction debt issuance. This dynamic poses challenges for the government's fiscal consolidation target to halve the FY10 deficit-to-GDP ratio by FY15 (CAO 2010).

LESSONS

- *The GoJ's broad contingent liability to natural disasters results from its responsibilities explicitly defined in Japanese laws and the implicit expectations of society, which can result in extraordinary fiscal costs, as evidenced by the GEJE.* The GoJ is expected not only to reconstruct assets, but to restore social and economic well-being following a major disaster. This role aligns with the Japanese values of solidarity and cooperation, but implies that the public finance system is highly exposed to disaster risks. The GEJE raised general account spending by nearly 16.6 percent in FY11—an earthquake striking Tokyo, for example, could stress the system much further. Quantitative analysis of the government's contingent liability to disasters would be an important first step toward management of its financial exposure to this type of event.

- *Local governments are at the frontlines of disaster response and reconstruction and thus the most aware of local needs, but local public finance has limited capacity to cope with large-scale disasters.* The liability of the central government was expanded following the GEJE (for example, under the Natural Disaster Victims Relief Law), and transfer schemes were designed to allow the central government to fund locally designed reconstruction plans. While the magnitude of the GEJE exceeded what might be reasonably expected for local

public finance to manage, it provides an opportunity to review and strengthen the effectiveness of local governments' disaster-financing mechanisms.

- *The GoJ's contingency budget allows it to quickly mobilize funding for an effective disaster response.*[14] The flexibility and immediate availability of the GoJ's contingency budget allowed it to approve funding within three days of the GEJE to finance immediate emergency relief. Although relief costs represent a very small portion of the overall amount spent on the disaster, they serve an essential function in mitigating additional fatalities and damages linked to a slow response effort.

- *Tax measures can be used effectively ex ante to incentivize investment in disaster prevention and ex post to facilitate cost-sharing of reconstruction by the population and private sector.* Japan has a series of laws that provide tax incentives for investment in earthquake mitigation. Although difficult to quantify, these incentives promote risk reduction and likely reduced losses from the GEJE in some areas. Following the event, the government immediately enacted tax relief measures for affected populations and industries, and it built tax incentives into its reconstruction policy. It also offered special tax deductions to individuals and corporations that contributed to the reconstruction and recovery effort, thus facilitating solidarity and cost-sharing by the unaffected population and private sector.

- *Financial demands placed on the government by major disasters exacerbate the underlying structural problems of the fiscal system.* The GEJE forced the government to issue additional debt and pass tax increases in an economic and fiscal environment in which these actions were not only unfavorable, but counter to fiscal management policy. The experience emphasized the imperative of having a robust fiscal system capable of absorbing large disaster shocks. For Japan to achieve prompt and enduring reconstruction, it should look beyond restoration, which brings the Japanese economy back to the predisaster state, and seek to strengthen the economy and society in a broader sense to prepare for the future.

- *A lack of ex ante financial planning for disasters can contribute to disagreements and possible delays around securing reconstruction funding.* Although Japanese law allows for the government to secure funding for disasters in broad terms, lack of a clear "blueprint" for how the government would finance reconstruction opened space for prolonged deliberation on appropriate measures. Alternative plans and road maps for flexibly financing reconstruction under different scenarios, both in terms of the type and scale of disaster and the economic and fiscal environment, could be designed to prevent this from occurring in the future.

RECOMMENDATIONS FOR DEVELOPING COUNTRIES

Japan's public finance responsively provided financing for an effective relief effort, but was stressed by the extensive burden of recovery and reconstruction funding requirements. In developing countries, where governments' fiscal options to finance disasters are likely more limited—for example, due to structural weaknesses such as lack of income support, inadequate financial resources, and lack of administrative capacity—fiscal impacts of these events can be even more substantial. The following recommendations could mitigate the impacts of disasters on governments' long-term fiscal balances and increase their financial response capacity in the aftermath of a disaster.

Treat disaster risks as a contingent liability of the government

Quantitatively assess the government's contingent liability in the event of natural disasters. Identify the government's explicit (that is, stated by law) and implicit (that is, socially and politically expected) contingent liabilities for disasters. Historical analysis, complemented with information from probabilistic risk models, can provide a sense of the government's recurrent financial needs as well as possible major losses from catastrophic events related to these contingent liabilities. In addition, where risks cannot be quantitatively assessed, they should be qualitatively identified and discussed. Clear definition of the government's contingent liability helps to protect public finance from an open-ended financial liability to disaster events.

Develop a disaster-risk-financing strategy as part of the government's broader fiscal risk management strategy. The disaster-risk-financing and insurance strategy should combine ex post and ex ante measures to optimize the timing, cost-efficiency, and effectiveness of disaster funding. For short-term postdisaster liquidity needs, the strategy should rely on ex ante budgetary and possibly market-based instruments, such as contingency budgets, reserve funds, and contingent credit. For the longer term, major reconstruction costs, a "blueprint" for mobilization of ex post financial resources (for example, debt issuance and tax increase) should complement the ex ante measures. Scenario analysis should be conducted to ensure the robustness of the strategy for disasters of varying type, magnitude, and location under different macroeconomic and fiscal conditions.

Understand the roles and financial responsibilities of the central and local governments in this process. Local governments should, to some extent, share financial responsibility for disasters affecting their territories. But local and central governments should agree together ex ante whether and how sharing of these financial responsibilities changes after severe disasters.

Reduce the contingent liability of the government in the long term

Use fiscal tools such as taxation and subsidization to encourage ex ante disaster risk management (DRM). The government could decrease residential and private sector dependence on postdisaster government aid by using tax and/or subsidy tools to encourage ex ante DRM. Although the relative power and ease of use of tax versus subsidy tools varies across countries, the government could achieve similar ends through either means by offering tax incentives or subsidies for investment in disaster prevention. It could also promote minimum levels of prevention by imposing tax penalties or fees for underinvestment in risk reduction and/or for risk-increasing actions.

Promote the development of private catastrophe risk insurance markets. The deepening of private catastrophe risk insurance markets shifts more of the burden of postdisaster recovery to specialized risk carriers. The government can encourage the development of functioning catastrophe risk markets by putting in place and enabling the legal and regulatory framework, developing risk market infrastructure, and facilitating risk-pooling mechanisms.

NOTES

Prepared by Motohiro Sato, Hitotsubashi University, and Laura Boudreau, World Bank.

1. As defined by the World Bank, a contingent liability is a spending obligation arising from past events that will be incurred in the future if uncertain discrete future events occur.

2. Indirect losses are losses that result from physical damage, such as business interruption, reduced tourism, reduced tax revenue, and so on.

3. Japan's fiscal year (FY) runs April 1 to March 31. The GEJE struck on March 11, 2011, toward the tail end of FY10.

4. FY10 GDP was ¥479.2 trillion and FY11 initial general account budget was ¥92.4 trillion (Ministry of Finance 2011).

5. This estimate includes the first and second supplementary budgets, which had already been approved at that time.

6. The level of payout for the agricultural insurance program remains uncertain due to the nuclear accident at Fukushima.

7. This amendment applies only to the GEJE. Cost sharing remains 50/50 for all other disaster events.

8. For the 23 years preceding 2010, Japan's general contingency budget was allocated ¥350 million; in 2010 this allocation was lowered to ¥300 billion, representing about 0.3 percent of the central government's initial general account budget for 2010.

9. Retroactive parliamentary approval is allowed for expenditures from the general contingency budget.

10. The *Contingency Reserve for Economic Crisis Response and Regional Revitalization* was introduced in the budget in FY10 in response to the worsening economic situation caused by the global financial crisis. The contingency budget had previously been used to support employment programs for college graduates as well as other economic support programs.

11. According to the IMF (2011), Japan's consumption tax (value added tax, or VAT) is the lowest of advanced economies with a VAT, and its personal income tax structure allows much room for deductions and provides low marginal rates for the middle class.

12. The GoJ has proposed to increase the consumption tax rate by 5 percent to fund increasing social expenditure costs until the mid-2010s as a part of its "unified reform of tax and social spending" initiative.

13. The Act for Special Measures for Securing Financial Resources Necessary to Implement Measures for Reconstruction Following the GEJE does stipulate, though, that reconstruction bonds must be redeemed by 2037, within the term of income tax increase (Article 71).

14. Equally as important, it was able to smoothly execute these funds for reconstruction due to pre-agreements with private sector firms. See chapter 20 of this series for additional information.

BIBLIOGRAPHY

CAO (Cabinet Office). 2010. *Fiscal Management Strategy.* Tokyo: CAO.

——. 2011a. *Disaster Management in Japan.* Tokyo: CAO.

——. 2011b. *Medium-Term Fiscal Framework (FY2012–FY2014).* Tokyo: CAO.

——. 2011c. *The Guidelines on Policy Promotion for the Revitalization of Japan.* Tokyo: CAO.

IMF (International Monetary Fund). 2011. *Japan: 2011 Article IV Consultation*. Rep. no. 11/181, Washington, DC.

Ministry of Finance. 2012. "Financial Budget for Each Fiscal Year." Tokyo. http://www.mof.go.jp/budget/budger_workflow/budget/index.html.

Ministry of Finance, Financial Bureau. 2011. "Highlights of FY2012 Government Debt Management." Tokyo.

National Tax Agency. 2011. *National Tax Agency Report 2011*. Tokyo: National Tax Agency.

Reconstruction Agency. 2011. *Basic Guidelines for Reconstruction in Response to the Great East Japan Earthquake*. Tokyo.

———. 2012. "Reconstruction Process on Track." Tokyo.

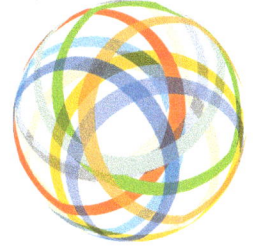

CHAPTER 32

Strategies for Managing Low-Probability, High-Impact Events

Every country should develop strategies for managing low-probability, high-impact extreme events—strategies that reflect their own as well as global experiences with megadisasters. These strategies should integrate structural and nonstructural measures tailored to local conditions. Forecasting and early warnings, land-use planning and regulation, hazard maps, education, and evacuation drills are all vital. Lessons from the Great East Japan Earthquake can help improve these nonstructural practices, which in Japan have been shaped by trial and error after experiences with many natural disasters. The international community can play an important role in developing knowledge-sharing mechanisms to help countries prepare for low-probability, high-impact extreme events.

FINDINGS

National strategies to address low-probability, high-impact extreme events

The Great East Japan Earthquake (GEJE) was the first disaster in Japan's modern history that exceeded all expectations and predictions. Its dimensions were almost beyond imagination (chapter 25). Its enormous impact prompted the government to seek a paradigm shift in disaster risk management (DRM), moving from structure-focused prevention to a strategy of mitigation by integrating structural and nonstructural measures.

Excessive reliance on structural measures proved to be ineffective, and even detrimental, when the forces of nature exceeded the structures' design limitations (chapter 1). In some towns, evacuation was delayed because people did not expect a tsunami to overtop an embankment as high as 10 meters or more. Some could not escape the tsunami in time because they had moved their homes to the lowlands along the coast to be closer to their source of income. They felt safe because high embankments had been built (chapter 11).

Addressing low-probability, high-impact extreme events requires an integrated DRM strategy, combining structural and nonstructural measures. Disasters should be categorized into two levels: level 1 consists of disaster events that occur with relatively high frequency (with a return period of around 100 years or less) and level 2 consists of events that rarely happen (with a return period of around 1,000 years or more). The GEJE was a level 2 event, as illustrated in figure 32.1. Level 1 events can be addressed mainly by disaster prevention structures, while level 2 events require an integrated DRM strategy.

Strategies for level 2 events should focus on saving lives. Measures to be used in an integrated manner to ensure immediate evacuation include installing disaster forecasting and early warning systems; land-use planning; designating and building of evacuation sites,

shelters, and other facilities; and installing structures to delay and weaken the force of waves. Education, practice drills, and mutual help mechanisms are extremely important. Urban and land-use planners need to consider mechanisms for speedy emergency evacuation and for sustaining social and economic activities. People's participation is the critical factor in the planning process.

During the GEJE catastrophic damage was inflicted when structures were overtopped by the tsunami, reached their breaking point, and suddenly collapsed. Structures should be resilient enough to hold up, or succumb gradually, even when the natural forces exceed their structural design limitation. Nonstructural measures such as land-use planning, forecasting and warning systems, evacuation drills, and public awareness-raising, should be designed with enough redundancy and flexibility to address different disaster scenarios.

Strategies should take into account the unexpected. In the GEJE, many plans did not specify the actions to be taken in the face of an unexpected event, contributing to catastrophic damage to facilities, communities, and socioeconomic systems.

Structural measures

Structural measures will continue to play a key role in managing low-probability, high-impact extreme events. Although many disaster prevention structures, such as tsunami defense dikes and gates, collapsed and were washed away in the GEJE, some withstood the waves even after they were overtopped, reducing the force of the tsunami and delaying its penetration inland (chapter 1). In a number of cases the dikes were not overtopped, and kept the hinterlands from being inundated. Postdisaster computer simulations for the Kamaishi Port indicated that the wave breakers around the port reduced the peak height of the tsunami by 40 percent: from 13.7 meters to 8 meters.

Damage by the tsunami of 10 meters or higher to structures and buildings was

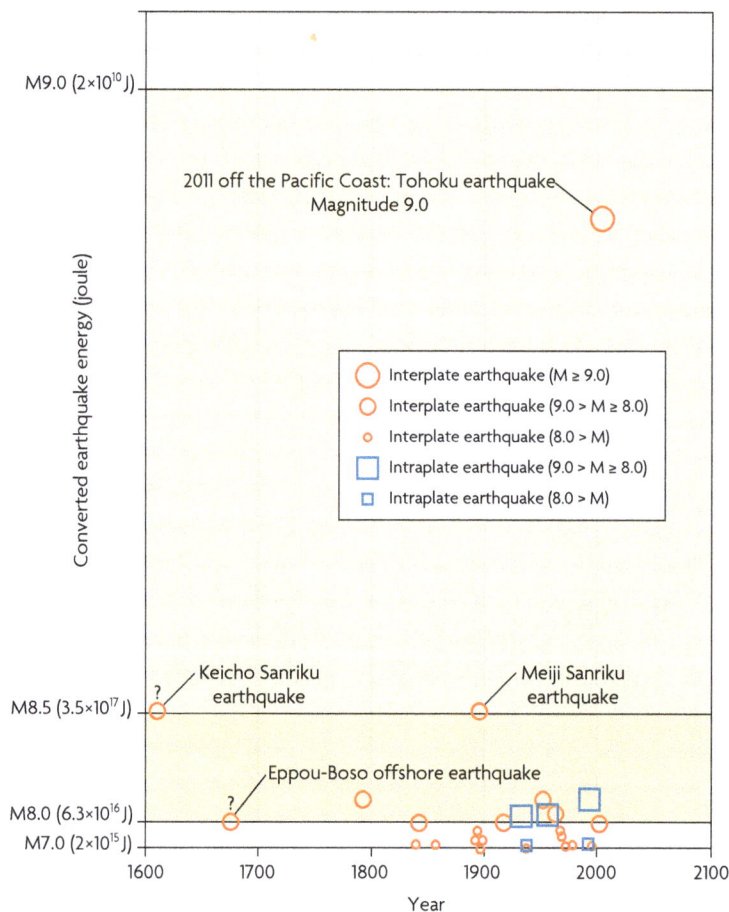

Figure 32.1 Magnitude of earthquakes in Japan

Source: Cabinet Office.

extensive and severe. Almost all buildings and structures made of wood were destroyed. Iron structures were left with only their skeletons. Most reinforced concrete buildings withstood the tsunami, although they suffered internal damage (chapter 2).

After the Indian Ocean tsunami and Hurricane Katrina, design standards for defensive structures, such as dikes and water gates, have been reevaluated. The conclusion is that using only preventive structures to defend against low-probability, extreme events is not an economically, environmentally, or socially viable option. For example, it is not realistic to try to protect hundreds or even thousands of kilometers of seacoast using embankments, even as high as 20 meters.

Tsunamis should be classified into two or more categories. Level 1 tsunamis may occur once in a 100 years; level 2 tsunamis are extreme events that may occur once in a 1,000 years or more. Disaster mitigation structures such as wave breakers and dikes should be designed to prevent inland penetration by level 1 tsunamis, saving lives and properties. Although these structures could be overtopped by a level 2 tsunami, they should be able to withstand complete collapse, thereby reducing the force of the tsunami and delaying its progress. In the case of level 2 tsunamis, the structure is not expected to achieve complete mechanical prevention, but rather to mitigate damage, in combination with other nonstructural measures.

Using infrastructure, such as highways and trunk roads, as defensive structures is also recommended. In the GEJE, coastal highways and trunk roads functioned not only as evacuation routes but also as temporary evacuation sites and even as dikes (chapter 3).

People in Kamaishi city's Katakishi District fled to the Sanriku Expressway, which had opened on March 6, 2011, just six days before the earthquake. The expressway, which was on a hill, first served as an evacuation area and then as a main road for delivering relief goods

and reconstruction materials. National routes running along the coast served as embankments preventing the tsunami from advancing inland.

Nonstructural measures

As Sanriku's coastal areas have been repeatedly hit by tsunamis, many towns and communities had developed both structural and nonstructural measures that mitigated the impact of the tsunami substantially.

In addition to information dissemination and evacuation measures, the following nonstructural approaches were found to be effective against extreme water disasters:

- Moving residential areas and public buildings to higher ground, while keeping commercial installations and activities based in the lowland coastal areas (chapter 12)

- Securing evacuation routes (such as roads and stairways) that connect public facilities (such as schools) to higher ground (chapter 8)

- Planting trees densely in coastal areas (chapter 13)

- Using tall concrete buildings (four to five stories or higher) as evacuation places

- Using highways and trunk roads as secondary protective embankments (chapter 4)

The Government of Japan enacted a new law—the Act on a Tsunami Resilient Community—to promote these nonstructural measures in the tsunami-affected municipalities (see chapter 12). The act requires restricting the construction of buildings in risk areas; introducing integrated tsunami mitigation plans comprising evacuation routes and facilities, hazard mapping, drills, and warning systems; relaxing the floor-space ratio of buildings to encourage the construction of taller buildings; reducing property taxes on designated evacuation sites; and relocating houses to higher ground.

Evacuation

Evacuation is the highest priority in low-probability, extremely high-impact events (chapter 11). A large number of casualties can be expected not only because of the scale of the event, but also because

- The lead time is shorter because of the sudden or unexpected occurrence of the event.

- Information networks and tools tend to malfunction when sensors and communication lines are destroyed, constraining people to react without accurate information.

- Evacuation options tend to be limited as the means of evacuation become fewer; for example, roads become impassable, traffic jams occur, and so on.

- People base their actions on past experiences with less-severe disasters, leading them to underestimate the time they have to evacuate and the severity of the consequences.

Raising awareness, education, and practice drills are the keys to ensuring faster, more complete evacuation in extreme events.

In Kamaishi City, where 1,000 people died out of a population of 40,000, the casualty rate among school children was low. Only 5 out of the 2,900 primary and junior high school students lost their lives. A survival rate of 99.8 percent for these school children is most impressive in a city where 1 in 40 lost their lives: the rate for school children was 20 times higher than for the general public. According to one headmaster, "repetitive drills, school education, and hazard maps" were the reasons for the high survival rate (chapter 8).

In Kamaishi city "a touch of disaster" is built into various lessons. In mathematics, for example, students may be asked, "If the speed of a tsunami is xx kilometers per hour when it hits land, how long will it take the tsunami to get from the coast to a house that is yy kilometers inland?" In a field exercise, students produced a tsunami hazard map on their own by visiting hazard and evacuation areas within the school district.

The students were also trained in key concepts, such as

- "Tsunami *tendenko*," that is, "Everybody should immediately evacuate without caring for anything or anybody else at tsunami onslaught."

- Do not believe in human assumptions of disasters, even one in a hazard map, as nature behaves differently from human assumptions.

- Do your maximum when encountering disasters. Always think and be prepared for the worst.

- Lead evacuation—you are saving others' lives by showing that you are evacuating for life and death.

Although more than 90 percent of students were out of school when the earthquake occurred on March 11 (whether they were walking home, playing outside, or in their homes), almost all of them headed for higher, safer areas on their own initiative and encouraged the others to run with them to safety. Having already discussed it in their homes, children and parents alike knew and trusted that they would all evacuate individually if a tsunami hit Kamaishi.

Keeping individual, community, and institutional memory alive between disasters is critical to successful evacuation. A number of monuments had been built in the coastal towns commemorating past events and citing lessons such as: "Run to a hill if you feel a strong shake or the sea suddenly withdraws." A nongovernmental organization (NGO) has called for the planting of cherry trees to delineate where the tsunami reached on March 11, so that future generations would remember the extent of the flooding.

The elderly, the disabled, and foreigners or outsiders to the locality needed extra help in evacuating. Sixty-five percent of those who died in the GEJE were more than 60 years old, which raised the issue of how senior citizens can be safely evacuated.

Hazard maps

Hazard maps are a useful tool for enhancing the preparedness of local governments, municipalities, and residents, but they can exacerbate the damage if not prepared or used properly. A number of cities and towns had produced and distributed hazard maps. In some of the towns they contributed to faster evacuation, but in others they actually provided misinformation since the tsunami was far larger than the hazard maps assumed. Casualties occurred because some of the designated evacuation sites and buildings where people had fled to were totally submerged. Many people who were living in nonflooding zones, according to the hazard map, had not evacuated when the tsunami hit (chapter 25).

Both level 1 and 2 events should be accounted for in hazard maps so that people will have enough information to deal with either category. Hazard maps should indicate all evacuation options. Just distributing these maps to citizens is not enough—evacuation practice drills should be conducted using these maps. Preparing hazard maps with people's participation will also help ensure effective evacuation.

Forecasts and warnings

Accurate forecasting and early warning systems are vital for safe and quick evacuation and disaster response. In the GEJE hundreds of thousands of people evacuated in response to the warning by the Japan Meteorological Agency (JMA) a few minutes after the earthquake. The Earthquake Early Warning System also enabled all the high-speed express trains, traveling at over 200 kilometers per hour, to come to a halt before the main tremor, which saved thousands of passengers. The emergency warning system announced the arrival of the main tremor nationwide on TV and other broadcasting systems, providing the public with a little lead time (a few to 10 seconds) to react (chapter 10).

Although the earthquake and tsunami warning system helped save many lives, there was room for improvement and some key lessons emerged. Because of the unprecedented size and complexity of the event, the JMA's first announcement underestimated the maximum tsunami height at 6 meters, while the actual height was more than 10 meters. Although the forecast was corrected 10–20 minutes later, the original estimate may have caused people to delay their evacuation, possibly leading to increased casualties. This occurred even though Japan is equipped with one of the most advanced forecasting and warning systems. The international community should invest not only in the installation of existing disaster forecasting and warning systems, but also in the development of new systems in combination with repetitive drills and practices. Advanced off-the-coast water pressure gauges and global positioning system (GPS)-based wave sensors have been effective in monitoring tsunami heights.

Addressing "chain-of-events" effects

The disaster unleashed a chain of events that affected people and organizations beyond Tohoku, including national, regional, and global economies. Following are a few examples of the chain of events observed in Japan:

- Earthquake and tsunami: nuclear accident: power shortage: economic stagnation: social unrest

- Earthquake and tsunami: dramatic increase in telecommunication activity: telecommunication system failures: interruption of social and economic activities (chapter 15)

- Earthquake and tsunami: damage to specific industries: interruption of parts supply: global slowdown of industrial activities (chapter 30)

Although it is impossible to foresee every eventuality, DRM strategies should include contingency measures for preventing the knock-on effects of low-probability, high-impact events (chapter 5). Providing for sufficient redundancy in various systems is one way of breaking the chain; business continuity planning is another (see chapter 9). Analyzing past examples of "chain-of-events" effects, and sharing them with the public, the business sector, and governments can help prevent them from recurring.

LESSONS

Overall strategy

- Use integrated disaster mitigation strategies, rather than structure-focused disaster prevention measures, to address low-probability, high-impact extreme events.

- Categorize tsunamis into level 1 events (fairly frequent disasters) and level 2 events (low-probability, high-impact extreme disasters). Level 1 can be addressed by preventive structures; level 2 requires integrated measures.

- For level 2 events, prepare strategies that focus on saving lives.

- Use resilient disaster mitigation systems, structural and nonstructural, in strategies to address level 2 events.

- Consider and discuss what should happen if an event exceeds expectations. This is critical in establishing effective, functional strategies.

Structural measures

- Structural measures can mitigate low-probability, high-impact extreme events if they are resilient and resistant to natural forces.

- Structural measures should be included in an integrated disaster mitigation strategy.

- Highways and trunk roads along the coast should be used as secondary protective embankments against tsunamis.

Nonstructural measures

In addition to information dissemination and evacuation, the following nonstructural measures have been effective against water-related megadisasters:

- Moving entire residential areas and public buildings to higher ground while keeping commercial enterprises and activities in the coastal areas

- Securing the evacuation routes (such as roads and stairways) that connect public facilities (such as schools) to higher ground

- Planting trees in coastal areas

- Using tall concrete buildings (of four to five stories or higher) as places for evacuation

Evacuation

- Drills, education, and awareness-raising are the keys to ensuring effective, more complete evacuation.

- Remember "tsunami *tendenko*," that is, everybody should evacuate immediately without waiting for anything or anyone else when the approach of the tsunami is assumed or feared.

- Prior discussion at home and in communities about evacuation helps ensure its success.

- Blind assumptions should not be made about any disaster, even those reflected in hazard maps, as nature behaves differently from human assumptions.

- Individual and institutional memory about past disasters should be kept alive to facilitate successful evacuation.

Hazard maps

- Hazard maps are a useful tool for enhancing the preparedness of local governments, municipalities, and individuals.

- Hazard maps should address both level 1 and 2 events.

- A hazard map functions well only in combination with awareness-raising, community education, and evacuation drills.

Forecasting and warning

- Forecasting and warning systems pay off.

- Tsunami and disaster warning networks should be built and used globally.

- The international community should promote and invest in the use and development of new technologies to improve the accuracy and timing of forecasts and warnings.

Addressing "chain-of-events" effects

- The indirect effects of extreme events travel far beyond the disaster-stricken areas; hence, building redundancy into systems helps break these chains of events.

- Probable chain-of-events effects should be considered in business continuity planning.

- Experiences of these effects should be evaluated and shared to help prepare for future events.

RECOMMENDATIONS FOR DEVELOPING COUNTRIES

Every country needs a national integrated DRM strategy. Many of the lessons from the GEJE are relevant for developing countries. Different combinations of structural and nonstructural measures may be used depending on a range of factors, such as socioeconomic conditions, budgetary constraints, geography, and the scale of the disasters. In the GEJE, DRM systems relied heavily on structural measures and could not prevent damages from the tsunami (figure 32.2d). The Japanese government is revising its tsunami DRM policies to better integrate structural and nonstructural measures (figure 32.2e). Level 1 tsunamis will be prevented by structural measures and level 2 tsunamis will be mitigated by both structural and nonstructural measures.

It is advisable to develop integrated measures for both level 1 and 2 events. For developing countries, greater reliance on nonstructural measures may be the most realistic approach even for level 1 events. But it is important to build structural measures to prevent loss of human lives and properties from frequent disasters. Disasters, especially high-impact events, tend to discourage people from investing for the future. Governments and communities should keep repeating the message that "prevention pays off," to avoid creating a vicious cycle between poverty and disasters.

Forecasting and early warning is fundamental. Developing countries can and should develop local networks for forecasting and warning about disasters. Countries can also join forces in building regional and international systems. For example, Sentinel Asia is a regional network for sharing satellite imagery and other observation data free upon requests by member countries.

Hazard maps are useful tools to help people save their own lives. Developing countries should take legislative, administrative, and financial measures to ensure that hazard maps are provided to all the disaster-prone localities. The international community should help countries to develop hazard maps that reflect the lessons described in this note. It would also be useful to create regional and global mechanisms to share good practices and examples of hazard maps.

Archiving disaster records and experiences in disaster databases is essential for designing viable DRM strategies. The government should

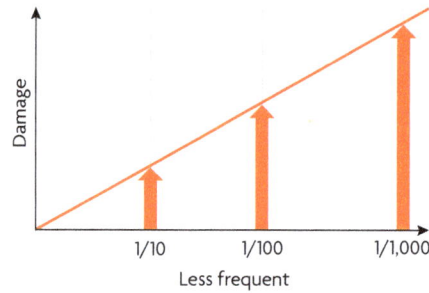

a. Disaster damage and frequency without countermeasures: Larger disasters occur less frequently than smaller disasters.

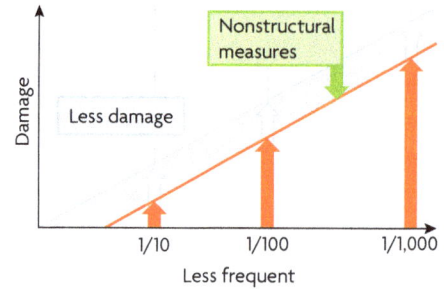

b. Disaster damage can be mitigated by nonstructural measures: Cases in cyclone DRM in Bangladesh and flood management before the early modern period in Japan.

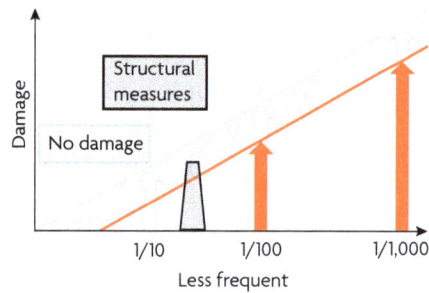

c. Structural measures can protect against frequent disasters: Cases in flood management in the early modern period in Japan

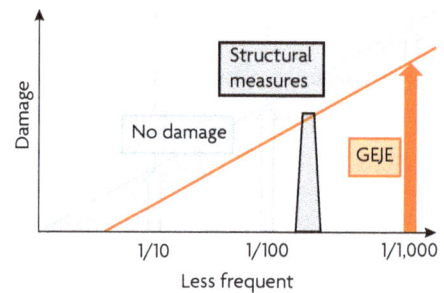

d. Structural measures protect against disasters that occur every few decades: Cases of tsunami management at the GEJE and current flood management in Japan.

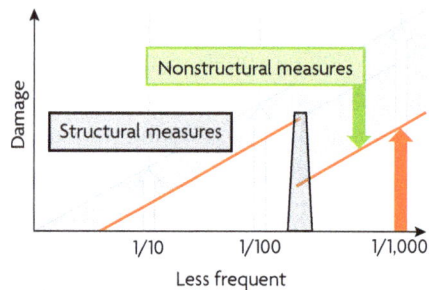

e. Tsunami damage will be mitigated by reconstructing resilient dikes and strengthening nonstructural measures.

Figure 32.2 DRM using structural and nonstructural measures

stress the importance of these less visible but critical activities and the people who are engage in them tirelessly. Regional data sharing would also benefit neighboring countries. Countries should put agreements in place to share hydrological, meteorological, geological, and other information.

Education, drills, and awareness raising are indispensable to avoid high death tolls in low-probability, high-impact extreme events, particularly in countries where physical defenses may be insufficient. The Japanese approaches to education, drills, and awareness raising have been developed over time through trial and error. But simply copying them exactly may not be advisable in other, often more challenging, circumstances. The first step is to evaluate, simulate, and test whether the Japanese measures are congruent with local social and cultural practices and behaviors.

Countries must learn from one another by sharing information and experience, since low-probability, high-impact extreme events happen infrequently in any given country. The international community could facilitate regular dialogues and information-sharing mechanisms, for example, through the United Nations. Regional cooperation mechanisms would serve not only to help disaster-affected countries but also to mitigate the negative interregional and international effects of megadisasters.

NOTE

Prepared by Kenzo Hiroki, International Centre for Water Hazard and Risk Management, Public Works Research Institute.

PART VII

Recovery and Relocation

CHAPTER 33

Relocation in the Tohoku Area

Relocation and new regulations for land use in at-risk areas are often proposed in the wake of mega-disasters such as the 2004 tsunami in the Indian Ocean, the 2008 earthquake in China's Sichuan Province, and the 2009 earthquake in Haiti. Since the 2011 Great East Japan Earthquake, the Japanese government has strengthened disaster risk management systems based on lessons learned from that event.

Steps to improve disaster risk management (DRM) in Japan have included new land-use regulations and relocation programs as well as better integration of structural and nonstructural DRM measures (see chapter 12). Local governments in the affected areas have regulated housing development in at-risk areas and are promoting relocation to higher ground. Because each local government has taken a different approach to such programs, levels of recovery vary across the affected areas. The recovery process is ongoing in Tohoku, where the Great East Japan Earthquake (GEJE)

struck with full force, but the actions taken to date offer many useful lessons.

One of those lessons is that relocation is effective in mitigating disaster damage but managing relocation projects—and consulting with affected communities—is challenging. It is difficult to achieve a consensus among community members on any rehabilitation plan. For example, while some prefer to rebuild their hometowns on the original sites, others want to move to safer areas.

Despite the challenge, governments must take a participatory approach and engage communities in the recovery process. Experts and civil society organizations (CSOs) can support local governments in efforts to consult community members. Meanwhile, a cross-sectoral approach—covering infrastructure, urban planning, disaster management, and economic activities—is best.

FINDINGS

Government initiatives

The Japanese government has enacted new laws and created new schemes for managing tsunami risks based on lessons learned from the GEJE. Government agencies are taking a holistic approach, integrating nonstructural and structural measures to develop safe communities (see chapter 12). Local governments have designated tsunami-risk areas where housing construction is prohibited. The national government has provided local governments with financial support for relocating affected people to safer areas, including for the development of residential sites and infrastructure (figure 33.1). Local governments have promoted relocation by building

consensus among community members, as they could not implement it by force. Given the limited financial capacity of local governments and the enormity of damages in the Tohoku region, the national government has covered the costs of relocation programs almost entirely, as an exceptional case.

Each of the local governments in Tohoku took its own approach to recovery by utilizing new as well as tested urban development schemes such as land pooling, readjustment, and redevelopment. Progress varies across cities in the affected areas. As of May 2013, local governments had started on the construction phase of 106 relocation programs (out of 328 planned) and 31 land pooling programs (out of 59 planned). Sendai City is promoting relocation projects as scheduled. On the other hand, affected people in some communities in the cities of Ishinomaki and Natori have not been willing to join recovery programs planned by city governments.

Relocation 1—the case of Sendai City

The Sendai City government is promoting the concept of "rebuilding better and safer." To this end, it has designated tsunami-risk areas where housing construction is prohibited and promoted relocation from risk areas to higher, safer ground. Also, the government is constructing public rentals for those who cannot afford to build new houses. Some 57,000 houses were damaged or destroyed by the earthquake tremors and tsunami waves that struck on March 11, 2011.

Tsunami-risk areas. In December 2011, the city government designated some 1,200 hectares as tsunami-risk areas, where some 1,500 families had lived before the GEJE. Residential development is prohibited in these risk areas. In the area, according to tsunami simulation, it is estimated that tsunami waves of the same scale as those that hit during the GEJE could cause floodwaters more than two meters deep even with countermeasures in place (map 33.1).

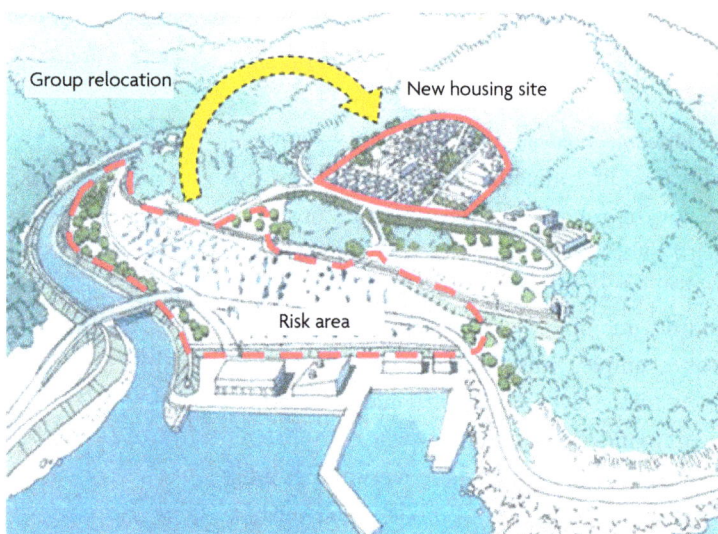

Figure 33.1 The relocation process

Source: Office of the Prime Minister.

The city government is promoting recovery programs for these 1,500 families, including

- *Individual relocation.* Affected people individually purchase land in safe areas and build as they see fit. The city government purchases the land they formerly occupied in at-risk areas and provides subsidies for moving and interest payments on new housing loans.

- *Group relocation.* Affected people build houses in new sites developed by the government. As noted previously, the city government purchases the land they formerly occupied and provides subsidies for moving and interest payments for new housing loans. The government is planning to complete the development of group relocation sites by March 2016.

- *Public housing.* The city government provides public rental apartments for those who cannot afford to build new houses.

As of May 2013, among families residing in at-risk areas, 49 percent chose group relocation, 27 percent individual relocation, and 23 percent public housing. Among the families who chose group relocation, 73 percent will rent land from the city government to construct houses. The city government is encouraging those affected to join individual or group relocation programs because public housing of standardized units cannot respond flexibly to people's needs (chapter 34).

Areas contiguous to at-risk areas. The city government is supporting some 2,300 families living adjacent to at-risk areas whose homes were flooded by the GEJE (map 33.1). In these areas, a tsunami of the same scale as that of the GEJE would produce floodwaters up to two meters deep, despite structural measures to counter the damage (according to simulations). The government provides subsidies to affected people in support of disaster management works—such as raising the level of the ground or constructing earth mounds on

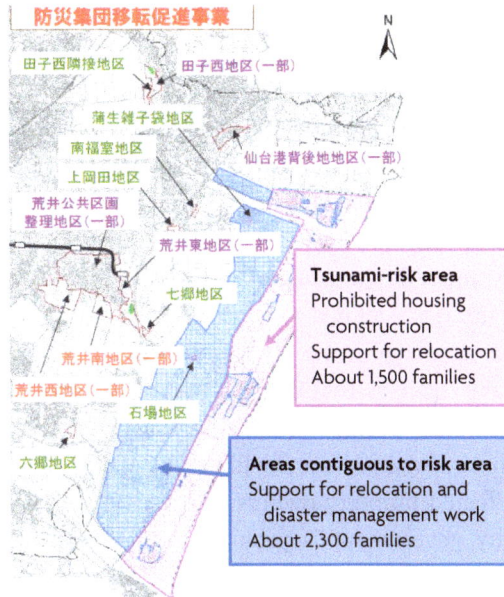

Map 33.1 Relocation project in Sendai City

Source: Sendai City.

housing lots. Alternatively, people may find a new plot and rebuild as they see fit. The city government provides subsidies for moving and interest payments for new housing loans.

Public housing. The city government is building 3,000 public rental units for affected people who cannot afford new houses. Before March 2015, some 1,620 units of public housing will be completed as conventional public works at 33 sites in the city. Meanwhile, the government has adopted the public-private partnership model and purchased 1,381 units from private companies. It will acquire land from private companies (who have negotiated with landowners), and these companies will construct housing and related facilities. The government selected 17 companies for this partnership in March 2013 by assessing the unit costs of land and housing, the certainty of land acquisition, the location, the support offered to communities, and other aspects of proposals.

Housing in hilly areas. Some 5,500 houses were severely damaged by landslides caused by the ground-shaking of the GEJE, and they must be rebuilt (chapter 2). The city government is planning to rebuild them through public works or by providing subsidies. The government will bear some 90 percent of the cost.

Relocation 2—Ogatsu District, Ishinomaki City

Most people in the Ogatsu District, Ishinomaki City, have been unwilling to join a relocation program planned by the city government, which is taking a similar approach to the one in Sendai City. In December 2012, the city government designated certain areas as high tsunami risk—defined as floodwaters at least two meters deep in the event of a tsunami of the same scale as that of the GEJE. The government has prohibited housing construction in these at-risk areas and will purchase land from affected property owners in these areas. The city government is promoting 47 relocation projects for some 7,000 families and is planning to construct 4,000 public rental units.

The city government and community members could not agree on a recovery plan in the downtown area of the Ogatsu District, affecting approximately 500 households. Because most downtown areas were flooded and designated as at-risk areas, the city government proposed relocation to higher ground. The move promised to protect community members against tsunamis but would have posed an enormous budgetary burden on the government, which would have had to develop new residential areas and associated infrastructure. Initially, a public survey indicated that many affected people wanted to rebuild their houses on the original site downtown. Though the city government announced a relocation plan to higher ground, only 12 percent of affected households indicated their willingness to relocate to the new site. Most decided not to join the relocation program, citing inconvenience because the government planned to relocate public services and various other facilities at scattered sites. Business and industrial areas, such as the fishing port and fishing industry facilities as well as government offices, tourism spots, and residential areas, were to be scattered across different locations. Meanwhile, right after the GEJE, some people moved from the Ogatsu District to the center of Ishinomaki City, where their workplaces were located, and never joined the relocation program.

Relocation programs risk disrupting communities already affected by demographic trends. The GEJE exacerbated the problem of an aging rural population that was already decreasing before the disaster. In general, younger people tend to move to urban areas and older people stay in their hometowns.

Land pooling on site—the case of Natori

The Natori City government has not been able to achieve consensus with community members on a rehabilitation program or to start a scheme in the Yuriage District. This district was the most seriously damaged in the city. More than half of the affected people chose not to participate in the program, moving on their own instead. Some 25 percent of community members showed a willingness to join the program in April 2013. The city government is currently reexamining the program to include relocation schemes.

The city government planned a land pooling (or land readjustment) scheme for 2,000 families on raised ground at the original sites. Because it obviates the need for relocation, this is a cost-effective approach. It saves the government the additional investment required for new infrastructure on yet-to-be-developed relocation sites.

A land pooling scheme is a development process based on consensus among all members in a community (figure 33.2). Such a scheme was used in reconstruction following the Great Hanshin-Awaji (Kobe) Earthquake in 1995 and has been widely used for urban development in Japan in normal times. Land parcels are assembled and after developing infrastructure and other public facilities in an assembled parcel, the land is returned to its original owners. Landowners equitably contribute a portion of their lands for developing infrastructure and public facilities. Thus, the reconstituted land lots are smaller than before the project started.

Management of nonresidential lands

How to use vast areas of nonresidential land designated as being at risk is becoming a crucial issue. For example, most people are no longer able to live in the coastal flatlands of the Minamisanriku region. Other uses for such land are also unclear. Most local governments plan to use it for three main purposes: (1) parks, (2) tourism (developing facilities such as souvenir shops and restaurants), and (3) fisheries (developing fish-processing and related facilities). Some local governments are planning to develop renewable energy projects to generate solar and wind power. The Miyagi Prefecture government, for example, is planning to develop solar power plants in devastated coastal areas. Local governments may also let the land revert to nature by abolishing infrastructure, such as roads and sewerage systems. However, cost-sharing and management mechanisms involving the private and public sectors have yet to be established. Governments need to establish budgetary and institutional mechanisms to manage these lands, secure financing to maintain them, and decide which institutions will do what.

Community participation

While communities have been extensively involved in the recovery process, local governments face practical issues on the ground. Many local governments have limited experience working with community members on planning and implementing projects. It is difficult for communities to achieve consensus on recovery plans because community members have different backgrounds and views. Some prefer reconstruction at the original site; others, relocation to higher ground. Disagreements arise, and compromises can be challenging. Each government has chosen a different method of reconstruction, and outcomes vary among cities.

Participation methods. Local governments and communities embarked on a variety of participatory activities. Some governments

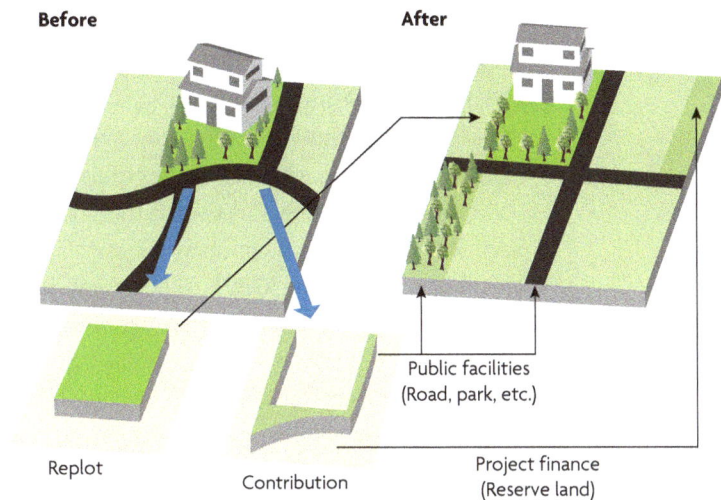

Figure 33.2 Land pooling scheme

Source: Ministry of Land, Infrastructure, Transport and Tourism.

organized workshops starting at the early concept phase, followed by careful consultation with communities to select proposals for recovery projects. Chapter 21 outlines the practice of Minamisanriku Town. The Tohoku Office of the Japan International Cooperation Agency (JICA) and Miyagi University began a joint pilot program to facilitate the reconstruction process in communities by assigning 10 community reconstruction facilitators (figure 33.3). They have organized and facilitated

Figure 33.3 Community rehabilitation facilitators

Source: Japan International Cooperation Agency (JICA).

consultation workshops to receive feedback on recovery plans from communities. In Nobiru District in Higashimatsushima City, communities took the initiative to recover by themselves (figure 33.4). They established the Nobiru Consultation Committee for Community Development in 2008, and conducted festivals and seminars before the GEJE. This committee created a working group for recovery in July 2012 to implement recovery activities by community members. The working group is formulating a development plan for relocation sites and is in charge of implementing the plan.

In some cases, even after consultations, community members' views were not properly reflected in recovery plans. This may have been because local governments, which had little experience with community participation and limited staff, had to formulate recovery plans within a limited time span. The Architectural Institute of Japan strongly recommends that local governments establish community-based consultation organizations with financial support to formulate community recovery plans.

Figure 33.4 Consultation at Nobiru District
Source: JICA.

Communication across various locations. Because many affected people live in temporary housing at various locations (chapter 22), it is difficult to establish and maintain communication, to share necessary information, and to receive feedback. After the GEJE, some local governments established information centers, while others used e-mail and websites in addition to conventional printed materials such as newsletters. Although information technology–based communication tools helped to promote communication among unintentionally separated community members to some degree, the digital divide between the young and the elderly (and between those who can afford the use of personal computers and those who cannot) has remained an issue, particularly for the aging rural population in the affected areas.

Community representation. People's opinions may not have been adequately reflected in municipalities' planning processes if the members of planning committees were not properly selected. Most local governments in Japan selected representatives from communities in a conventional way, choosing those who were the heads of community organizations or commercial associations, predominantly middle-aged or elderly males. But people's views on recovery depend on various factors, such as occupation, generation, gender, and scale of damage. Local governments should conduct a stakeholder analysis to select community representatives and seek various ways of receiving feedback from communities, such as web-based surveys and workshops.

How experts can help

Experts and CSOs are expected to play a supporting role in formulating recovery plans. Many experts in architecture, civil engineering, and urban planning voluntarily provided support to communities in the wake of the GEJE. Local governments should also play such a role, but they often face difficulties supporting communities because of excessive workloads in times of crisis, limited capacity, and limited experience. Outside experts and CSOs can (1) formulate and make presentations of alternate recovery plans, (2) facilitate discussions and assist in building consensus in communities, and (3) bridge the gap between the government and communities by moderating discussions. Chapter 21 explains the case of Minamisanriku Town, where university researchers supported the formulation of a recovery plan. In Ogatsu District, Ishinomaki City, Tohoku and other universities have supported community organizations and the Ishinomaki City government in promoting recovery programs. Researchers and students have proposed rehabilitation projects such as public housing and community centers.

Three types of relationships involving outside experts are particularly important:

- *Relationships with communities.* Experts should conduct a survey of community members' opinions, closely consult with community members, and build consensus. They should not only compile community opinions and propose plans but also incorporate their own professional views into the plans based on their understanding of the potential demands and requests.

- *Relationships with various experts.* It is important not only to work with experts on physical infrastructure, such as in civil engineering, architecture, and urban planning, but also with experts and CSOs in social welfare, education, and health. To restore daily life to its predisaster rhythms, a range of community activities must be carefully examined. Experts and CSOs need to cooperate with each other and help formulate recovery plans through coordination and teamwork.

- *Relationships with local governments.* The relationship between experts and local governments should be established to ensure a win-win situation. By collaborating with experts, governments can decrease the burden of consulting with communities because experts can play roles of facilitators and consensus-builders. And experts can have their ideas realized on the ground. Because it is difficult to create such relationships in a short time frame immediately after a disaster, it is important to establish them beforehand, as was demonstrated in Nobiru District.

Government coordination efforts

Coordination among various sectors is crucial for recovery. A wide range of reconstruction projects (roads, tsunami dikes, schools, and houses) have been simultaneously implemented. The concerned organizations require close collaboration to work efficiently. An integrated approach, such as a project combining roads with dikes, is best. Without coordination, a public works department may build an embankment to protect low-lying areas at a site that has been marked out for housing by the urban development department. The Sendai City government established the Steering Committee for Disaster Recovery in May 2011 to make decisions and coordinate recovery policies, plans, and programs. The mayor chairs the committee, which consists of 23 heads of organizations that are part of the city government. Thirty-six committee meetings have been held over the past two years.

In formulating urban plans, local governments should consider various components, including DRM, quality of life, economic activities, and environmental impacts. Relocation to higher ground can provide affected people with safer housing but may interfere with livelihoods or cause adverse environmental impact through development work. Land-use regulations are usually required to implement recovery plans.

LESSONS

- *Disasters exacerbate the existing problems of an aging and dwindling rural population.* The GEJE has aggravated demographic issues that were serious even before the event. If the local government cannot formulate recovery programs in a timely manner, communities can easily be disrupted. The elderly may refuse to relocate. Meanwhile, younger people may be unwilling to return to their hometowns, instead moving elsewhere to restart their lives.

- *Relocation is effective, but implementing relocations can be challenging.* Reconstructing towns on higher ground is regarded as an ideal approach for mitigating disaster damage. But some cities have faced difficulties owing to out-migration and the management of low-lying lands. People decide whether to participate in relocation projects by examining their prospects for earning a living at the new sites, how long it will take to relocate, and the convenience of the new sites. As cases show, local governments cannot "sell" some projects; community members have failed to reach consensus on some plans formulated by local governments. In Ogatsu District, Ishinomaki City, the affected population preferred to rebuild the town in the original area—in opposition to a relocation plan. On the other hand, the people of Natori City preferred to move to higher ground, even as the local government was recommending rehabilitation of homes at the original site.

- *Community participation is key to promoting recovery, but local governments face practical issues.* Community participation in the consultation process is needed to respond to a wide range of needs. Local governments are required to organize various events, such as workshops, but such tasks are a burden for cities damaged by the disaster, where government agencies are overtaxed by recovery works. Where affected people live in transition shelters, communication with them poses an additional challenge. Experts from outside the disaster-struck areas as well as CSOs can support consultation processes and communicate with the affected population.

RECOMMENDATIONS FOR DEVELOPING COUNTRIES

Governments should examine various recovery schemes such as relocation to safer areas, reconstruction at original sites, and land pooling. When planning a recovery scheme, it is crucial to consider community needs. But there is a trade-off between speed and quality in the recovery process. A government can rehabilitate towns promptly by taking a top-down approach. Community consultation requires more time. Following the Sichuan earthquake in 2008, Chinese governmental organizations took a top-down approach with limited consultation with affected communities, rebuilding houses and infrastructure at a rapid pace. But tall residential buildings inconvenienced affected people who had lived in rural villages before the disaster. People and local governments have also had to share the unnecessarily high costs of operation and maintenance of the new facilities and housing.

Local governments should establish a participatory mechanism because community participation is essential in promoting recovery. One lesson from the humanitarian response systems used after the Indian Ocean tsunami in 2004 is the importance of striving to understand local contexts and working with and through local structures. Experts and CSOs are expected to play a role in assisting recovery, for example, by organizing and facilitating workshops or consultation meetings and working with the government and other experts.

A cross-sectoral approach is required to rehabilitate people's daily lives. Organizations should harmonize recovery plans among all sectors concerned, such as roads, DRM, and urban planning. Coordination among local governments, the ministries of the central government, and reconstruction agencies is crucial for effective planning and implementation.

Local governments should lead recovery, but support from the national government is essential. Because local governments can more closely respond to the varied needs of affected people on the ground, they should take the principal responsibility for recovery planning and implementation. The national government should support local government efforts by creating legislation and new project schemes, providing subsidies, and providing technical support (such as conducting tsunami simulations and dispatching technical staff).

NOTE

Prepared by Michio Ubaura, Tohoku University; Akihiko Hokugo, Kobe University; Mikio Ishiwatari, World Bank; and the International Recovery Platform.

BIBLIOGRAPHY

Architectural Institute of Japan. 2012. *Recommendation on Urban Development in Rehabilitating from the GEJE* (in Japanese). Tokyo. http://www.aij.or.jp/scripts/request/document/20121115.pdf.

Nian, S. B. 2013. "The Vision of Social Governance: Implication of Wenchuan Experience on the Post Disaster Reconstruction on Lushan." In *Proc. Forum on Post-Disaster Revival and International Disaster Reduction*, 25. May 14–15, 2013, Sichuan University.

Sendai City. 2013. *7th Report on Recovery in Sendai* (in Japanese). http://www.city.sendai.jp/shinsai/report/report7.pdf.

Nobiru Town Planning Committee (in Japanese). http://mm.higashimatsushima.net/matsumng/introduction.do?id=00007.

TEC (Tsunami Evaluation Coalition). 2006. *Joint Evaluation of the International Response to the Indian Ocean Tsunami: Synthesis Report*, Key Messages. London: TEC.

Ubaura, M. 2012a. "Are We Planning 'Mistakenly Excessively Defended Cities and Towns'?" Symposium on "Learning from Great East Japan Earthquake," 603–06. March 1–2, 2012, Tokyo.

———. 2012b. "Land Use Plan and Regulation for Reconstruction of the Affected Area and its Problems." Eighth APRU Research Symposium on Multi-hazards around the Pacific Rim, 121–22. September 20–22, 2012, Sendai.

CHAPTER 34

Reconstruction in the Tohoku Area

It is best when those who have lost their homes to a disaster can assume responsibility for rebuilding their dwellings, so that they match their needs. In planning and administering disaster-recovery assistance, governments should endeavor to harness people's natural interest in rebuilding as they see fit while also providing special support for the vulnerable, such as low-income households and the elderly.

In addition to providing financial support for individual rebuilding efforts, local governments in the Tohoku region are constructing public rental housing and housing complexes for those who cannot afford to rebuild by themselves. Because completing large tracts of public housing in a short time is a difficult task, local governments responsible for reconstruction works should seek assistance from other organizations and experts and from public-private partnerships (PPPs). The needs of the affected population change as reconstruction progresses and as people age. The most vulnerable groups, notably people of low-income and the elderly, depend on public housing, and

effective government assistance is particularly important to them. In particular, measures are required to prevent so-called solitary deaths (*kodokushi*) (chapter 22).

FINDINGS

Rebuilding schemes

In the aftermath of the Great East Japan Earthquake (GEJE), local governments have been promoting the reconstruction of permanent housing through two schemes: (1) self-reconstruction and (2) public housing. Wherever possible, local governments should encourage affected community members to assume responsibility for rebuilding their lost dwellings. This approach is desirable because it allows people to rebuild in a way that suits their needs and because it lightens the load on government. Some groups, however, such as low-income households and the elderly, cannot rebuild on their own because of financial constraints. Local governments are providing these people with public housing.

Self-reconstruction

In accordance with the Act on Support for Reconstructing the Livelihoods of Disaster Victims, Japan's national government provides up to ¥3 million to people who lost their houses. Because this amount is not enough to rebuild a house, local governments provided additional financial support. Some members of the affected population had to supplement these resources with other sources of financing, such as housing loans.

There are significant regional differences in the amounts provided by the government, which affects the speed of recovery in each prefecture. The percentage of people receiving financial support is higher in the Miyagi and Fukushima prefectures than in the Iwate Prefecture. There, people could not rebuild until development work on higher ground was completed. As of October 2012, the national government had provided a total of ¥248.2 billion to 183,264 households. For people affected by the nuclear accident, the government has not yet established a scheme for reconstructing houses (chapter 36).

Public housing

Public housing provides a safety net to people who have lost their homes. For vulnerable people who cannot afford to rebuild homes lost in the GEJE, local governments are constructing public rental housing complexes (figure 34.1). Anyone who lost a home is eligible to apply for public housing. Some who would not otherwise be vulnerable but who are still repaying

Figure 34.1 Public housing in Sendai City
Source: Sendai City.

loans on houses lost during the tsunamis cannot contemplate financing new housing (chapter 32). Others worried about risks such as land subsidence, future tsunami risks, and contamination from radiation may resist rebuilding on their original sites. According to surveys, the number of people choosing public housing is increasing in Kamaishi City, while the number of those choosing self-reconstruction is decreasing.

The Iwate, Miyagi, and Fukushima prefectures plan to complete their public housing by March 2016. Some 6,000 units are planned in the Iwate Prefecture; 15,000 units in the Miyagi Prefecture; and 3,700 units in the Fukushima Prefecture—a total of 24,700 units. Governments have acquired land for an additional 12,804 units and began construction of 2,152 units in the Iwate and Miyagi Prefectures in May 2013.

Public housing is not necessarily the best option for affected communities. People accustomed to living in spacious houses—in farming villages, fishing villages, and rural towns, in particular—may find it difficult to adapt to small public housing units. Public housing is also not always suited to the lifestyles of the people living in it. Housing units are usually built according to a uniform design. People cannot change the floor plans or furnishings. The locations of these housing units is usually determined by lottery, and people cannot freely choose their units. In Sendai City, the city government prioritizes groups consisting of more than five families that choose to join a community.

Project management

Constructing large tracts of public housing quickly is a difficult task. Governments are trying to accelerate the process through various measures, notably parnerships—such as PPPs—and with the help of a variety of outside organizations and experts. The public housing construction process after the GEJE has included three main steps (figure 34.2). The first step is to collect the information needed to understand where the best options lie, starting with existing land-use planning and relocation/reconstruction planning documents, surveys of citizens, community consultations, and workshops. The second step relates to design and construction. Once the type of housing has been selected, the design phase begins (with the selection of designers, builders, materials, and the selection of a management approach), followed by construction of the units and monitoring of progress and costs.

Local governments in the affected areas of Japan are considering the needs of community members while erecting public housing. Governments are trying to ensure that residents of public housing are not isolated from their communities. Through housing preference surveys, governments provide the most vulnerable people with detailed information on housing schemes in an effort to better understand their needs.

Local governments are trying to hasten the process of construction through flexible land acquisition, standardized materials, and PPPs. They must coordinate the construction process with overall recovery plans, community needs, and project management procedures. Local governments have contracted with private companies to manage reconstruction works. Some governments have applied for design-build schemes to decrease project management workloads (see chapter 33 for a PPP case in Sendai City).

Local governments require assistance with project management because they are stretched thin by various recovery tasks. In the Miyagi Prefecture, municipalities will construct all public housing, and the prefecture will reduce the municipal burden by taking over project management. The Iwate and Fukushima prefectures are undertaking the construction of some of the public housing. As of May 2013, the government housing agency was also building 2,143 units of public housing at 30 sites in three prefectures. When completed, the units will be transferred to local governments.

Local governments should plan the operation and maintenance of public housing units

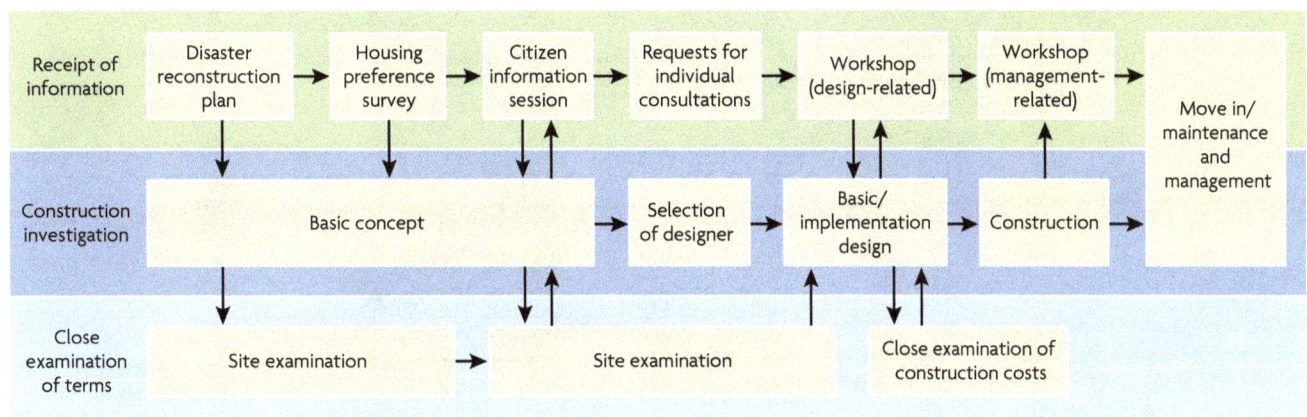

Figure 34.2 Construction of public housing: A three-step process

at the design stage. Although the national government finances the construction of public housing, local governments must operate and maintain it. Ishinomaki City had 1,700 public housing units before the GEJE; it is now building another 4,000 units. Minamisanriku Town plans to construct 1,000 units, and Ofunato City 900 units. Local governments must find ways to operate and maintain the new units efficiently (figure 34.3). They must also plan for the dismantling or reuse of empty units (for example, to make way for other public facilities).

A variety of services is typically required, including nursing care for senior citizens. In public housing, where people cannot expect help from large families or fellow community members, governments must provide these services. The Kobe City government provided services at transition shelters following the Great Hanshin-Awaji (Kobe) Earthquake in 1995, and some local governments in areas devastated by the GEJE have also begun to do so (chapter 22).

The situation on the ground: The case of Shichigahama

Shichigahama is a small scenic town with a population of about 20,000. It is located about 15 kilometers (km) from Sendai City, the largest city in Tohoku. Most houses are built within a circle about 5 km in diameter. Some 46 percent of the city was flooded by the tsunamis, and 1,323 homes were completely or partially destroyed. As of February 2013, 222 public housing units were under construction, accounting for approximately 3 percent of the total 6,540 households in the town. (This figure is smaller than the 6.9 percent in Ishinomaki City that plans to construct 4,000 units.) Housing complexes are to be developed on five sites that suffered severe damage. In Shobuda,

Figure 34.3 Managing public housing

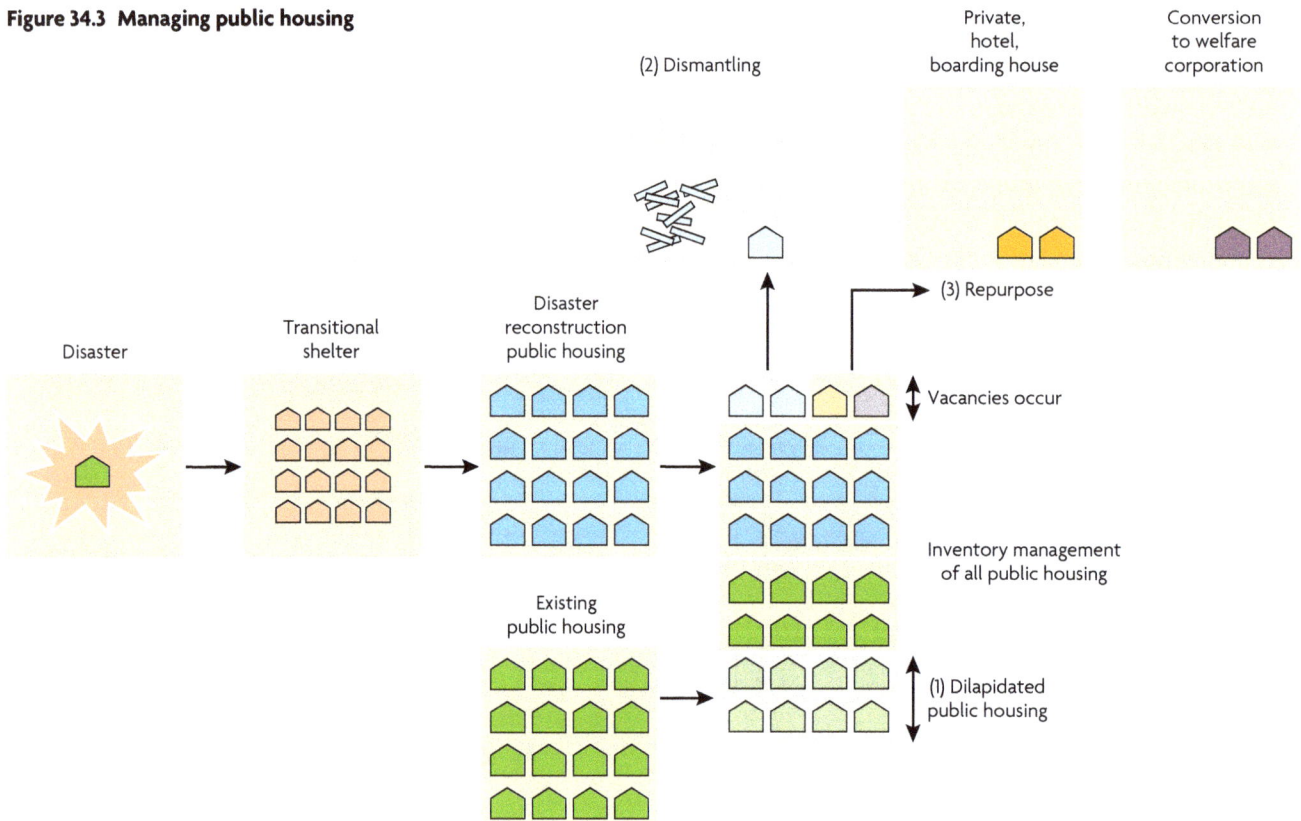

the most severely damaged district, more than 100 homes are being built, but in other sites the number is lower.

After close consultation with the affected households, the number of planned units was decreased. The town government tried to decrease the number of public housing units, citing difficulties in operation and maintenance. The government conducted two preference surveys in July 2011 and February 2012, among 1,000 affected families, to decide the proper number of public housing units and the scale of other recovery schemes. While nearly one-third of affected people indicated their willingness to live in public housing in the first survey, the number decreased to less than one-fourth in the second survey. Between the first and second surveys, the town government explained the details of self-reconstruction, group relocation, and public housing in a series of interactive consultations. Through the consultation process, some became convinced of the advantages of self-reconstruction and group relocation. Others continued to prefer public housing.

Local governments can sometimes better manage the process of housing reconstruction with outside assistance. It is important to prepare for the particular needs of vulnerable groups staying in these units. Most people living in public housing are more than 70 years old and need barrier-free, easy-access structures. In Shichigahama, experts and researchers assisted the town government with planning, designing, project management, and public consultation. The town government asked the Miyagi Prefecture government to take over the tasks of designing buildings and selecting contractors. The town government also organized and coordinated meetings among designers, contractors, and government organizations.

The challenge is to respond to people's needs. Local governments should take into account community feedback regarding the design of public housing units. Ideally, every designer should organize workshops to gather feedback to then use in the design process. At these workshops, participants could discuss a wide range of issues, such as the management of public spaces and formulation of new communities.

LESSONS

- *Reconstructing permanent housing.* An essential task of government is to help people affected by natural disasters, particularly the most vulnerable groups, to reconstruct permanent housing. A large segment of the affected population can bear the responsibility of rebuilding homes to match their needs with financial support from the government. Others, however, are unable to reconstruct their own homes for one reason or another. Governments should be prepared to assist these vulnerable groups, including the elderly and low-income households.

- *Local governments should strive to identify the best way to manage the process of housing reconstruction.* Completing a large number of public housing units within a short time frame is a difficult task. Local governments should adopt PPPs and seek assistance from other agencies and organizations, domestic and foreign.

- *Close communication between the government and affected communities is an essential aspect of an effective response.* Communicating with the elderly in public housing can be especially challenging, and plans should be made to meet that challenge.

RECOMMENDATIONS FOR DEVELOPING COUNTRIES

Governments should establish support mechanisms for housing reconstruction, particularly for vulnerable and low-income groups.

Wherever possible, people affected by disasters should be permitted—even encouraged—to assume responsibility for rebuilding their own homes, with financial support from governments. Following the 2001 Gujarat earthquake in India and the 2005 earthquake in Pakistan, governments supported the reconstruction of core housing units with a cost of $2,000–$4,000 per family that could be expanded to meet family needs. For low-income and other vulnerable groups, governments must create social safety nets. In the aftermath of the GEJE, local governments have constructed rental units. In India and Pakistan, governments provided additional financial support to low-income groups.

Support from experts and private sector involvement are useful. Experts can help local governments effectively consult with affected communities and advise on project implementation. Local governments are well advised to take advantage of the private sector's experience with project management. The Sendai City government has purchased housing where private companies acquire the land and manage construction (chapter 33). Governments should establish PPPs to promote greater housing reconstruction within a limited timeframe.

Local governments should formulate plans to operate and maintain public housing. While the central government should provide financial support for construction, local governments and the affected population will have to operate and maintain public housing. Local governments should consider operation and maintenance at the design stage.

NOTE

Prepared by Yoshimitsu Shiozaki, Kobe University; Yasuaki Onoda, Tohoku University; Mikio Ishiwatari, World Bank; and the International Recovery Platform.

REFERENCES

Indrasafitri, D. 2012. "Yasuaki Onoda: Exiting His Comfort Zone." *Jakarta Post,* February 10. http://m.thejakartapost.com/news/2012/02/10/yasuaki-onoda-exiting-his-comfort-zone.html.

Onoda, Y. 2011. "From the Front Lines of Reconstruction Planning (2) Kameishi City, Iwate Prefecture: Toward Creative Reconstruction Plan Formulation." *Gekkan Jichiken* [Local Government Studies Monthly] 53 (626): 5–58.

Onoda, Y., and S. Fukuya. 2011. "Students' Skills Help to Forge a New Tohoku." *The Japan Times,* September 25, 7–8.

Onoda, Y., and Y. Kato. 2011. "Using the Wisdom of Architects in Reconstruction: The Activities of ArchiAid (A Reconstruction Support Network of Architects in the Great East Japan Earthquake)" [in Japanese]. *Chiiki kaihatsu* [Local Development] 564: 15–21.

Trohanis, Z., and G. Read. 2011. "Housing Reconstruction in Urban and Rural Areas." EAP DRM Knowledge Notes, World Bank, Washington, DC.

CHAPTER 35

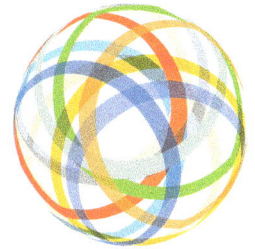

Cultural Heritage and Preservation

A country's cultural heritage is fundamental to its national and community pride and for social cohesion; historical monuments are regarded as national and community treasures. Because these properties are deeply connected to people's lives and communities' histories, their disappearance is equivalent to losing part of a nation's identity. The Japanese government and cultural heritage experts have recognized historical records and cultural properties as national and community treasures. Following the Great East Japan Earthquake, the Agency for Cultural Affairs (ACA) and expert groups began rehabilitating and preserving cultural properties. Various organizations rescued and preserved a wide range of historical records and cultural properties damaged by the tsunami waves and earthquake tremors.

Over the last several years, natural disasters have caused enormous losses to cultural heritage sites around the world, such as the damage done to the historical citadel of Bam in the Islamic Republic of Iran by the 2003 earthquake to the Prambanan Temple Compounds in Indonesia by the 2006 earthquake to historic churches in L'Aquila in Italy by the 2009 earthquake.

In Japan, earthquakes and tsunamis have damaged an enormous number of cultural properties—for example, 744 designated cultural properties were damaged by the Great East Japan Earthquake (GEJE). Temples that

the government had designated as national treasures collapsed or cracked. The Matsushima Islands, which were designated by the national government as "special places of scenic beauty," were also damaged. The stone walls of the Edo Castle in Tokyo, designated as a special historic site, also collapsed.

To mitigate potential disaster damage to cultural and historical heritage, governments, museums, experts, researchers, property owners, and communities should devise collaborative mechanisms before disaster strikes. For example, databases with detailed information on heritage properties (including images saved

323

as electronic files) are useful when rescuing properties following a disaster—and they become valuable records should the original properties be lost.

FINDINGS

Preservation measures implemented by governments and experts

Historical records and cultural properties are symbols of people's lives. After the GEJE, the commissioner of the Agency for Cultural Affairs (ACA) delivered a public appeal highlighting the importance of rescuing nondesignated cultural artifacts. Large volumes of historical records are stored in local communities, including old documents; antique works of art; and folk craft articles used in the agricultural, fishing, and forestry industries as well as those used in securing food, clothing, and shelter. The quality and quantity of old documents created during Japan's Edo period, from the 16th to the 19th century, are greater than those of the same period kept in other countries.

After the Great Hanshin-Awaji Earthquake (Kobe earthquake) in 1995, the government recognized the need to preserve cultural properties. In 2004, the Cabinet Office formed the Committee to Protect Cultural Heritage Properties from Disasters. The committee's report highlights that public awareness of the need to transfer historical heritage properties to future generations is essential to preserving these properties. Based on these recommendations, the government decided to protect cultural properties in local communities as well as those designated as significant cultural properties at the national level.

National government schemes after the GEJE

The ACA started two schemes following the GEJE: (1) "cultural property rescue" to preserve historical materials and art objects, and (2) "cultural property doctor" to preserve historical buildings. Donations of some ¥270 million were received from the public for these activities during the year following the disaster.

Cultural property rescue. This scheme was aimed at preserving or rescuing historical materials and art objects such as documents, paintings, sculptures, and crafts. The ACA formulated a rescue committee that consisted of research organizations, museums, libraries, the private sector, and civil society organizations (CSOs) throughout the country. Some 4,900 experts had participated in the scheme as of March 2012. In Miyagi Prefecture, expert teams rescued dozens of properties at 58 locations consisting of museums, schools, private houses, temples, and shrines. For example, at the Ishinomaki Cultural Museum, which was severely damaged by tsunami waves, these experts fumigated, cleaned, dried, or rehabilitated folklore materials, arts, crafts, unearthed human bones, and historical maps. They then transported and stored these artifacts at other museums, universities, and private warehouses in Sendai and Tokyo. The experts rescued statues of the Buddha and scriptures from damaged temples. The scheme also covered zoological and botanical specimens at natural history museums. Experts rehabilitated a stuffed specimen of a whale measuring some 10 meters from a maritime museum in Rikuzentakata City, moving it to the National Museum of Nature and Science in Tokyo (figure 35.1). The Japan Self-Defense Forces, a

Figure 35.1 Rehabilitating a whale specimen

Source: National Museum of Nature and Science.

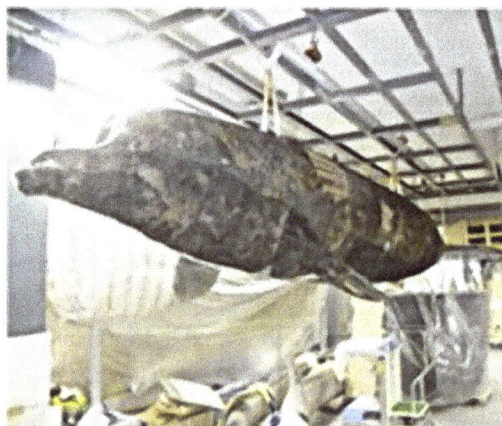

significant presence in response works following the GEJE (chapter 14), helped to transport these heavy materials.

Before the GEJE, the ACA was not well prepared for megadisasters. Rescue activities began only after some 20 days had passed. Damage to historical records from saltwater and mold had already set in. Experts were selected on an ad hoc basis. The organizations to which the experts belonged had to cover travel costs at the initial stage because the ACA did not have any extra travel budget for disasters. The procedures for requesting and dispatching experts were confusing and the recovery processes complicated. To be most effective, preservation projects must begin immediately after a disaster.

Architects and building experts participated in the scheme, "Doctors for Buildings: Diagnosing and Treating Damage to Historic Buildings." They assessed damages and provided technical advice for preserving and rehabilitating historical buildings. In total, 467 experts conducted activities on 3,936 buildings in 198 municipalities before March 2012. In the second year, the experts provided building owners with detailed advice on methods and cost estimates for preserving and rehabilitating works.

The activities of nongovernmental organizations

Experts in historic preservation established Shiryo Networks, a nongovernmental organization, following the Kobe Earthquake, to protect historical records from disasters. These experts included university researchers, graduate students, curators, local government staff, and private experts in the restoration of cultural assets. They worked with community members who had an interest in preserving the historical culture of their own communities. Groups have been established throughout Japan; currently some 20 are working at the prefectural level. The networks not only meet when disaster strikes, but they also take preventive measures. Shiryo Net in Kobe, which accumulated experience in the Kobe Earthquake, now functions as a national secretariat to coordinate networks at the prefectural level.

The case of the Miyagi prefecture

In the Miyagi Prefecture, experts prepared for disaster by establishing networks of government entities, local communities, and property owners. They started identifying historical documents and recording materials by taking photographs during normal times, which helped them preserve and rehabilitate cultural properties following the GEJE.

Predisaster: Activities begun eight years before the GEJE

Experts began to rescue damaged historical records following an earthquake in 2003, saving some 200,000 historical records in five towns affected by the disaster. These included some 100,000 records from the family of Yonosuke Saito, the second-largest landowner in Japan before World War II. The Saito family had donated these records to Tohoku University.

The Miyagi Shiryo Net was established in 2004 to coordinate all organizations concerned. Shiryo Net has been promoting the preservation of historical records kept in communities in partnership with governments, owners, and local residents. In 2007, the organization gained the status of a nonprofit organization in Japan. When the Iwate-Miyagi Nairiku Earthquake struck in June 2008, it began collecting information on the day of the earthquake and began rescuing damaged historical records from the affected areas two weeks later. The Shiryo Net calls for preserving damaged historical records through various media, including television news and newspapers.

Preservation in Miyagi involves two main activities:

- *The first (undertaken in normal times) is to identify historical records stored in communities*. After creating a primary list and

Symbol of reconstruction: Storehouse of Eiichi Homma family, Ishinomaki City, Miyagi Prefecture

The tsunami destroyed Mr. Eiichi Homma's house, a landmark in Ishinomaki City, Miyagi Prefecture. The two-story storehouse, built in 1897, held old documents and other historical records. Although the storehouse was flooded, the historical documents were stored on the second floor and were spared the tsunami waves. About 50 cardboard boxes of materials were temporarily moved to the Tohoku History Museum in Sendai City on April 8, 2011.

Mr. Homma was initially planning on demolishing the storehouse but decided to repair and preserve it based on experts' advice. An expert team led by Mr. Toshiro Sato, an architect in Fukushima, conducted a survey, and found that the building had suffered no major structural damage and could be preserved with minor repairs.

The building serves as a symbol of reconstruction; the town association's emblem hangs on one of its walls. Local nongovernmental organizations raised funds for its repairs. A workshop was held at the site on September 24, 2011, to promote understanding of the significance of preserving storehouses. Donations of more than ¥3 million were received from other areas. The repair work started on March 1, 2012.

Workshop at Homma family storehouse

Source: Daisuke Sato.

getting an overview of the situation through a literature survey, experts formulate detailed information on various objectives in collaboration with government agencies and residents in the target communities. By 2003, the network had completed primary lists in 61 of the 73 local governments in the Miyagi prefecture. Also, researchers surveyed detailed documentation kept by 415 families and organizations.

- *The other activity is to photograph all documents.* When large sets of documents are found, it is highly likely that they will not only include information about the owner's ancestors, but also about the history of the community as a whole. This type of activity not only aims to collect materials based on a particular research area, but also to archive all historical records. In Miyagi, 52 individual surveys were conducted and some 350,000 digital images were captured. In Ogatsu and Kitakami districts in Ishinomaki City, some 30,000 old documents were destroyed by tsunami waves, but their images had been captured before the disaster in more than 70,000 electronic files.

During and following disasters: rescuing historical records

The Miyagi Shiryo Net contributed greatly to the rescue and preservation of cultural property during the earthquake that damaged the building of the secretariat of Shiryo Net at Tohoku University in Sendai City on March 11, 2011. The secretariat moved to another facility on campus on March 15 and resumed its activities.

Gathering damage information. The Miyagi Shiryo Net could not begin restoration until the end of March because of damage to transportation networks and a shortage of gasoline. During this period, it collected damage information from property owners, local residents, and government officials, making use of the networks of key stakeholders that it had established during the eight years prior to the disaster. It had collected information on more than 500 damaged historical records by the end of March. Because the coastal areas were the most severely damaged, the Shiryo Net prioritized activities in those areas based on aerial photos published online in mid-March.

Temporary movement out of the affected areas. The Miyagi Shiryo Net conducted the first survey of damage in Ishinomaki City, Miyagi Prefecture, on April 3. It began its rescue of historical records at the home of Eiichi Homma in Kadonowaki District on April 8 (figure 35.2 and box 35.1). As of February 15, 2013, 86 sets of historical records had been temporarily moved out of the disaster area. Of these, 64 were moved to Sendai City and are still being processed. Of these, 50 sets, or about 80 percent, were from owners living in tsunami-affected areas. In 2013, nearly two years after the disaster, some records are still being moved out of the tsunami-affected areas.

Emergency processing of damaged documents. Documents were damaged by seawater, sand, sludge, and other substances brought in by tsunami waves. Universities and research institutions elsewhere in the country helped to repair and preserve these documents. Volunteers, university students, and local people helped clean off sand and sludge, rinsing items to remove salt, and then drying them out. The Nara National Research Institute for Cultural Properties, which excavates and preserves the ancient capital of Nara, announced in April 2011 that it would help repair and dry out documents. Miyagi Shiryo Net sent old documents from the Kimura family in Onagawa Town in the Miyagi prefecture (box 35.2) and large quantities of other damaged items to the institute. Also, Miyagi Shiryo Net took photos of damaged historical records that were undergoing emergency processing. As of February 15, 2013, about 150,000 digital images had been taken. Only 10 of the 64 collections of materials have been completely processed and returned to their owners. It may take years to finish processing all the records.

LESSONS

- *It is important to prepare for disasters by conducting collaborative activities with local communities during normal times.* The owners of historical records, local residents, government officials, and experts should be involved in creating mechanisms

Figure 35.2 Rescue activities at the home of Eiichi Homma in Ishinomaki City, Miyagi Prefecture
Source: Shuichi Saito.

Kimura family documents washed ashore, Onagawa City, Miyagi Prefecture

The Kimura family in Onagawa City in the Miyagi Prefecture is an old family whose members served as Okimoiri (village heads) during the Edo period. Old documents, designated as the cultural properties of Onagawa City, were stored in three tea chests in the family's storehouse. The Kimura family house was destroyed by the tsunami, and the tea chests were washed away. But about a month after the disaster, someone found one of the chests in the Tsukahama District on the opposite shore of the bay and delivered it to the Onagawa City office. Miyagi Shiryo Net retrieved these historical materials from the city office on May 12. Although two months had passed since the tsunami, the historical materials were still completely drenched. The documents were sent to the Nara National Research Institute for Cultural Properties. In August, after the documents had been successfully dried, volunteers removed salt from the materials. Experts continue to repair them.

One owner of the documents, who published an opinion in a local newspaper in December 2011, said "Having lost my parents in the tsunami, as well as my home and all of my possessions, I was able to see some hope in the survival of these old documents."

Rescued chest of Kimura family documents

Source: Daisuke Sato.

Collecting and preserving disaster materials

Records on and materials related to disasters are useful in understanding disasters, for transfering knowledge to coming generations, and for preparing for future disasters. Following the Kobe earthquake in 1995, the Hyogo Prefecture preserved and collected disaster-related materials. Volunteers, libraries, and local governments were involved in these activities. "Earthquake materials" were not limited to conventional disaster data on earthquakes, such as disaster scale and damage, but also included information on the recovery progress of the affected population, governments, CSOs, and others.

With help from collection experts, the Hyogo Prefecture government collected a wide range of materials—such as books, memos of personal experiences and town meetings, leaflets, and wall posters—during the preservation and rehabilitation process. These cover information on (1) the earthquake itself, (2) the damage it caused, (3) the response to the event, (4) the daily lives of affected people, and (5) the process of developing reconstruction plans and projects. The materials consisted of printed information, images, and voice recordings. Printed information included (1) books, photos, newspapers, newsletters, maps, and so on; (2) private leaflets, fliers, wall posters, internal company memos, newsletters, volunteer information and diaries, and records of personal experiences; (3) research reports, survey reports, and policy proposals; (4) lecture notes and seminar and symposium products; and (5) statistical data. Images included television images, media photos, videos, 8 mm film, and other photos as well as electronic materials on CD-ROM or other media and microfilm.

for preservation. Without systems for preserving historical records, records in private collections are at a high risk of disappearing during disasters.

- *Digital copies should be made of original historical records.* These copies are a crucial contribution to the preservation and rehabilitation process when original records are lost to a disaster.

- *The national government plays a critical role and needs to be prepared for disasters.* Rescue and repair schemes functioned well to preserve cultural properties after the GEJE disaster. It took some 20 days for these rescue activities to begin, however, and damage from seawater and mold had already set in. Ideally, preservation work

should begin immediately after a disaster. Experts were selected on an ad hoc basis, and the ACA did not have an adequate travel budget. Procedures for requesting and dispatching experts were confusing.

- *Museums should produce a database of properties.* Information on properties is crucial for conducting preservation work after disasters. At a museum in Rikuzentakata City, it was quite difficult for experts to address the property and materials they encountered because staff had died in the disaster and all of the information was lost.

- *Governments should embrace the importance of preserving cultural heritage.* Protecting and preserving cultural properties and historical buildings are often considered low priorities in disaster management. The disaster risk management plans of governments rarely cover the preservation of cultural heritage.

RECOMMENDATIONS FOR DEVELOPING COUNTRIES

The national government should prepare for disasters by creating systems to preserve cultural assets. These systems need a permanent secretariat, a roster of experts, budgetary arrangements, and procedures for dispatching and requesting experts. Retrofitting is effective in protecting historical buildings from earthquakes (chapter 2).

Museums should make individual preparations for disasters. Each should develop a database of properties so that preservation work can proceed smoothly after a disaster strikes. Also, each museum should develop a priority list of properties and identify areas for their safekeeping. In Turkey, for example, authorities developed a digital inventory of cultural heritage buildings in Istanbul (with support from the World Bank) to be used in devising countermeasures.

Governments at all levels should include the preservation of cultural properties in disaster risk management plans. Preservation should be recognized as an integral part of rehabilitation.

Community leaders should understand and embrace their community's historical culture and develop the ability to conduct basic preservation efforts for the sake of cultural heritage.

Researchers, government organizations, private sector actors, and communities should be involved in establishing networks to preserve cultural assets and properties during normal times. Cooperation with international organizations is also useful. Following the Indian Ocean tsunami in 2004, the National Committee of the International Council on Monuments and Sites (ICOMOS) played a significant role in the recovery of cultural sites in Sri Lanka and successfully advocated for the importance of doing so by including cultural heritage values in postdisaster recovery plans.

NOTE

Prepared by Daisuke Sato, Tohoku University; Hiroshi Okumura, Kobe University; Kazuko Sasaki, Kobe University; Mikio Ishiwatari, World Bank; and the International Recovery Platform.

BIBLIOGRAPHY

Asahi Shinbun. 2011. "Old Documents and Diaries Restored at the Nara Research Institute, Supports Affected Areas with a Massive Dryer, Great East Japan Earthquake" (in Japanese). April 21.

Disaster Reduction and Human Renovation Institution. 2005. *Report from the Committee to Investigate the Disclosure of Earthquake Materials* (in Japanese). Kobe.

Great Hanshin-Awaji Earthquake Memorial Association. 2001. *Report from the Committee of Classification and Disclosure Standards for Earthquake Materials* (in Japanese). Kobe.

Hirakawa, A. 2005. "Disaster Preparedness Measures Shift from Post-Disaster Preservation to Pre-Disaster Planning" (in Japanese). *Rekishi Hyoron,* 666.

Historical Science Society of Japan, ed. 2012. *The Role of Historical Science in the Face of Great Earthquakes and Nuclear Disaster* (in Japanese). Tokyo: Aoki Shoten.

Hyogo Prefecture. 1995. "Great Hanshin-Awaji Earthquake Reconstruction Plan" (in Japanese). http://web.pref.hyogo.lg.jp/wd33/wd33_000000043.html.

Miyagi Shiryo Net. 2007. *Miyagi Network for Preserving Historical Materials* (in Japanese). Research report from the Project on Measures to Protect Cultural Properties from Disasters conducted on Behalf of the Agency of Cultural Affairs from 2005 to 2006, Sendai.

Movable Cultural Property Rescue Manual Editorial Committee, ed. 2012. *Movable Cultural Property Rescue Manual* (in Japanese). Tokyo: Kubapro.

Okumura, H. 2012. *Major Earthquakes and the Preservation of Historical Records: From the Great Hanshin-Awaji Earthquake to the Great East Japan Earthquake* (in Japanese). Tokyo: Yoshikawa Kobunkan.

Rescue Committee of Cultural Properties Damaged by the Tohoku-oki Earthquake, Committee Secretariat. 2012. *Activity Report on FY 2011* (in Japanese). Tokyo.

Shiryo Net. 1999. "Kobe and Heike in History" (in Japanese). Kobe Shimbun Press Center, Kobe.

Tohoku History Museum. 2012. *Damaged Cultural Property, Toward Recovery—Report of Cultural Property Rescue* (in Japanese). http://www.thm.pref.miyagi.jp/topics/detail.php?data_id=385.

UNESCO (United Nations Educational, Scientific and Cultural Organization). 2010. *Managing Disaster Risks for World Heritage.* Paris: UNESCO.

CHAPTER 36

The Recovery Process in Fukushima

The recovery process following the nuclear accident at the Fukushima Daiichi Nuclear Power Station on March 11, 2011, presented challenges different from those faced in the recovery of areas damaged by the tsunami waves and tremors of the Great East Japan Earthquake. The nuclear accident that followed the waves and tremors left communities concerned with the serious effects of radiation exposure, relocation, the dissolution of families, the disruption of livelihoods and lifestyles, and the contamination of vast areas.

Following the nuclear accident, people in Fukushima were removed to municipalities and prefectures outside their home communities. There they faced difficulties finding housing, jobs, and schooling in unfamiliar places. Many families separated in the process of seeking employment and uncontaminated places to live. To date, those affected do not have a clear vision of when they can return to their original communities. Even in areas where living restrictions have been lifted, there are few job opportunities, educational opportunities, and medical and other social services. In addition, the fear of radiation has not yet dissipated. Many displaced people continue to reside in transition shelters, perpetuating the possibility of conflict between the host community and temporary residents. Many of those who lived outside of evacuation zones left the Fukushima area voluntarily. These include mothers and children, for whom support programs are generally inadequate.

The challenges of the recovery process in Fukushima are similar to those encountered following disasters in other parts of the world. After some natural disasters, such as volcano eruptions, people cannot return to their original communities because of prolonged events or drastic changes in geographical features. Also, in complex emergencies—such as those spurred by armed conflicts in developing countries—refugees and internally displaced persons are often forced to stay in unfamiliar environments outside their hometowns or countries for a long period without future prospects or hope.

To address the effects of such disruption in Japan, the government, private sector, universities, and civil society have worked with affected communities in Fukushima to support their daily needs and future interests. The purpose of this note is to outline what these organizations did and how they did it, and to recommend what other responders can do in the face of similar events in the future. In the case of Fukushima during the Great East Japan Earthquake (GEJE), it was found that restoring livelihoods, caring for children, rehabilitating communities, and communicating risks were crucial to the recovery process.

FINDINGS

How did the accident affect the people of Fukushima?

Although more than two years have passed, the people of Fukushima are still struggling with the effects of the nuclear accident that prompted a physical and mental health crisis for area residents. Many suffered the stress of sudden displacement and prolonged evacuation, fears over the possible health effects of continued exposure to low-level radiation, an exodus from the area, dissolving communities, and conflicts with host communities.

Prolonged evacuation. An enormous number of people are still in transitional shelters and other locations that are not their homes. As of September 2012, 1,121 people had died in Fukushima from physical and mental exhaustion caused by the accident. This number includes 35 who died in the six months between March and September 2012, more than a year after the accident. Meanwhile, the risk of death among the elderly increased during evacuation. For example, the mortality rate among residents evacuated from nursing homes following the accident was 2.7 times higher than before it.

Some 160,000 people, approximately 8 percent of the total population of the prefecture, left their hometowns for transition shelters in the wake of the GEJE (as of December 2012). Of this group, 111,000 are from restricted areas, mainly Futaba County, where the nuclear station is located, while 49,000 people evacuated voluntarily. Approximately 60,000 are residing outside the Fukushima Prefecture, most of them elsewhere in Japan but some overseas. More than 60 percent of the people feel isolated from other people in the country.

The Tokyo Electric Power Company (TEPCO) has been paying compensation to those affected by the accident. TEPCO began with a lump-sum payment of ¥1 million per family moved from the evacuation zones in April 2011, before the final compensation payment. As of June 2013, TEPCO had paid some ¥2.5 trillion in compensation to affected people and companies. The company is also paying a ¥100,000 payment every month to each evacuee as compensation for nonphysical damages (pain and suffering, stress and strain), payable until the evacuee is able to return to his or her hometown. But glitches in the payment process have been observed. Various organizations have requested that compensation payments be accelerated. The Japanese government established the Nuclear Damage Liability Facilitation Fund in September 2011 to support TEPCO in paying compensation to affected people, and it has formulated compensation guidelines.

Fear of radiation. More than 80 percent of the population in Fukushima City fear radiation; according to a city government survey, that fear was growing even a year after the accident. People, in particular families with children, are trying to avoid radiation risks by checking the contamination level of foods, hanging laundry inside houses to dry, buying bottled water, and avoiding highly contaminated areas.

The fear of radiation has disrupted families in the restricted areas because that fear is unequally distributed in terms of generation, gender, and living conditions. According to a questionnaire conducted by researchers at Fukushima University in September 2011, 98 percent of evacuees from Futaba County experienced separation from family members. Most typically, fathers continue to stay in Fukushima for livelihoods while mothers and children live outside Fukushima to avoid radiation risks.

A dwindling and aging population. The accident and subsequent evacuation have exacerbated preexisting problems, such as an aging and declining population. Between the accident and January 2013, the population of the prefecture decreased by 60,000 to 1.96 million. In 2012, 13,348 residents left Fukushima, of which 3,009 were children younger than 14 and 4,030 were aged 25–44. In addition, because almost half of the younger generation shows no intention of returning to their hometowns, it is expected that the population will continue to dwindle.

Even in areas where aerial radiation levels are relatively low, some displaced persons do not intend to return home because of the lack of public services. Although the mayor of Hirono Town lifted evacuation orders in March 2012, and the town government rehabilitated most infrastructure, only one-fourth of the town's residents had returned as of June 2013. The government has supported radiation decontamination and debris removal and has rebuilt infrastructure; nevertheless, services vital for daily life—such as those supplied by local shops and medical facilities—have not fully resumed. Meanwhile, people are worried about another accident at the nuclear stations.

Fracturing of communities. Evacuees are dispersed both within and outside of Fukushima, and it is critical for the municipality government to address their various needs in a timely way. They are making every effort to maintain community bonds. To help people affected by the GEJE stay in contact and maintain community bonds, the local government in the town of Tomioka has compiled a telephone directory of evacuees from which residents can obtain contact information for their hometown neighbors and friends. The government also established an information system utilizing electronic tablets for social workers. The accumulated welfare and medical information promises to help save the lives of residents.

Conflicts between communities. In March 2013, Fukushima's governor spoke of conflicts between evacuees and host communities. It was reported that evacuees in transition shelters were being harassed by graffiti painted on vehicles or public buildings saying "evacuee, go home." Conflicts between evacuees and their host communities have accompanied government support programs and compensation payments from TEPCO. People in host communities, such as Iwaki City, who suffered from the tsunami waves and earthquake tremors but were not compensated resent the evacuees of the nuclear accident from Futaba County who are being compensated.

Since the evacuees use hospitals, roads, and other public facilities, people in host communities do not receive public services in the same way they did before the accident. Housing shortages have arisen in host cities as affected persons receiving government financial support rent private houses. The evacuees pay local taxes to their original municipalities, not to the host municipalities that are providing them public services.

Governments and other organizations provide limited support to promote the peaceful coexistence of evacuees and their host communities. Iwaki City, which suffered from earthquakes and tsunamis and hosts some 24,000 evacuees from other affected municipalities, has asked the national government to provide it with additional financial support to strengthen the public services it provides to evacuees.

Recovery progress. The progress of recovery in Fukushima is lagging behind other

Figure 36.1 content:

Key policy and planning measures	Great East Japan Earthquake 11 March 2011	Specific policy and planning measures for Fukushima Reconstruction
Reconstruction Design Council	1 month	
Seven Principles for Reconstruction Framework	2 months	
Basic Act for Reconstruction	3 months	
Reconstruction Headquarters Recommendation by Design Council	4 months	
Basic Guidelines for Reconstruction		
	5 months	Special Act on Evacuees from Nuclear Accident
Prefecture and Municipality Recovery Plans	6 months	Consultation Committee on Fukushima Reconstruction
	7 months	
	8 months	
	9 months	
Law for Special Zone for Reconstruction	10 months	Fukushima Prefecture Recovery Plan
Reconstruction Agency	11 months	
Reconstruction Grant Projects	1 year	**Law for Special Measures for Rebirth of Fukushima**
	13 months	
	14 months	
	15 months	Law for Support to People Affected by Nuclear Accident
	16 months	
	17 months	Basic guidelines on Fukushima reconstruction
		National Grand Design
	18 months	Consultation Committee on Relocating Communities for Long-term Evacuees
	2 years	Recovery plans in 4 municipalities
	10 years	

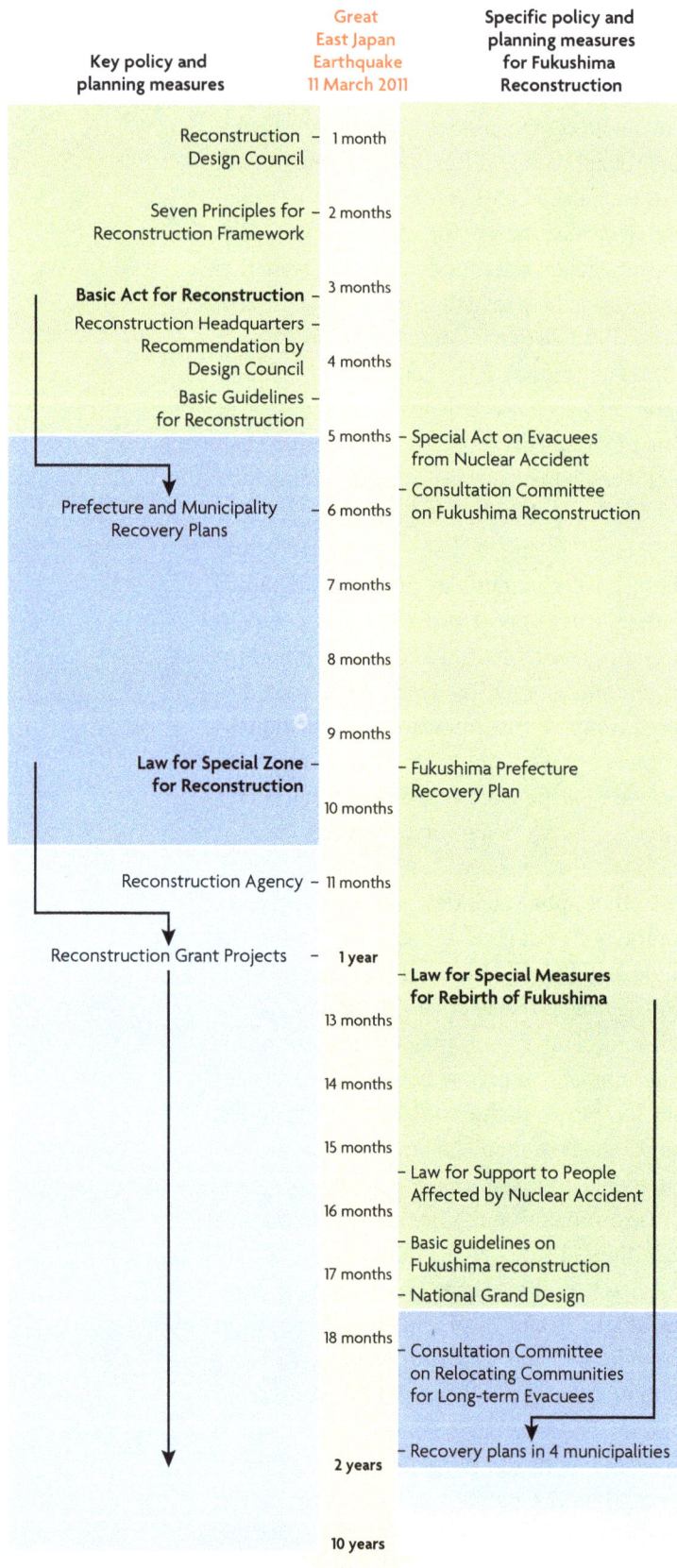

Figure 36.1 Chronology of key policy and planning measures for recovery from the GEJE, in general and in Fukushima

disaster-affected areas (figure 36.1). Municipalities damaged by tsunamis produced recovery plans within a year of the GEJE (chapter 21), while municipalities in Fukushima needed two years because it remained unclear when people could return to their hometowns and when the towns would be free of radiation. Some plans in Fukushima still do not indicate when people can return to their communities.

Legislation has been enacted to support affected people, and the government has formulated reconstruction guidelines. Various consultation processes among mayors, ministers, governors, and other key stakeholders to promote recovery are ongoing. Yet compensation and subsidies cannot restore original lifestyles, which were deeply connected with the culture, natural environment, and human relations of communities.

Support for recovery from government and other sources

The national and prefectural governments created a framework to support recovery. It includes laws, guidelines, plans, consultation processes, and budget allocations. Universities and civil society organizations have provided additional support for the affected population.

Government support

Legislation and planning. A law for "special measures for the rebirth of Fukushima" was enacted in March 2012, one year after the accident. The cabinet adopted basic guidelines for Fukushima reconstruction in July 2012. These guidelines aimed to promote reconstruction and revitalization following the nuclear accident in a holistic way. The Fukushima prefectural government formulated a recovery plan in December 2011. Its basic concepts were to (1) build a safe, secure, and sustainable society free of nuclear power; (2) revitalize Fukushima by bringing together everyone who loves and cares about it; and (3) rehabilitate towns so they can be a source of pride again. The four municipalities of Namie, Okuma, Futaba, and Tomioka—where people cannot return—formulated recovery plans from September 2012 to March 2013. Because it is not known when people will be able to return to their home communities, the plans cover relocation to other municipalities, but they do not include detailed rehabilitation of the original communities.

Budget. The Fukushima government allocated a budget of ¥1.5764 trillion in the financial year 2013 to fund activities detailed in the recovery plan. The government allocated a budget for flagship programs as shown in table 36.1. Further, the government allocated a supplementary budget of ¥256.9 billion.

Consultation. Various consultation processes among mayors, ministers, governors, and other key stakeholders are ongoing. In August 2011, the national government created the Consultation Committee on Fukushima Reconstruction to examine recovery activities. Chaired by the reconstruction minister, the committee consists of concerned ministers and mayors of affected municipalities. The mayors of affected municipalities, Fukushima's governor, and ministers have met periodically since March 2012 to exchange views.

The national government revised its evacuation zoning regulations in April 2012, announcing that the residents of four municipalities could not return home for a few years (map 36.1). With this step, relocating people to other municipalities became a crucial issue. In September 2012, the national and Fukushima Prefecture governments jointly created the Consultation Committee on Relocating Communities for Long-Term Evacuees in September 2012. The committee examines (1) the period, scale, and other issues of relocation; and (2) public housing for those affected. The committee consists of the mayors of the receiving municipalities, the mayors of the four affected municipalities, and the reconstruction minister.

Table 36.1 Budget of the Fukushima Prefecture for flagship programs in FY 2013

SAFE DAILY LIVING		397.1 (¥ BILLION)
	Rehabilitating the environment	281.2
	Supporting the rehabilitation of daily lives	72.2
	Protecting health	22.1
	Raising children	21.6
Job opportunities		*155.4*
	Rehabilitating agriculture	32.6
	Rehabilitating small and medium enterprises	114.4
	Promoting renewable energy	4
	Developing the medical industry	4.4
Urbanism and networking		*89.6*
	Networking	1.1
	Rehabilitating tourism	0.7
	Rehabilitating cities	58
	Rehabilitating transport	29.8
Total		642.1

Source: Fukushima Prefecture.

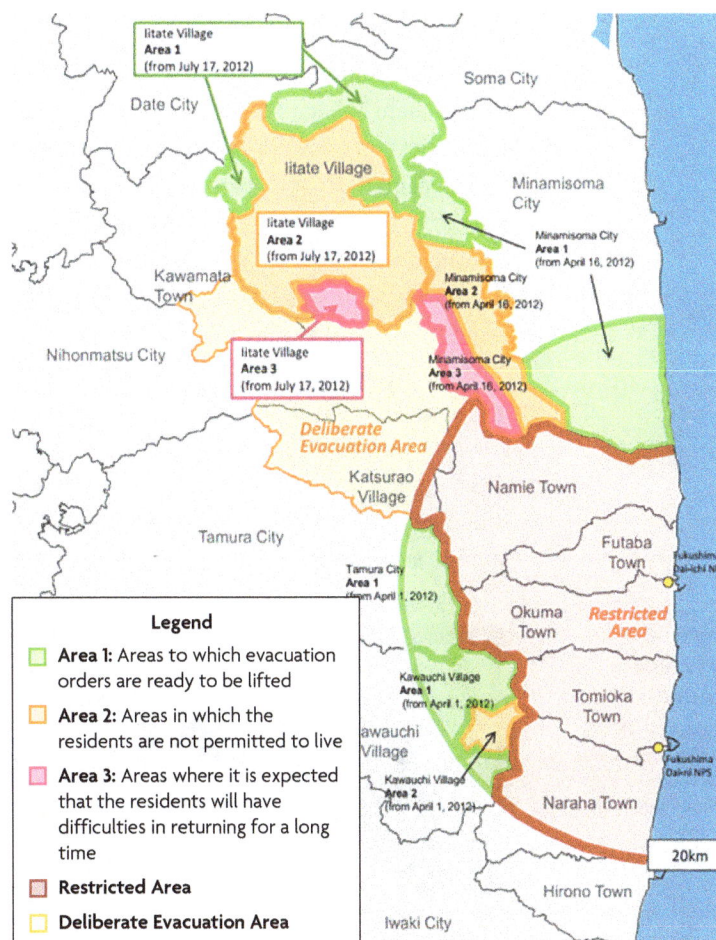

Legend

- **Area 1:** Areas to which evacuation orders are ready to be lifted
- **Area 2:** Areas in which the residents are not permitted to live
- **Area 3:** Areas where it is expected that the residents will have difficulties in returning for a long time
- **Restricted Area**
- **Deliberate Evacuation Area**

Map 36.1 Rearrangement of evacuation zoning

Source: Ministry of Economy, Trade and Industry.

Support from Fukushima University

Fukushima University, the only national university in the prefecture, established the Fukushima Future Center for Regional Revitalization (FURE) in April 2011. The purpose of the center is to conduct research on scientific changes and damage caused by the GEJE and the ensuing nuclear accident. In addition, FURE assists in the rehabilitation and revitalization of Fukushima by supporting the formulation of action plans and implementing recovery in consultation with affected communities. The center consists of four support groups focused on (1) children and youth, (2) community support, (3) industrial restoration, and (4) environment and energy. It signed agreements with eight municipalities in January 2012 to support the formulation of recovery plans, to provide advice on decontamination, and to conduct surveys of the affected population. The center created two satellite offices in Kawauchi Village and Minamisoma City to support the affected population in those areas.

The center also supported the revision of a recovery master plan for the Fukushima Prefecture. One year after the disaster, the prefectural government began major work on a "Fukushima Master Plan" that sets goals from 2013 through 2021. In the review process, it was crucial to hear the voices of residents; however, most residents were hesitant to voice their opinions or did not know what they should ask for. The university carried out an opinion survey on the master plan to ascertain the public's wishes. They gathered opinions of more than 1,200 residents and compiled their recommendations for Fukushima's governor.

The university has supported the Odagaisama Center (loosely translated as the Tomioka center for mutual support in rebuilding town life) at a temporary housing complex in Koriyama City. This center provides support to the inhabitants and is a place for evacuees to communicate and interact (figure 36.2). It also operates a traditional handicraft workshop, which provides training opportunities for residents who have lost their jobs due to the accident. The center's activities are based on self-governance by residents exposed to risks caused by the separation from their original communities and families. As of August 2012, 30 percent of the residents of Tomioka Town had left Fukushima, 18 percent lived in transition housing units, and 52 percent lived in rented apartments in Fukushima.

Critical areas of support

This section examines four issues identified by experts at workshops. They are consistent with known features of disaster recovery efforts in developing countries: (1) rehabilitation of the community, (2) communication of risks, (3) caring for children, and (4) restoration of livelihoods.

Rehabilitation of communities in temporary towns or migrant communities. Because decontamination of radioactive areas takes a long time, at the end of 2011, the municipal governments of Namie, Okuma, Futaba, and Tomioka began planning "temporary towns," or migrant communities, for those ousted from their original municipalities. Municipal governments and public facilities as well as residents were relocated to these temporary towns.

Local governments have encountered difficulties in planning these towns. The host municipalities cannot prepare for the towns without knowing the number of evacuees. Yet municipalities cannot calculate the numbers

**Figure 36.2
Odagaisama
Center**

Source: © Fukushima University. Used with permission. Further permission required for reuse.

because people will not declare their willingness to move in the absence of a detailed plan. Among the affected population of Namie Town, 19.5 percent have said they want to live in the temporary towns, while almost half say, "I don't know," because of the limited information available to them.

Among the several issues to resolve are the following: (1) conflicts over resources, job opportunities, and other issues with host communities; (2) difficulties forming networks with other members of the home community; (3) difficulty selecting sites for temporary towns through participatory planning; (4) difficulty defining the respective roles of the original local government, the local host government, the prefectural government, and the national government; and (5) the possibility that temporary towns will become permanent.

Affected people and municipalities are examining two methods for developing temporary towns: the concentrated community and the distributed community.

Concentrated community. A concentrated community is one in which affected people will live in a specified area separate from the host community. Most housing, municipal facilities, and other public functions will be newly constructed. People in Namie Town call that planned town "Little Namie," referring to it in the same way that Japanese communities in foreign countries are often referred to by names such as "Little Tokyo." These facilities are intended to be used only for a few years. After evacuees return to their home community, operation and maintenance problems may emerge. Such concentrated communities also have a huge impact on town planning in the host communities because it is difficult to find enough space for them.

Distributed community. A distributed community is one in which affected people blend into the host communities. They use the existing facilities in the host municipalities, and new schools and facilities are unlikely to be built. Having evacuees live alongside residents

of the host cities may minimize conflict, but it makes it more difficult for them to maintain bonds with home.

In December 2012, the Iitate Village office agreed with the Fukushima City government to develop a "temporary town" in Fukushima City. Some 3,800 people (more than half the population) and village offices have already moved from Iitate Village to Fukushima City. Public housing, a junior high school, a kindergarten, agricultural facilities, and other necessary infrastructure have been constructed. Because this is the first case of an agreement about a temporary town, the details of programs (such as cost sharing and responsibilities between the two municipal governments) are to be determined later.

Rehabilitating credibility: Community-based monitoring and communication of risks. To produce a database that is accountable, various stakeholders (such as landowners, consumers, farmers, governments, and community-based organizations) are involved in monitoring radiation. Accurate and scientific data on radiation are essential for rehabilitating agriculture by reducing radiation risks, conducting decontamination works, and resuming agricultural activities. The government organizations and academia lost credibility due to inappropriate communication with the public during the accident (chapter 27). Uncertain information that flooded the Internet also accelerated the decreasing confidence in these organizations.

FURE is conducting community-based monitoring with farmers and agricultural cooperatives to accumulate reliable data and to produce detailed maps (100 meter mesh) useful for rehabilitating agricultural activities and daily lives (figures 36.3 and 36.4). Through joint monitoring by experts and farmers on a community scale, stakeholders share important information. By taking part in the measurement process, farmers and residents gain confidence in the process. Contamination

**Figure 36.3
Farmers
monitoring
radiation**

Source: © Fukushima
University. Used with
permission. Further
permission required
for reuse.

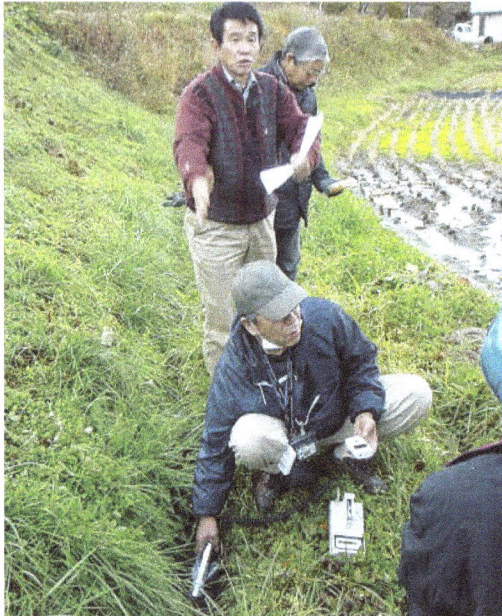

**Figure 36.4
Radiation map**

Source: © Fukushima
University. Used with
permission. Further
permission required
for reuse.

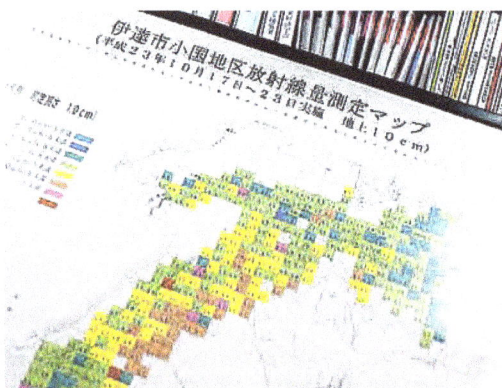

Chernobyl accident. University researchers teach farmers that radioactivity is transmitted to plant bodies. To revitalize agriculture and recover food safety, the university is working with various stakeholders in four stages:

Stage 1: Investigating the actual situation of radioactive contamination

- Mapping radiation of farmlands and residential lands

Stage 2: Countermeasures at the production stage

- Testing cultivation in paddies and upland fields

- Clarifying the mechanisms of cesium absorption and transfer and evaluating the effects of potassium and zeolite

- Providing guidance on farming appropriate to the contamination level of fields

Stage 3: Countermeasures at the marketing stage

- Expanding and improving radiation measurement systems and facilities

- Raising the capacity of measurement technicians

- Advising on the development of measurement systems

Stage 4: Countermeasures at the consumption stage

- Providing opportunities for communication between producers and consumers, such as *Fukko* ("revitalization") Marche (figure 36.5)

- Surveying consumers

Rehabilitation for the future: Care for the vulnerable, especially children. Communities cannot be sustained without children and young people. Some 90 percent of people in Fukushima City worry about their children's future. Because children are especially vulnerable to radiation, concerned governments and

maps (which government agencies have produced by monitoring radiation by airplane, vehicle, and monitoring posts) are not helpful in rehabilitating farmers' daily lives and agricultural activities. This is because their large scale can provide an overview of contamination but not information on the distribution of contamination on a community scale.

Fukushima University is helping farmers promote safe agricultural products and processed foods. The university conducts tests of paddy rice to measure the absorption and transfer mechanisms of cesium 134 and 137 from soil and water. Researchers have also applied their experience and knowledge of agriculture revitalization following the

organizations are focusing on efforts to care for them, such as giving them emotional support, counseling, and education (chapter 19).

Students have had trouble adjusting to their new schools. The temporary schools lack facilities and accessibility, and students are separated from their friends from home. As a result, children at transition shelters are not receiving the same quality of education that they received before the disasters. Approximately 30,000 children under 18 years of age lived at transition shelters as of October 2012. Of this number, some 17,000 had moved outside Fukushima.

Based on a survey of thyroid health, the Fukushima Prefecture and the national ministry of the environment found no significant differences between children living within or outside Fukushima. After the Chernobyl accident, the greater scale of iodine 131 contamination over cesium 134 and 137 contamination caused many cases of thyroid cancer among young people. To avoid overdoses, countermeasures such as evacuation, examination of food, and restrictions on food distribution immediately after nuclear accidents are a priority.

Parents and children who remain in Fukushima face difficulties in their daily lives, stemming from stresses caused by the accident. A Fukushima Prefecture survey found that 13 percent of children in evacuation areas suffered from mental health problems as compared with 9.5 percent in other areas in Japan. The prefecture provided mental care to a high-risk group composed of 7 percent of the children.

A Fukushima University survey found that parents in areas where radiation levels are higher suffer from more stress than parents in other areas. Also, parents suffer more stress as children grow. Children are showing signs of stress, such as fear, anxiety, and regression. They are restricted from playing outside, and their physical inactivity causes obesity. In 2012, the obesity ratios of the children in Fukushima Prefecture were the highest in Japan.

As most students attend schools near their new residences and outside Fukushima, it is becoming increasingly difficult for families to return to their original towns. Access to schools, relationships with new friends, and fear of radiation are bottlenecks preventing them from returning to their hometowns. The Hirono primary school in Hirono Town resumed classes in August 2012, but only 20 percent of students came back to the school.

The prefecture has introduced a variety of measures that target children. The government provides free medical care for children under 18 years of age and will be conducting lifelong medical examinations of the thyroid glands of children who were under 18 at the time of the accident.

Fukushima University has initiated a "Children Campus" program at the university. Children who lived in different transition shelters can gather at the university, play with university students, and attend classes. The program also provides recreational space that children cannot find at the transition shelters (figure 36.6).

Figure 36.5 *Fukko* **("revitalization") Marche**

Source: © Fukushima University. Used with permission. Further permission required for reuse.

Figure 36.6 Supporting children affected by the disaster

Source: © Fukushima University. Used with permission. Further permission required for reuse.

Rehabilitating jobs: Livelihood restoration. Before the accident, the main job opportunities in the affected areas were in the electric power industry, including the nuclear stations. Some people who worked in these areas kept their jobs because the industries remained, or because they found opportunities at other stations. In 2012, the ratio of job offers, including temporary and permanent ones, to job seekers in the Fukushima Prefecture was 1.18, the highest of all the prefectures in Japan. Significant increases were seen in construction jobs (including decontamination) and jobs nursing the elderly.

There is, however, a mismatch between demand for labor and the preferences of the labor force. People prefer permanent jobs in a service industry, whereas most available jobs are for temporary manual labor. The ratio of permanent job offers to job seekers is just 0.72. The number of job offers in the construction sector was 3,616, but only 1,037 were accepted. In the manufacturing industry, by contrast, 6,249 people were looking for jobs, but there were only 3,776 offers. Job seekers who receive compensation from TEPCO tend to be more selective about salary and job conditions.

In highly contaminated areas, the situation is more severe than in other areas in the prefecture. While the ratio of job offers to job seekers is very high, at 2.42 in Futaba County, job seekers have not returned because of fear of radiation or the inconveniences of daily life. Some local businesses, such as retail stores, cannot resume their services because of a labor shortage and a dearth of customers.

In the case of Kawauchi Village, from which all residents were evacuated, the village government has been encouraging people to return home following the lifting of living restrictions in January 2012. The lack of job opportunities is one of the main challenges to returning. Some 250 people lost their jobs following the accident. The village office induced three companies to set up operations in Kawauchi Village, generating half the number of jobs lost. But for the reasons mentioned above, job offers are not being filled.

The government provides private companies with high subsidies that cover 75 percent of the construction costs of factories. As of December 2012, it had decided to provide 291 private companies with subsidies, which are expected to create more than 4,000 jobs. In 2012, 102 new factories were built, 2.4 times the number in 2010. These factories provided more than 2,200 jobs.

Students of Fukushima University have organized *Fukko* Marche in Fukushima City and Tokyo and are working with farmers associations, women's groups, nongovernmental organizations (NGOs), and local agencies to promote agro-related industries (figure 36.6). While Fukushima is famous for its agricultural produce (such as fruits, vegetables, mushrooms, rice, and rice wines), their sales were affected by rumors after the nuclear accident. The *Fukko* Marche offers demonstrations of radiation measurements of agricultural products.

Women who had been engaged in agribusinesses in the Abukuma region established a women's organization called the *Ka-tyan no Chikara* ("power of moms") project in October 2011, seven months after the earthquake, in Fukushima City, where they were temporarily staying. A major part of the Abukuma region in the central and eastern part of the Fukushima Prefecture is now a restricted area for agricultural production because of the nuclear accident. The objectives of *Ka-tyan no Chikara* are to contribute to the recovery of the nuclear-affected Abukuma region, build a network among women from the region, create employment, demonstrate the safety of products from Fukushima, and build and sustain communities that include evacuees and residents. The members of the network produce various kinds of processed agricultural foods, such as rice cakes, pickles, sweets, and lunch boxes. With support from Fukushima University and other agencies, all of their products are examined for radioactivity and are sold with a certification that guarantees safe levels of radioactivity.

LESSONS

- *Nuclear disaster can divide a society.* The affected population of Fukushima has been divided by differences in radiation exposure, risk perception, age, and income. Following adjustments to evacuation zoning, some affected people have begun to return to their hometowns. More than 20,000 people in four municipalities, however, will not be able to return to their communities for at least five years because of high levels of radiation. People who live outside the evacuation zones and who have voluntarily evacuated out of fear of radiation receive less government support than evacuees from the evacuation zones. Some groups, in particular families with children, are seriously concerned about radiation and have moved outside the prefecture, while others stay on. In general, younger people tend to move away and start new lives, while older people seek to return to their home communities. People with higher incomes are more likely than poorer people to voluntarily relocate.

- *Prolonged evacuation causes conflict between communities.* Conflicts have emerged between evacuees and host communities. Municipalities in the prefecture that suffered from the earthquakes and tsunamis are hosting evacuees from areas affected by the nuclear accident. Because the evacuees occupy housing and use public services (such as health, education, and transport facilities) in the host communities, natives encounter shortages, leading to resentment, which is exacerbated by the fact that

evacuees from the nuclear accident are being compensated by TEPCO.

- *Developing "temporary towns" is an enormous challenge.* Developing temporary sites for evacuees in other municipalities is more complicated than the normal practice of building resettlement shelters in the disaster-affected area. It is necessary to clarify responsibilities and cost-sharing arrangements among the affected and host municipalities and with the national and prefectural governments. The question of how to use the facilities and buildings of the temporary towns after evacuees return to their hometowns will have to be studied and resolved.

- *People face an uncertain future.* Those affected by the GEJE have mixed feelings. They wish to return home, but cannot, nor can they lead their daily lives as they did before the accident. In addition to radiation, various factors, such as a lack of employment opportunities, make people hesitant to go home. Ways must be found to narrow the huge gap between job offers and job hunters. Government financial incentives for private companies and entrepreneurs can create good jobs. In addition to decontamination, adequate social services (such as education, health, and transportation) will be required to induce evacuees to return home.

- *Radiation monitoring requires participation* from various stakeholders (such as communities, governments, and academia) to produce a database that is accountable. It will be necessary to measure radiation levels on individual farming plots and to set up a reliable monitoring system. Merely providing risk information on radiation is not enough to prevent rumors or to overcome their influence.

- *Providing support at various locations is another challenge.* People affected by the nuclear accident have evacuated to other areas in and around Fukushima, and communities and families have been separated, which complicates efforts to reach them for the purposes of providing support and assistance and reaching consensus on recovery plans.

RECOMMENDATIONS FOR DEVELOPING COUNTRIES

Provide national government support to host municipalities. Conflicts between refugees and internally displaced persons, on the one hand, and their host communities, on the other, can be avoided by lessening the burden on the latter. The host communities may well face difficulties in sharing public services and resources with displaced persons. In prolonged situations of displacement, conflict can become severe. The presence in Pakistan of 3 million Afghan refugees over the past 30 years has had profound social, economic, and environmental impacts for the host country. National governments should support activities to promote coexistence between the displaced persons and host communities. The Refugee Affected and Hosting Areas programme of the United Nations Development Programme aims to ensure refugees' peaceful coexistence with local communities. The program helps host communities train human resources, distribute food and water, and build facilities such as farm roads, water supply and sanitation infrastructure, and medical stations.

Make care for children a priority because children are particularly vulnerable to disasters. Mental health and education programs are needed for affected children. Before the accident in Fukushima, children in rural areas enjoyed living in spacious houses and playing outside. Children are very sensitive to the anxiety and uneasiness of their parents. They sometimes regard their parents' anxiety and uneasiness as their own problem and tend to be hesitant about expressing their feelings. Programs for parents, such as counseling, should accompany activities organized for children.

Provide community-based monitoring. Communities can gain valuable information by monitoring disaster situations that can be put to use in managing disasters—for example, in understanding risks and the importance of evacuation plans. In Sri Lanka, community members are monitoring rainfall using simple equipment and warning other members when rainfall reaches the point where it could cause floods or landslides (chapter 10).

Collaborate with a wide range of stakeholders. Academic institutions can offer help with risk communication, job creation, and technical knowledge. Businesses can also play a crucial role in job creation. Private staffing agencies can reduce the government's burden by hiring affected people (chapter 24). To create livelihoods for refugees, it is vital to engage profit-oriented and commercial institutions and companies. In South Africa, the private temporary recruitment agency matched skilled refugees to labor markets that had a shortage of local talent.

Create jobs. Many evacuees from the GEJE are hesitant to return home because job opportunities are scarce. Government support and financial schemes that provide funds to the private sector for starting businesses in these areas are required. In addition to creating job opportunities for heads of families, spouses also need jobs.

Have municipalities prepare evacuation plans, especially those in which crucial facilities, such as nuclear stations, are located. These municipalities should prepare for serious accidents, knowing that residents may respond to such accidents by migrating. They should raise public awareness, prepare contingency plans, and establish partnerships with the national government (chapter 11). Also, municipalities should discuss arrangements for transition shelters with neighboring municipalities in the event of accidents.

NOTE

Prepared by Mikio Ishiwatari, World Bank; and Satoru Mimura, Hideki Ishii, Kenji Ohse, and Akira Takagi, Fukushima University.

BIBLIOGRAPHY

Fukushima City. 2012. "Survey on People's View on Radiation" (in Japanese). http://www.city.fukushima.fukushima.jp/soshiki/7/kouchou12090501.html.

Fukushima Governor. 2013. "Comments on Recovery Plan in Areas Lifting up of Restricted Area" [in Japanese]. http://www.cms.pref.fukushima.jp/download/1/tokusohou-keikakuikenn.pdf.

Fukushima Prefecture. 2013. "Panel on Survey on People's Health" (in Japanese). http://www.cms.pref.fukushima.jp.

Hokkaido Shimbun. 2013. "Two Years from 3.11: Difficult Recovery" (in Japanese). February 23. http://www.hokkaido-np.co.jp/cont/tooifukkou/188947.html.

Kahokushinpo. 2013. "Two Years from GEJE" (in Japanese). March 5.

McCurry, J. 2012. "Japan's Tohoku Earthquake: 1 Year on." *Lancet* 379: 880–81.

——. 2013. "Fukushima Residents Still Struggling 2 Years after Disaster." *Lancet* 381: 791–92.

Nomura, S., S. Gilmour, M. Tsubokur, D. Yoneoka, A. Sugimoto, et al. 2013. "Mortality Risk amongst Nursing Home Residents Evacuated after the Fukushima Nuclear Accident: A Retrospective Cohort Study." *PLoS One* 8 (3): e60192.

Tokyoshinbun. 2013. "All People Died because of the Accident: Prolonged Evacuation in Fukushima" (in Japanese). March 13.

Tsutui, Y., M. Tominaga, M. Takahara, and R. Takatani. 2012. "Stress on Parents and Children" (in Japanese). http://www.fukushima-u.ac.jp/press/H24/pdf/34_03.pdf.

UNDP (United Nations Development Programme) Pakistan. 2013. "Project Brief: Refugee Affected and Hosting Areas Programme." http://www.undp.org/content/dam/pakistan/docs/CPRU/Project%20Briefs/UNDP-PK-CPRU-RahaPB-2013.pdf.

UNHCR (United Nations High Commissioner for Refugees). 2012. "Livelihood Programming in UNHCR: Operational Guidelines." http://www.unhcr.org/4fbdf17c9.html.

CHAPTER 37

Spreading the Word: Raising Capacity for Disaster Risk Management in Developing Countries

The ultimate objective of the Learning from Megadisasters project is to share Japan's knowledge of disaster risk management (DRM) and postdisaster reconstruction with other countries vulnerable to disasters and to help mainstream DRM policies in those countries. The first phase of the project produced the chapters that make up the bulk of this volume and formed a community of practice (CoP) capable of helping developing countries draw real benefits from the lessons encapsulated in the chapters. The key feature of the second phase of the project is the development of on-demand capacity-building programs for high-profile countries for which the lessons from the project have particular relevance. Successful implementation in those countries will pave the way for adoption of similar approaches in other countries.

The capacity-development program is focused on seven pilot countries: Armenia, Indonesia, Kenya, the Kyrgyz Republic, Maldives, Sri Lanka, and Uganda. An additional capacity-building exercise was held in Sierra Leone in October 2013. The goal of these programs is to widen the dissemination and application of the knowledge generated in Phase 1 and to identify steps and measures to enhance disaster preparation and responsiveness in pilot countries based on the Japanese experience.

At the outset of Phase 2, the pilot countries identified the most important lessons learned from Japan's experience. Country-specific capacity-building programs were designed around those needs, comprised of workshops, learning sessions organized by the World Bank Institute's (WBI) Global Development Learning Network (GDLN), face-to-face meetings, a study tour in Tohoku, and follow-up efforts to identify plans for next steps, including further actions and activities for each country, such as technical assistance, studies, and—especially—the addition of specific topics and measures to existing or planned operations. The latter, of course, is the definition of mainstreaming.

The program in each country was tailor-made to the needs and interests each country had identified. Part of the tailoring process involved the identification of specific groups of stakeholders whose roles in the disaster risk management (DRM) process made them good candidates for capacity building. Specific activities were designed with the clients and in consultation with knowledgeable staff within the Bank and in partner institutions, including those engaged in ongoing or planned development operations, thereby amplifying the effect of the capacity-building program. Some examples of program activities follow:

- In Kenya, together with United Nations Development Programme (UNDP), two workshops have targeted national government officials and provincial and local officials in an effort to advance the policy goal of decentralizing DRM in the country.

- In Uganda, preparatory meetings have been held with the Office of the Prime Minister and other governmental departments. Because the country recently adopted comprehensive DRM regulations but lacks the institutional capacity to implement the program, capacity-development activities focus on strengthening pertinent institutions through measures adapted to the Uganda contest. Two main training activities have been organized. The first, aimed at national government officials and several provincial representatives, focuses on specific tools and measures (including structural ones). The second—for members of parliament and chaired by the DRM minister—aimed to raise awareness among decision makers.

- In Indonesia, Sri Lanka, and the Kyrgyz Republic, several workshops and video conferences have been organized in cooperation with ministries, academic institutions, members of the United Nations system, and other important DRM actors.

- In Armenia, the activities carried out under the capacity-development program targeted the academic and research community and technical experts at the national level. The focus was on structural measures and building codes, which is understandable because the country is at high risk for earthquakes and is in the process of developing structural and nonstructural measures to mitigate earthquake damage.

- In Sierra Leone, representatives of the national and provincial governments gathered to learn about the effects of natural hazards on development and the contribution that effective DRM can make to development. The measures and tools that the Japanese have designed and implemented were presented to demonstrate that a culture of prevention helps to sustain growth.

A DYNAMIC COMMUNITY OF PRACTICE

An important aim of Phase 2 is to engage clients and experts through the community of practice (CoP) described below and to build capacity to address specific country needs and interests.

To build developing countries' capacity in DRM, the WBI and its partners in the project (through the platforms offered by the Tokyo Distance Learning Center [TDLC]) have designed a program to exchange and share knowledge and to deliver it through blended-learning events and opportunities. With additional content provided by units of the World Bank Global Facility for Disaster Reduction and Recovery [GFDRR], the Social Development Department, and the East Asia and Pacific Region) and other organizations

(Japan's Ministry of Finance and the Ministry of Land, Infrastructure, Transport and Tourism [MLIT]; the Asian Disaster Reduction Center; the Japan International Cooperation Agency [JICA]; the International Recovery Platform [IRP]; and local research and academic institutions, such as Tohoku University and Fukushima University), the Government of Japan and the World Bank developed the Learning from Megadisasters CoP.

Designed in Phase 1 of the project and launched in October 2012 (at the World Bank's annual meeting, held that year in Japan), the CoP provides a virtual classroom environment, allowing participants to register, access reading materials, view presentations offered at video conferences and webinars, and engage with other participants in live discussions as well as through facilitated e-discussions and blogs. As of April 2014, the CoP serves as a venue where more than 1,000 DRM experts and practitioners from 83 countries share views, best practices, and documents; make suggestions; and engage in discussions, all in furtherance of the goal of disseminating the knowledge and lessons assembled in Phase 1.

The ultimate goal of the CoP is to foster a more responsive and effective DRM culture in developing countries by sharing best practices in DRM (notably from the earthquake and tsunami of March 11, 2011), building the capacity of DRM practitioners, promoting DRM as a critical component in development strategies and policies, and providing continuous DRM education (through webinars, discussion forums, and blogs).

The CoP has carefully tested its interaction model. Before launching the community, the team piloted it with a selected group of DRM experts to assess the design, understand how to define roles and responsibilities, test the activities, refine the communication strategy, understand what types of professionals the community would attract, and learn what motivates those professionals to be active members. After launching the community, the team polled members before updating the community's design and activities.

In March 2013, the Learning from Megadisasters CoP won an award as the best "Collaboration 4 Development" (C4D) CoP. Member engagement was identified as the leading key to success.

STUDY TOUR IN TOKYO AND SENDAI

Policy makers and practitioners from five targeted countries (Indonesia, Kenya, Maldives, Sri Lanka, and Uganda) were invited to Japan for a study tour designed to help these countries mainstream DRM into their development policies and operations. The tour, held in Tokyo and Sendai from June 24–27, 2013, brought the visitors into contact with Japanese organizations, academic institutions, and other key organizations. They visited Sendai City to learn about Japanese disaster management systems and absorb the lessons of the Great East Japan Earthquake (GEJE).

The delegation first visited the Japan Meteorological Agency, which is responsible for monitoring earthquakes, tsunamis, and various weather events, and for issuing warnings and alarms.

An international technical workshop on DRM and postdisaster reconstruction was organized by the World Bank, IRP, and JICA, and was held at the Tokyo Development Learning Center. At the workshop, the delegates from Indonesia, Kenya, Maldives, Sri Lanka, and Uganda joined Japanese and foreign government officials and practitioners of DRM.

Participants invited from five countries then traveled to Sendai to visit disaster sites, including Arahama Elementary School, where the principal at the time of the GEJE explained how students evacuated from the school and were later rescued. The mayor of Sendai City, Ms. Emiko Okuyama, explained the current progress of recovery and remaining challenges in her city. Sendai will be hosting the UN World Conference on Disaster Risk Reduction in 2015.

The delegation then moved to Tohoku University for a policy dialogue with academics specializing in various areas of DRM and reconstruction. They heard presentations on humanitarian logistics management, ground motion characteristics and vibration damage, mechanisms of destruction of coastal levees, geographic information system (GIS) and geodesign as a disaster reconstruction planning tool, forward creative reconstruction, and medical management of large-scale disasters.

The delegation also visited the Cabinet Office in Tokyo to meet with Mr. Yoshitami Kameoka, Parliamentary Secretary for Disaster Management, who stressed the importance of international cooperation in DRM to counter the increasing incidence of extreme weather events.

While in Tokyo, the delegation paid a visit to the MLIT to meet with officials and the technical staff of the Water and Disaster Management Bureau and the Policy Bureau. The key message delivered by the Japanese hosts was the importance of taking advance action to prevent and mitigate disasters before disaster strikes so that the impact of disasters can be minimized.

The delegation concluded its visit at the MLIT's Disaster Management Center.

SENDAI POLICY DIALOGUE AND THE BIRTH OF A NEW DRM HUB

Government ministers from around the world met in Sendai, Japan, for the Sendai Dialogue on October 9 and 10, 2012—a special event on managing disaster risk co-hosted by the Government of Japan and the World Bank. Part of the International Monetary Fund–World Bank Group annual meetings program, the dialogue brought together delegates to the annual meetings, disaster experts, and other stakeholders to build a global consensus on the need to better prepare for disasters around the world.

The Sendai Dialogue was an occasion to express solidarity with the people of Japan and a unique opportunity to learn from the Japanese experience with DRM. The event also drew on the experiences of other countries that have faced large-scale disasters. A report prepared by the partners for the event argues that the practice of DRM is a defining characteristic of resilient societies and should therefore be integrated—or mainstreamed—into all aspects of development. Natural hazards need not turn into disasters, the report urges. By investing in DRM rather than merely responding to disasters, lives, property, and the expense of rebuilding can be saved.

A joint statement at the conclusion of the Sendai Dialogue highlighted that Japanese know-how and expertise should be utilized to help vulnerable developing countries build their resilience to disasters, and that knowledge and partnerships should be expanded to support DRM policies and programs. It was then agreed to establish a DRM hub in Japan to facilitate the connection between Japanese centers of excellence in government, civil society, the private sector, and academia with the international development community, giving special attention to some particularly vulnerable countries.

CONCLUSION

The global cost of natural hazards in 2011 alone was estimated at $380 billion—resources that could have been used in productive activities to boost economies, reduce poverty, and raise the quality of life. No region or country is exempt from natural disasters, and no country can prevent them from occurring. But all can prepare by learning as much as possible about the risks and consequences of devastating events and by making informed decisions to better manage both. Disaster management is increasingly important as the global economy becomes more interconnected, as environmental conditions shift, and as population densities rise in urban areas around the world. As was shown by the GEJE of March 11, 2011, proactive approaches to risk management can reduce the loss of human life and avert economic and financial setbacks. To be maximally effective and to contribute to stability and growth over the long term, the management of risks from natural disasters should be mainstreamed into all aspects of development planning in all sectors of the economy.

Index

Iwate prefecture
    breakwaters in, 28, 28*f*
    debris in, 15–16
    dike systems in, 26, 26*b*
    evacuation centers destroyed in, 56
    floodgates in, 27
    green belt damage in, 118, 118*m*
    hazard map of, 16, 17*m*
    hospitals damaged in, 57, 57*f*
    livelihood and job recovery in, 216, 216*f*
    local disaster management plan for, 74–75
    NGO operations in, 129
    population numbers in, 113
    public housing in, 318, 319
    reconstruction planning in, 184
    subsidence in, 45
    telecommunications damage in, 134, 135*m*
    tsunami monuments in, 100*b*
    volunteer efforts in, 130, 130*f*

J

JA Kyosai, 257, 258, 260–261, 260*b*, 264, 265
Japan. *See also* disaster risk management (DRM);
    Government of Japan (GoJ); Great East Japan
    Earthquake (GEJE); Tohoku Tsunami; *specific*
    cities, municipalities, and prefectures
    building codes in, 9, 33–35
    business continuity plans in, 85
    coastal geography of, 25, 117
    deaths caused by disasters in, 249, 250*f*
    exports and domestic production affected by
        GEJE, 2, 17
    Learning from Megadisasters project,
        sponsorship of, 2, 4
    local government structure in, 110
    magnitude of earthquakes in, 298, 298*f*
    relief goods delivery system in, 143–144, 144*f*
    sovereign credit rating of, 290, 290*f*
"Japan as One" Work Project, 16, 213–214
Japanese Disaster Relief Act (1947), 194
Japanese Earthquake Reinsurance Co. (JER), 258,
    259, 265–266
Japanese government bonds (JGBs), 289–290,
    291–292, 291*f*
Japanese Red Cross Society (JRCS), 128, 129, 130,
    197
Japan International Cooperation Agency (JICA),
    39, 40–41*b*, 41, 279
Japan Meteorological Agency (JMA)
    earthquake-warning system, 10–11, 45, 95–96,
        95*f*, 301
    magnitude calculation by, 92, 98*n*1
    tsunami-warning system, 92–95, 92*f*, 93*b*, 94*m*,
        94*f*
Japan Self-Defense Forces (JSDF), 126, 126*f*, 130,
    151, 234, 324–325
Japan Society of Material Cycles and Waste
    Management (JSMCWM), 205

Japan Women's Network for Disaster
    Reconstruction and Gender, 165–166
JER (Japanese Earthquake Reinsurance Co.), 258,
    259, 265–266
JGBs (Japanese government bonds), 289–290,
    291–292, 291*f*
JICA (Japan International Cooperation Agency),
    39, 40–41*b*, 41, 279
jichikai (neighborhood associations), 6, 14, 67–68,
    162
JMA. *See* Japan Meteorological Agency
job creation. *See* livelihood and job creation
Jokoji Temple, 58, 58*f*
JRCS (Japanese Red Cross Society), 128, 129, 130,
    197
JSDF (Japan Self-Defense Forces), 126, 126*f*, 130,
    151, 234, 324–325
JSMCWM (Japan Society of Material Cycles and
    Waste Management), 205

K

Kamaishi City
    disaster awareness in, 300
    evacuation centers in, 244, 244*f*
    hazard map of, 242*m*
    neighborhood associations in, 67–68, 68*f*
Kamaishi-Higashi Junior High School, DRM
    education at, 78, 78*b*, 78*f*
"Kamaishi Miracle," 5, 11, 78, 78*b*
Kamaishi–Yamada Road, 50
Kanto Earthquake (1923), 212
Ka- tyan no Chikara ("power of moms") project,
    340–341
Kenya, capacity-building programs in, 346
Kesennuma City
    DRM education in, 78–79
    neighborhood associations in, 67, 67*f*
Kimura family documents, 327, 327*b*
Kitakami City, livelihood and job recovery in, 214
Kitakami River flooding, 28, 29*m*
Kobe Earthquake (1995)
    building collapses in, 34, 34*f*
    community centers opened following, 197*b*
    disaster waste from, 203, 204
    emergency response in, 69
    government spending on, 285
    lessons learned from, 72, 75, 198
    reconstruction following, 37, 212
    temporary housing following, 14, 200
kodokushi (solitary death), 198, 200, 212
Kuji Port breakwater project, 253, 254*m*
Kyrgyz Republic, capacity-building programs in,
    346
Kyu-Kitakami River flooding, 28, 29*m*

L

land banks, 201
land pooling schemes, 310, 311*f*
landslides, 37–38, 39, 45, 97*b*

land-use planning, 18, 110, 112–114, 113*f*, 185–187, 186*b*
large enterprises, business continuity plans among, 85–86
laws and regulations
 for building permits, 40, 40*f*
 for disaster risk management, 3, 71–74
 for land use, 18, 110, 112–113
 for reconstruction, 182–183, 183*b*, 187–189
 for relocation, 11
Learning from Megadisasters project, 2, 4, 345, 347
level 1 and 2 disasters, 29, 29*f*, 30–31, 298, 299, 302
liquefaction, 9, 36, 37, 37*f*, 41*n*1, 46
liquefied petroleum gas (LPG) facilities, 57–58, 57–58*f*
livelihood and job creation, 211–219
 financial support for, 16
 gender considerations in, 163–164
 government initiatives for, 213–214
 historical background, 211–212
 humanitarian assistance in, 214*b*
 lessons and recommendations, 217–219
 NGO and private sector initiatives, 214–215, 219
 out-migration, prevention of, 112
 in radiation-affected areas, 340–341
 results and challenges, 216–217, 216–217*f*
 in transitional shelters, 197–198
Local Autonomy Act (1947), 16
local disaster management plans, 73–75, 76
local government partnerships, 149–153
 for debris management, 209–210
 disaster relief agreements, 151, 152, 158
 evolution of, 150–151
 lessons and recommendations, 152–153
 twinning arrangements, 12–13, 151–152, 151*t*, 189
logistics management, in relief goods distribution, 144, 145–148
low-probability, high-impact events, 297–304
 chain-of-event effects of, 301–302, 303
 evacuation in, 300–301, 302
 forecasts and warnings for, 301, 303
 hazard maps for, 301, 303
 lessons and recommendations, 302–304
 national strategies to address, 297–298
 nonstructural measures for, 299, 302, 303, 304*f*
 structural measures for, 298–299, 302, 303, 304*f*
LPG (liquefied petroleum gas) facilities, 57–58, 57–58*f*

M

marine earth stations, 137
maternal care, 163
Maule Earthquake (2010), 264–265, 264*t*
medium enterprises, business continuity plans among, 85–86
Meiji-Sanriku Tsunami (1896), 5, 21*n*2, 26*b*, 27, 99–100
men, relief and recovery needs of, 164, 164*b*. *See also* gender considerations

METI (Ministry of Economy, Trade and Industry), 273–274, 279
MHLW (Ministry of Health, Labour and Welfare), 165
MIC (Ministry of Internal Affairs and Communications), 137, 138
Michi-noeki (road stations), 50–51, 51*f*, 51*t*, 52
migrant communities, 20–21, 336–337, 341
migration of residents, 112, 187, 187*m*, 187*f*
Minamisanriku
 disaster management facilities in, 50–51
 evacuation centers in, 157
 population decrease in, 187, 187*f*
 public housing in, 320
 reconstruction planning in, 184, 184*f*, 188*m*, 188–189
 relocation programs in, 185, 311–312, 313
 shopping villages in, 215, 215*f*
 temporary housing in, 198
Ministry of Economy, Trade and Industry (METI), 273–274, 279
Ministry of Health, Labour and Welfare (MHLW), 165
Ministry of Internal Affairs and Communications (MIC), 137, 138
Ministry of Land, Infrastructure, Transport and Tourism (MLIT)
 cost-benefit analyses conducted by, 252, 252–253*f*
 emergency response efforts of, 11–12
 on floodgates and inland lock gates, 30
 hazard map Web portal, 234, 234*m*, 243
 on hydrometeorological disasters, 43–44, 45, 46
 on infrastructure rehabilitation, 174, 175
 on liquefaction, 37
 on livelihood and job recovery, 213
 regulatory impact analyses conducted by, 254
 road station development, 50
 subsidence mapping, 45
 TEC-FORCE, establishment of, 127
 on tsunami countermeasures, 113
Miyagi prefecture
 building damage in, 36
 cultural heritage and preservation in, 325–327, 326–327*b*, 327*f*
 debris in, 16
 dike systems in, 26
 green belt damage in, 118, 118*m*
 hazard map of, 16, 17*m*
 hospitals damaged in, 57, 57*f*
 livelihood and job recovery in, 216, 216*f*
 NGO operations in, 129
 public housing in, 318, 319
 reconstruction planning in, 184, 184*f*
 telecommunications damage in, 134, 135*m*
 tsunami run-up in, 28, 29*m*
 volunteer efforts in, 130, 130*f*

Miyako City
dike systems in, 26*b*
hazard map of, 229*m*
tsunami monuments in, 100*b*
Miyako Road, 50
MLIT. *See* Ministry of Land, Infrastructure,
Transport and Tourism
mobile-phone communications, 133–136, 134*f*,
135*m*, 136*f*
mobilization. *See* coordination and mobilization of
emergency response
moment magnitude, calculation of, 92, 98*n*1
monuments, preservation of, 19
multifunctional infrastructure, 49–53, 254–255
multihazard disaster risk management, 6, 6*f*
municipal buildings, relocation of, 12, 12*f*, 55–56,
56*f*, 149–150
Muteki Ltd., 260*b*

N

Nagoya City, partnership with Rikuzentakata City,
150–151
Naraha City, partnership with Aizu-Misato City,
158
National Government Defrayment Act for
Reconstruction of Disaster Stricken Public
Facilities (1951), 173–174
national monuments, preservation of, 19
National Mutual Insurance Federation of
Agricultural Cooperatives, 257, 258, 260–261,
260*b*, 264, 265
National Research Institute for Earth Science and
Disaster Prevention (NIED), 239
Natori City
green belt damage in, 119, 119*f*
relocation programs in, 310
natural disasters. *See specific names and types of
disasters*
Natural Disaster Victims Relief Law (1998), 284,
286
Natural Hazards, UnNatural Disasters (United
Nations–World Bank), 256
neighborhood associations, 6, 14, 67–68, 162
networked relocation, 198
newsletters for evacuation centers, 158*b*
NGOs. *See* nongovernmental organizations
Nidec-Shimpo Corp., strategies for minimizing
supply chain disruptions, 278–279
NIED (National Research Institute for Earth
Science and Disaster Prevention), 239
Nikkei Index, 272, 273*f*
Nissan, strategies for minimizing supply chain
disruptions, 278
Nobiru Consultation Committee for Community
Development, 312, 312*f*
Nokia, business continuity plan developed by, 84*b*

nongovernmental organizations (NGOs)
community participation of, 65
historic preservation by, 325
livelihood and job creation by, 214–215
mobilization of, 128–129
in reconstruction planning, 15
nonprofit organizations (NPOs)
community participation of, 65
mobilization of, 128–129
nonresidential lands, management, 311
nonstructural measures, 63–121
business continuity plans, 83–89. *See also*
business continuity plans
community participation, 65–69. *See also*
community-based disaster risk management
cost-effectiveness, assessment of, 254
disaster management plans, 71–76. *See also*
disaster management plans
education sector, 77–82. *See also* education
sector
evacuation, 99–108. *See also* evacuation
findings and recommendations, 10–11
green belts, 117–121. *See also* green belts
for low-probability, high-impact events, 299,
302, 303, 304*f*
reconstruction planning, 109–115. *See also*
reconstruction planning
structural measures, integration with, 5, 10, 25,
30–31
warning systems, 91–98. *See also* warning
systems
NPOs. *See* nonprofit organizations
nuclear power stations. *See also* Fukushima
Daiichi Nuclear Power Station; Fukushima
Daini Nuclear Power Station; radiation
contamination
damage to, 58–60, 59*b*
evacuation preparedness and, 106
near epicenter, 58, 59*m*

O

Odagaisama Center, 336, 336*f*
Ofunato City, public housing in, 320
Ofunato Junior High School, as evacuation center,
157, 157*f*
oil refineries, damage to, 57–58
Okawa Elementary School, student and teacher
casualties at, 101*b*
Okirai Elementary School, evacuation bridge at,
101*b*
Omoto Elementary School, evacuation stairway at,
51, 51*f*, 101*b*
Onagawa Nuclear Power Station, 59*m*, 59*b*
OpenStreetMap (OSM), 237–238
Operation Toothcomb, 174–175, 175*m*, 175*f*
OSM (OpenStreetMap), 237–238
Otsuchi Town, municipal building damage in, 12,
12*f*, 55–56, 56*f*, 150, 150*f*

transitional shelters, 193–202. *See also* evacuation
   centers
   advantages and disadvantages of, 200, 200*t*
   community building and emotional care in, 197,
      197*b*
   evolution of, 198–199
   framework for, 194
   lessons and recommendations, 14, 199–201
   livelihood support in, 197–198
   private rental apartments, 196–197
   public housing, 19, 197, 309, 318–320*f*, 318–321
   "roof first" concept for, 199, 199*b*
   support systems in, 197–198
   temporary housing, 195–196, 195–196*f*, 198–199,
      198–199*f*
   transportation needs in, 197
   types and characteristics of, 194–197, 194*f*, 195*m*
transportation needs, in transitional shelters, 197
trash. *See* debris and waste management
trench-type earthquakes, 224, 225*m*
TSE (Tokyo Stock Exchange), 272
Tsunami and Storm Surge Hazard Map Guidelines
   (GoJ), 229
tsunamis, 25–32. *See also* dike systems; disaster
   risk management (DRM); *specific tsunamis*
   breakwaters for, 8, 26, 28, 28–29*f*, 29
   comparison of damage from, 250, 250*f*
   dams for, 43, 44
   evacuation routes for schools, 51, 51*f*, 101*b*
   floodgates for, 27, 27*f*, 29–30
   inland lock gates for, 29–30
   lessons and recommendations, 30–31
   sediment deposits from, 204–205
   simulations of, 185, 185*m*, 185*f*, 229
   two-level categorization of, 29, 29*f*, 30–31, 31*n*1,
      299, 302
   warning systems for, 91–95, 92*f*, 93*b*, 94*m*, 94*f*
Turkey, cultural heritage and preservation in, 328
twinning arrangements, 12–13, 151–152, 151*t*, 189
Twitter, 14, 137, 238

U

Uganda, capacity-building programs in, 346
UKG (Union of Kansai Governments), 151–152
UNDP (United Nations Development
   Programme), 342, 346
unemployment. *See* livelihood and job creation
UNEP (United Nations Environment Programme),
   209*b*
UNICEF (United Nations Children's Fund), 166
Union of Kansai Governments (UKG), 151–152
United Nations Children's Fund (UNICEF), 166
United Nations Development Programme
   (UNDP), 342, 346

United Nations Environment Programme (UNEP),
   209*b*
United States, aid and assistance from, 130
Unosumai Elementary School, DRM education at,
   78, 78*b*, 78*f*
urban planning, 18, 110, 112–114, 113*f*, 185–187, 186*b*
Ushahidi, 238*m*, 238–239
utilities. *See* public utilities

V

vehicular evacuations, 10, 102, 105, 106
very small aperture terminals (VSATs), 136
violence, domestic, 163
visitors, evacuation measures for, 103
voice messaging services, 135–136, 136*f*
volunteers
   fire corps, 65, 67, 69, 91
   mobilization of, 130, 130*f*
VSATs (very small aperture terminals), 136

W

Wakabayashi ward, neighborhood associations in,
   68, 69*f*
warning systems, 91–98
   for earthquakes, 10–11, 45, 95–96, 95–96*f*
   evacuation, coordination with, 10
   for landslides, 97*b*
   lessons and recommendations, 96–97
   for low-probability, high-impact events, 301, 303
   short message service alerts, 95
   on television, 95, 137
   for tsunamis, 91–95, 92*f*, 93*b*, 94*m*, 94*f*
waste management. *See* debris and waste
   management
water supply systems, 12, 173, 176, 177*f*
welfare shelters, 157, 164–165
women. *See also* gender considerations
   on disaster prevention councils, 165
   empowerment of, 165–166
   maternal care for, 163
   privacy and security concerns of, 162–163
   socioeconomic status of, 162
   workload and livelihood concerns, 163–164
World Bank
   capacity-building programs, 345, 346
   community of practice, formation of, 346–347
   Learning from Megadisasters project,
      sponsorship of, 2, 4
   Sendai Dialogue event, 348
World Travel and Tourism Council (WTCC), 272
WTCC (World Travel and Tourism Council), 272

Y

Yoshihama village, relocation of, 111, 114

www.ingramcontent.com/pod-product-compliance
Lightning Source LLC
Chambersburg PA
CBHW080226270326
41926CB00020B/4154